Complete vehicle

Michael Trzesniowski

Complete vehicle

Michael Trzesniowski
Pankl Racing Systems AG
Kapfenberg, Austria

ISBN 978-3-658-39669-5   ISBN 978-3-658-39667-1   (eBook)
https://doi.org/10.1007/978-3-658-39667-1

Preface

Now the manual series is already in its 2nd edition, and the family of these books is growing. While there were already five volumes at the beginning, another volume has been added for this edition. The authoritative idea that individual special volumes can go into depth without space problems has thus proven itself. Since the publication of the first edition, further findings have been added, which have found their way into the corresponding chapters, or new chapters have been added.

That the contents nevertheless fit together and complement each other as if in a single book – one of the great strengths of the original book *Race Car Technology* – is ensured by the editor in a comparable way to how the project manager keeps an eye on the overall function in a major design project.

The racing car technology handbook series is dedicated to the racing vehicle from conception, design and calculation to operation and its (further) development.

The first volume, "Basic Course in Race Car Technology", thus offers not only current considerations but also a historical overview of motorsport, racing operations, such as the rescue chain, and a comprehensive overview of the technology used in racing cars as a general introduction to the subject. For more than fifteen years, the author has been concerned with the driving dynamics and chassis tuning of production passenger cars.

Volume two, "Complete Vehicle", starts with the chronological design process and therefore begins with concept considerations, considers safety aspects and the design of the driver's environment, describes aerodynamic influences and then looks at the frame and body work design.

Volume three "Powertrain" deals with all forms of drive systems and their energy storage and continues in the sense of load flow via start-up elements and transmissions up to the side shafts. Electrical systems and electronic driving aids have also found their appropriate place in this volume.

Volume four, "Chassis", is devoted exclusively to the decisive subsystem and its components that determine driving behaviour. Tyres and wheels, wheel-guiding parts, springs and dampers, steering and brakes are covered.

Volume five, "Data Analysis, Tuning and Development", deals with the phase that follows once the vehicle has been designed and built. The development and tuning of a racing vehicle require a much different approach than its construction and key tools – such as data acquisition and analysis, simulation and testing – are therefore presented. The subject of data acquisition and analysis is profoundly presented by an author who is confronted with this activity on a daily basis.

For volume six, "Practical Course in Vehicle Dynamics", authors have been recruited who have decades of experience as race engineers and race drivers on the race track. In their work, they describe the practical tuning of racing vehicles, underpin what they present with examples of calculations and thus also build a bridge to the theoretical considerations in the other volumes.

I wish all readers that they will find "their" volume in the abundance offered and that they will get essential impulses for their studies, profession and/or leisure time from reading it, be it because they are designing a vehicle, building one, operating and improving one or because they are analysing one with a thirst for knowledge.

Kapfenberg, Austria   Michael Trzesniowski
Spring 2019

Greeting

27.05.2019
EC Todsen

Edition: Racing Car Technology Handbook: Six Volumes

Motorsport continues to inspire. For as long as there have been cars, drivers have been pushing their racing cars to their technical and physical limits, engaging in gripping and exciting competitions. But the competition doesn't just take place on the race track. The foundation for success is laid in the development departments and design offices. In-depth knowledge of vehicle technology and development methodologies, along with thorough and timely project management, creative problem-solving skills and unconditional team play, determine victory or defeat.

Motorsport continues to be a model and guide for technological progress – be it in lightweight design, material selection or aerodynamics. Chassis and tire technology also benefit immensely, and new safety concepts are often based on experience from the racetrack. However, the influence of motorsport is particularly evident in the powertrain: In addition to the impressive increase in performance and efficiency of the classic internal combustion engine drive system, the key future technologies of hybrid and purely electric drive have also successfully arrived in motorsport competition and are continuing this successfully and with public appeal in a partly completely new setting. It remains very exciting to observe which attractive innovations digital networking solutions and autonomous driving systems will generate in motorsport. I recommend that young engineers in particular acquire the tools for their future careers in motorsport. What you learn in motorsport sticks. Formula Student already offers an ideal environment to start with.

I am very pleased that the book series *Handbuch Rennwagentechnik* has been so well received and that the second edition has been published within two years. This shows that the competencies addressed are clearly presented in this work and conveyed in an understandable way.

This book series has deservedly become a well-known and valued reference work among experts. The work brings students closer to the fascination of motorsport and racing enthusiast laymen to a deeper technical understanding.

I wish you much success on and off the race track!

Prof. Dr.-Ing. Peter Gutzmer

Deputy Chairman of the Executive Board and Chief Technology Officer, Schaeffler AG, Herzogenaurach, Germany

Abbreviations, Formula Symbols and Units

Equations given in the text are generally quantity equations. The quantities can be used in any units, preferably in the SI units (meter-kilogram-second system). The unit of the quantity to be calculated then results from the selected units of the variables. Sometimes, the numerical value equations commonly used in practice are also given. With these, the equation is only correct if it is calculated with the specified units. The unit of the result variable is therefore also given in the text.

Geometric *Points*

M	*Centre point*
V	*Vehicle centre of gravity*
W	*Centre of tyre resp. wheel contact*

Indices

If more than one index occurs, they are separated by a comma. The order of indices is this:

For forces, the first index indicates the location or point at which the force is applied and the second index indicates the direction of the force, e.g. $F_{w, z}$ ... wheel contact force (vertical force at the tyre contact point). The vehicle fixed coordinate system used is defined in the glossary.

Additional specifications, such as front, rear, driven, etc., follow as further indices.

0	*Zero-point position or starting point. Ambient*
1	*To the top/in jounce/in compression/in*
2	*To the bottom/in rebound/out*

(continued)

b	*Bending*
c	*Chassis, frame*
C	*Coolant*
co	*Cornering*
dr	*Drag*
e	*Effective*
f	*Front, forward*
i	*Inner wheel, inner*
critical	*Critical*
L	*Aerodynamic*
l	*Left, left-hand side*
M	*Engine resp. motor*
m or med	*Middle, mean*
max	*Maximum or permissible maximum value*
min	*Minimum*
n	*Rated value*
o	*Outer wheel, outer*
r	*Rear, aft*
Rd	*Rod, linkage resp.*
Ro	*Roll*
rs	*Right, right-hand side*
rsl	*Resulting*
S	*Anti-roll bar, stabilizer*
Sp	*Spring*
t	*Total, nominal value resp.*
ts	*Torsional*
T	*Tyre*
tc	*Turning circle*
V	*Overall vehicle*
X or x	*Longitudinal direction in general*
Y or y	*Lateral direction*
Z or z	*Vertical direction*

Distances in mm

a to p	*Distances and length (in general)*
b_f or b_r	*Track width, front or rear*
B_t	*Overall width of the vehicle*

(continued)

c	Chord length of a wing
d or D	Diameter, in general
D_S	Track circle diameter (front)
$D_{S,r}$	Track circle diameter, rear
D_{tc}	Turning circle diameter, wall to wall
f	Camber (wing)
h or H	Height, in general
H_t	Overall height of the vehicle
h_V	Height of the vehicle centre of gravity
l	Wheelbase
l_f or l_r	Distance of vehicle centre of gravity to middle of front or rear axle
L_t	Overall length of the vehicle
r	Effective control arm length or force lever in general
R	Path radius
s	Travel or stroke, in general
s	Span of a wing
s_t	Total wheel travel
t	(Wall) thickness
t	Maximum airfoil thickness

Angle in ° or rad

α	Angle of attack
α_f or α_r	Slip angle of front or rear tyre
β	Angle, in general
β	Sideslip angle (attitude angle)
δ	(Wheel) Steer angle
δ_i or δ_o	Actual steer angle, inner or outer wheel

Masses, Weights in kg

m	Mass, weight or load in general
$m_{V,f}$ or $m_{V,r}$	Axle load, front or rear
$m_{V,t}$	Gross vehicle weight

Forces in N

$F_{L,X}$	Aerodynamic drag
$F_{L,Z}$	Aerodynamic downforce
$F_{W,X,a}$ or $F_{W,X,A}$	Accelerating force in the centre of tyre contact of one wheel (a) or both wheels of an axle (A)
$F_{V,Y}$	Lateral force at vehicle centre of gravity
$F_{W,Y}$	Lateral force at wheel contact point
$F_{W,Z}$	Corner weight (wheel load)
$F_{V,Z,t}$	Cross vehicle weight

Torques and Moments in Nm

M_b	Bending moment

Spring rates in N/mm

c_f or c_r	Rate of the body supporting spring at parallel springing, related to the centre of tire contact of one axle side, front or rear

Dimensionless Key Figures

c_A	Downforce coefficient
c_W	Drag coefficient
i_m	Axleload ratio front/rear
k_A	Area ratio of a diffuser
S	Safety factor

η		Total efficiency of geartrain and final drive
Φ_{Sp}		Ratio of wheel-related spring rates front/rear
Λ		Aspect ratio of a wing
$\mu_{W,Y}$		Coefficient of static friction in lateral direction

Other Sizes

Δ	Change, difference	
ρ	Density	kg/m^3
σ	Stress	N/mm^2
ρ_L	Density of air	kg/m^3
A	Area, cross-section area	m^2
a_x	Longitudinal acceleration in general	m/s^2
a_y	Lateral acceleration	m/s^2
E	Modulus of elasticity, Young's modulus	N/mm^2
g	Acceleration due to gravity	m/s^2
I	Area moment of inertia	mm^4
$J_{V,Z}$	Mass moment of inertia of vehicle around the Z axis	kgm^2
k	Heat transfer coefficient	$W/(m^2\,K)$
P	Power	W
P_e	Effective power of engine	kW
p_0	Ambient pressure (air pressure)	bar[1]
Q	Heat flow	W
R_e	Yield strength	N/mm^2
R_L	Gas constant of air	$kJ/(kgK)$
R_m	Ultimate tensile strength	N/mm^2
T	Thermodynamic temperature	K
T	Time	s
v_L	Air-flow velocity	m/s
v_V or v_X	Longitudinal velocity of vehicle	m/s or km/h
W	Section modulus under bending	m^3
W	Work	J

---

[1] 1 bar = 100 kPa. Although the valid SI unit for pressure is Pascal (Pa), the book uses the unit bar, which is more "handy" in practice.

Other Abbreviations

FVW	*Fibre composite material*
CFRP	*Carbon fibre reinforced plastic (CFRP)*

Contents

Introduction

1

1.1 Types of Race Cars

Motor sports are defined as all sports involving motor-driven land or water vehicles (automobile, motorcycle, motorboat sports). Automobile sports include road racing, rally and touring car racing, auto- and rallycross and veteran car racing; motorcycle sports include road racing, enduro racing, speedway and ice speedway; motorboat sports include motorboat racing (regattas on a circuit of 1500–2000 m in length marked by turning buoys in several heats) and offshore sports, and in a broader sense also water skiing. In the following, multi-track competition craft will be the focus of consideration, Fig. 1.1.

It is not possible to present a generally valid classification of the competition vehicles solely according to competitions or vehicle types. The starting fields and the technical specifications of individual regulations are too varied. However, regardless of the type of competition, some typical vehicles can be categorized according to technical aspects, Fig. 1.2. Cup vehicles based on road vehicles, rally vehicles based on road vehicles,

M. Trzesniowski, *Complete vehicle*,
https://doi.org/10.1007/978-3-658-39667-1_1

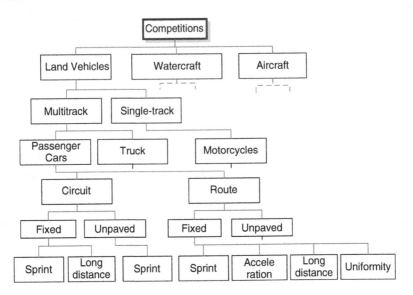

Fig. 1.1 Classification of motor sports (selection)

two-seater sports prototypes built for racing purposes only, single-seater racing vehicles (monoposti) with open cockpits and open (free-standing) wheels, and touring cars.

Individual competitions can be assigned to these vehicles:

Cup cars: Caterham Hankook, Clio Cup, Ford Fiesta Cup, GTM series, Lupo Cup, Mini Challenge, Polo Cup, Porsche Cup, Porsche Super Cup, Seat Leon SC, Yaris Cup, etc.

Rally cars: hill climb rally, national championships, HJS rally, world championship, etc.

Sports prototypes: 24 Hours of Le Mans, WEC (World Endurance Championship), ALMS (American Le Mans Series), DPi (Daytona Prototype international), FIA GT, Radical Race Cup, rhino's GT Series, Sebring, etc.

Formula cars: A1 GP Series, F3 Euro Series, Formula 1, Formula 2000 (former Easter), Formula 3, Formula BMW (former ADAC), Formula 4, Formula Ford, Formula König (series finished), Formula Opel, Formula Renault, Formula Renault EM, Formula Renault V6, Formula Student, Formula V, Lista Formula Junior, Recaro F3 Cup, etc.

Touring cars: 24 Hours Nürburgring, ADAC Procar, Castrol Haugg Cup, Divinol Cup, DTM (German Touring Car Masters), FIA ETCC, FIA WTCC, Langstrecke Nürburgring (VLN – Long distance Nürburgring), STT, etc.

A further subdivision is provided by the FIA International Sporting Code (Appendix J Article 251, see Annex). According to this, multi-track competition vehicles are divided into categories and groups. A distinction is made between Category I ("homologated production cars"), Category II ("racing cars") and Category III (trucks). In detail, a distinction is made between:

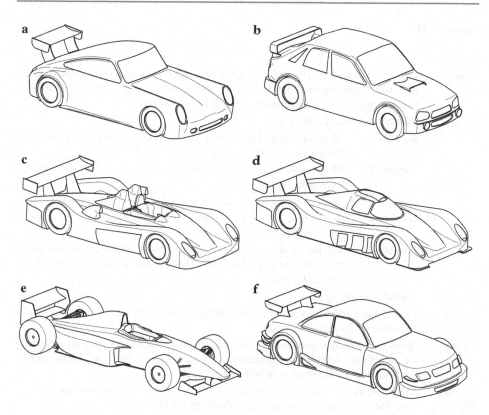

Fig. 1.2 Typical racecars. (**a**) Cup vehicle, (**b**) Rally vehicle, (**c**) Open sports prototype, (**d**) Closed sports prototype, (**e**) Formula car, (**f**) Touring car

Category I

Group A:	Four-seater *touring cars* with standard bodywork, production of at least 2500 units per year. WRC (world rally car) are also included
Group B:	Two-seater GT vehicles (*grand touring cars*). These are road-going racing cars, production volume at least 200 units per year
Group N:	*Production cars*. These are four-seater production cars with minor modifications, production volume at least 2500 units per year
Group SP:	Super production cars. Production volume at least 2500 units per year
Group T2:	*Series cross-country cars*

Category II

Group CN:	Production sports cars. These are two-seater prototypes with a near-series engine with a maximum displacement of 3000 cm^3. Fuel tank capacity less than 100 l. minimum weight depends on engine capacity, e.g. 625 kg for 3000 cm^3	
Group D:	International formula racing cars	
	Formula 1:	Monoposto with V6 engine, capacity 1600 cm^3, with turbocharger, energy recovery systems, minimum weight 690 kg
	Formula 3:	Monoposto, engine derived from a large series unit, displacement not exceeding 2000 cm^3
	Formula 3000:	Monoposto, engine capacity up to 3000 cm^3, minimum weight 625 kg
	Formula 4	4-cylinder 1.6-l engine, minimum weight 570 kg, entry-level series with cost limits
	Formula E	Electrically driven formula cars, max. Power 200 kW, minimum mass with driver 800 kg (of which 200 kg battery), 18″ wheels with profiled standard tyres
Group E:	Formula-free racing cars	
Group GT1:	Grand touring sports cars. These are road-going vehicles with open or closed cockpits, two-seaters with max. Two doors	
Group GT2:	Series *grand* touring sports cars. These are road-going vehicles with naturally aspirated engines of max. 8000 cm^3 or supercharged engines with max. 4000 cm^3 displacement. Air restrictors are mandatory for both engine types	
Group GT3:	Cup grand touring sports cars. The vehicles belonging to this group are listed by the FIA. These cars are homologated individually by the FIA	
Group SR	Sports car. Two-seater, pure racing vehicles with open or closed cockpit. In the latter case with two doors. Minimum weight 750 kg (SR2) and 900 kg (SR1). Engines: SR1: Naturally aspirated petrol engines up to 6000 cm^3, supercharged petrol engines up to 4000 cm^3 and supercharged diesel engines up to 5500 cm^3; SR2: Naturally aspirated petrol engines up to 4500 cm^3, supercharged petrol engines up to 2700 cm^3. Fuel tank capacity 90 l. front headlights and rear tail lights mandatory	
Group T1:	Off-road prototypes (*modified cross-country cars*)	

Category III

Group F:	*Racing trucks*
Group T4:	Raid rally trucks (*cross-country trucks*)

For the famous 24-hour race in Le Mans, the organizer ACO (see appendix) issues its own regulations. There are several vehicle categories whose engines all have an air restrictor:

Le Mans prototype:	LMH (Le Mans Hypercar, open or closed), LMDh (Le Mans Daytona Hypercar, open or closed)
LM grand Tourismo:	LM GTE pro (professional), LM GTE am (amateur)

There are also other special vehicles for other competitions, e.g. dragsters for acceleration races or autocross and rallycross vehicles.

As can be seen from the above, if one wants to make a general, rough classification of racing vehicles that applies to all competitions, the only classification that remains is that of vehicles with free-standing wheels and those with enclosed wheels.

Basically, a classification of racing vehicles is not necessary for their construction. The design of a racing vehicle is technically oriented primarily to the intended use, but only within the limits set by various regulations. Nevertheless, no regulations are presented in detail in this book. A regulation has among other things the task to ensure a competitive equality ("rules of the game") and is often changed. It is therefore primarily essential for these specifications to be easily measurable or verifiable. However, many other regulations are generally worthy of attention for the designer in that they have been created by accidents and incidents in the past and thus represent an enormous wealth of experience. Certain passages can therefore be found in almost all regulations. In the present work, we will only refer to individual statements in the regulations if they are relevant to safety or to the understanding of a chosen solution. When designing a vehicle, the currently (!) valid regulations must be consulted in any case, if one wants to avoid that the new "wonder car" cuts a bad figure already at its first public appearance, because it does not pass the technical inspection. The FIA regulations can be read in detail or downloaded from the Internet [1].

Figures 1.3, 1.4, 1.5, 1.6, 1.7, 1.8, 1.9, 1.10, 1.11, and 1.12 show, in loose order, some examples of different racing vehicles.

Fig. 1.3 Indy Car. High speed vehicle for oval courses

Fig. 1.4 FIA GT vehicle

Fig. 1.5 Touring car. Vehicle based on series parts

Fig. 1.6 Formula 1 car. Monoposto with open wheels and open cockpit

Fig. 1.7 LMP1 vehicle. Long-distance vehicle with open, two-seat cockpit and enclosed wheels

Fig. 1.8 Cart. Monoposto, no moving parts in the wheel suspension, no differential, asymmetrical construction

Fig. 1.9 Cup vehicle close to series production

Fig. 1.10 Rally car. Rally vehicles move on paved and unpaved roads

Fig. 1.11 Racing motorcycle

Fig. 1.12 Raid truck. Commercial vehicles are also used on the circuit and, like the vehicle pictured, off-road for competitions

1.2   Comparison Racing Versus Mass Production

Racing vehicles are practically as old as the vehicles themselves. As soon as man invented a vehicle, he also raced it. At that time, racing and everyday vehicles were identical in construction. However, throughout history, vehicles were built specifically for racing. These racing vehicles have only one purpose, to win races. This means to drive through a certain distance as fast as possible within the limits of the rules and regulations, and to complete necessary and permitted maintenance and/or repair work just as quickly. The

vehicle must therefore be capable of high performance (see Chap. 2 *Concept*) and be easy
and quick to repair. Table 1.1 lists some differences between utility vehicles and racing
vehicles.

Simple solutions are sought for racing vehicles. The vehicle must also be adjustable to
different track and weather conditions by relatively simple means. The former concern, for

Table 1.1 Differences in requirements between road vehicles and racing vehicles

Request	Road vehicle	Racing car	Rem.
Security	High, selling point	Caused by rules	(a)
Comfort	High, selling point	Unimportant; partly even undesirable	(b)
Styling	Important, sales argument	Rather unimportant, importance for sponsors and private drivers	
Lifetime	10–15 years	1–3 years, depending on category and owner	
Costs	Important, economic efficiency	Rather unimportant, depending on owner	
Dates	Important, but not fixed	Important and immovable	(c)
Legal requirements	Many, country specific	A set of rules for each category	
Planning period	3–5 years and more	Often less than a year	
Production volume	Particularly high	Single pieces, small series	
Maintenance	Rather seldom, due to legal reasons	Very often, practically before and after every run	
Repair	In specialist workshops under relatively little time pressure with all the tools and machines available	At the race track under high time pressure and only with the means which are allowed or available	
User	General public, not specially trained	Selected group of people, mostly professional users	
Design speed	Partly up to max. 250 km/h; average operating speed considerably lower	Over 350 km/h; highest operating speeds aimed at	
Night-time capability	Important: Lighting, instrument lighting	Only for rally cars and endurance racing cars	
Winter suitability	Important: Starting behaviour, heating, ventilation, tyres, snow chains …	Not required except for rally cars	

Remarks:
(a) Proof of safety-relevant characteristics required by the regulations – tendency increasing
(b) A racing driver wants to "feel the car", i.e. a softly upholstered seat, for example, prevents the
driver of a formula car from experiencing the limit
(c) The start of production (SOP) may be postponed, the race weekend not

example, brake cooling and wear, balance between drag and downforce, the latter ambient temperatures and precipitation.

The different requirements inevitably result in different working conditions for those involved in motorsport compared to similar positions in series development. Unconventional working hours, more direct responsibility and absolute dedication are expected of them [2]. Decisions often have to be made quickly and are sometimes not fully comprehensible to engineers from production manufacturers because they come from experience and emotion [2]. Nevertheless, many large automotive companies have more or less direct links with motorsport. The reason lies, among other things, in the marketing benefits of involvement in motorsport. For example, the sales figures of the then DaimlerChrysler AG increased from 21.3 to 36.3% since the Formula 1 team was officially called McLaren-Mercedes [2]. The advertising value from mere participation in Formula 1 is estimated at EUR $1.8-10^9$ for the energy drink manufacturer Red Bull between 2000 and 2014 [3].

Depending on the formula, motorsport also offers the opportunity to use and test new materials and systems. The usual constraints of series development, such as cost pressure, restrictions on existing or specific production facilities, often stand in the way of the introduction of new techniques and materials. And it is not uncommon for racing to be the driving force behind a development that later finds its way into series-production vehicles, Table 1.2. The frequently asked question about the influence of motorsport on progress in series development cannot therefore be answered simply. The motorsport departments of even large automotive groups are usually organizationally and geographically detached from the influence of the factory. The work is carried out by specialists and

Table 1.2 Motorsport as a pacemaker for new materials and technologies [4]

Year	Company/ vehicle	Comment
1895	Michelin	Vehicle with pneumatic tyres in the Paris-Bordeaux-Paris race
1899	Dürrkopp	Development of a small sports car with an aluminium body to reduce weight.
1900	Maybach/ Daimler	The Daimler delivered to Jellinek, which he then named Mercedes after his daughter, had an engine made largely of aluminium and magnesium and a honeycomb radiator made of brass
1934	Auto union	Crankcase and cylinder heads made of cast aluminium for the 16-cylinder engine
1962	Porsche	Titanium for the connecting rods of the formula 1 engine
1963	Porsche 904 GTS	First German production vehicle with GFRP bodywork
1967	Porsche 910/8	Use of an aluminium grid tube frame with partial secondary function of the tubes as an oil line
1971	Porsche 917	Use of magnesium for tubular space frames
1981	Hercules/ McLaren/lotus	Load-bearing structure of formula 1 cars made of carbon-fiber-reinforced plastics (CFRP) for the first time

the direct influence of the namesake manufacturer is primarily financial. Other racing companies are in any case small manufacturers who work completely independently of large car companies. The designs are special designs, which do not aim at a large number of pieces. It is rare, but it does happen, that the same personnel drives series and motorsport projects. The engine subsystem group provides successful examples of this. Probably only because many racing engines were and are initially derived from existing production engines. Despite this minor direct influence of motorsport on series production, indirect influence and adoption of techniques cannot be denied.

The supporting structure of two production sports cars has clearly borrowed from racing.

For example, the Porsche GT and the Mercedes McLaren feature CFRP frames. The production volume of such CFRP hollow section frames could also be increased so that it becomes economically interesting at least for niche vehicles [5].

The core package process for casting metal parts was initially only used for special models and racing. In the meantime, it has also been further developed for large series.

The properties demanded of lubricating oils by racing engines are now being put to good use in passenger car start-stop systems [6].

Even in the field of commercial vehicles, which by their nature are cost- and use-oriented, examples can be listed. In 2014, Volvo introduced the I-Shift, a dual-clutch transmission for heavy trucks, and in the same year MAN introduced its TGX engine, a unit with top-down cooling.

In [7], the same trends in racing and production vehicle development are also noted:

- Multi-valve engines with increasing market shares,
- Turbocharged engines also with increasing market shares,
- Torque/power increases,
- Displacement decreases: power density improves,
- More torque/power is achieved from smaller, lighter engines,
- Compression ratio increases,
- Rated speed decreases,
- Mean effective pressures rising,
- Electronics have a high priority on a broad front: engines, transmissions, brakes, etc. and, more recently, the holistic networking of systems to form an overall system,
- Since the introduction of limited fuel quantity and limited fuel mass flow in some racing series (e.g. WEC, Formula 1), the focus in racing has also been on the economy of the powertrain.

In general, it can be stated that the benefit of motorsport for series development probably depends to a large extent on the regulations and on the organisation of the company. Regulations that prohibit the use of systems used in production vehicles prevent progress caused by motorsport. On the other hand, the simultaneous development of production and motorsport power units by a team equally improves reliability and racing suitability [8]. It is also noticed that with the increasing demand for reliability, e.g. for endurance racing, the solutions are getting much closer to the series solutions, which naturally results in a greater benefit for both sides.

The most recent example of how a transfer between motorsport and series development was deliberately forced is provided by the 2006 Le Mans winner: Audi R10 TDI. For marketing reasons and for the simple reason that there has been very little experience with diesel racing engines to date, series development provided significant input in the design of the racing engine [9].

The fact that not all solutions can be adopted directly from racing for series production is often due to the fact that the development goals are not the same. In the case of a series product, the optimisation of the function is also in the foreground at the beginning and is later replaced by the search for the best design (strength and material consumption), but ultimately the economic production and operation are in the foreground. The required suitability for everyday use also demands simple and safe operation of series products without special training.

For the future, a new field of activity is likely to arise on the racing side, which is already commonplace in series development, namely through the idea of environmental protection. Competitions that focus on minimal fuel consumption already exist. Competitions in which vehicles with alternative drive systems compete are under discussion. Energy recovery systems (KERS – *kinetic energy recovery systems*), with which braking energy can be used for subsequent acceleration, have already been developed and are in use in some major racing series (Formula 1, WEC). Such systems for the recuperation of braking energy are particularly interesting for vehicles with electric drives, as they help to reduce the energy requirements of the batteries. In racing series, where the efficiency of the powertrain is decisive for victory or defeat, combustion engines, electric motor-generator units and energy storage systems benefit equally from the development goal of "reducing losses". Hybrid powertrains, i.e. powertrains in which combustion engines and electric machines work together, initially even benefited from the experience gained from passenger car series development in Formula 1 and the WEC. Now that individual teams are tackling the development of their own drive concepts in Formula E, the performance limits of such systems including energy storage (battery, supercaps, . . .) will certainly be pushed. The vehemence and determination with which developments are driven forward in motorsport is certainly invaluable for the series.

1.3   Development Process

If a new vehicle is to be created, the entire product development process must be run through, Fig. 1.13. The concept phase is characterized by the search for the ideal solution, which is dictated by the regulations and restricted by time and cost specifications. During

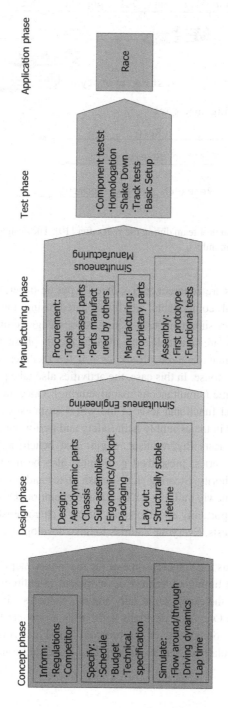

Fig. 1.13   Sequence of a vehicle project. The procedure describes the complete product development process of a racing car in the essential steps

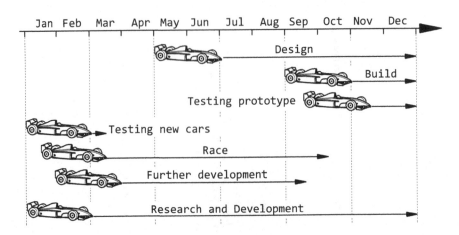

Fig. 1.14 Activities of a Formula 1 team during the year, after [10]. The design and construction of a new vehicle takes barely six months

the design phase, the ideas are concretized, i.e. literally put on paper, and mathematically validated. (Safety) critical components are investigated using the FE method.[1] All components are worked on simultaneously (simultaneous engineering). As soon as the manufacturing documents are available, the manufacturing phase begins. The newly designed parts are either manufactured off-site or – if the facilities and their utilization allow it – manufactured in-house. In this case, the activities also take place simultaneously in order to save project time (simultaneous manufacturing). Once all parts are available, assembly can begin. Initial functional tests of the wheel suspension, gearshift, steering, brakes, etc. are carried out in the assembly hall. Safety and performance relevant parts are also subjected to special tests (hydropulser, engine test bench, axle test bench, …). Depending on the racing series, prescribed parts must also be homologated, i.e. type-tested by a motorsport authority. Steering shafts, crash boxes and roll cages belong to this category. Now the new or newly built car can be put into operation. Drivers, mechanics and race engineers have the opportunity to get to know the car's characteristics and to eliminate weak points during track tests. Everyone's aim is to have developed a competitive vehicle by the time the race starts.

But even for already existing vehicles some of the mentioned steps have to be taken for further development. The timing of vehicle development is entirely determined by the event calendar and – with the exception of rally and raid vehicles – by the warm season.

Every year, a Formula One team develops and builds a new car that contains hardly more than 5–10% of the components of the previous car (Fig. 1.14). This results from optimizations and the regulations may have been changed significantly compared to the

---

[1] See glossar.

Table 1.3 Developmental key data of some racing classes

Class	Total budget [€]	Test kilometers per team	Test days per year and team	Vehicle mileage	Number of races
Formula 1	47 to 400 ·10^6 [2, 11]	19,000 [12]	65	3000 km [12] Engine: 400 km until 2003 From 2005: 1200 km	17
LMP1	44·10^6 [13]				
Formula 3 [14]	300–350,000		25	Engine: 1 season (with 1 revision)	20 on 10 weekends
Formula Renault [14]	350,000	Up to 12,000	20–25	Engine: 3 revisions per season	10–13
Formula A-Lista junior [14]	30–50,000			Engine: 1 season	

previous year. A Formula 1 car consists of more than 3500 components [10]. Table 1.3 summarises typical key data of various vehicle developments.

The main assemblies of a multi-track racing vehicle are found in all types and essentially there is no difference in construction when considering vehicles with free-standing and enclosed wheels, Figs. 1.15 and 1.16. Attached to the rear end of the chassis is the engine, to which in turn is attached the gearbox. Both form the load-bearing (stressed) structure of the rear of the vehicle, which houses the chassis at the rear. To the side of the cockpit are heat exchangers for engine cooling and possibly intercooling. The front end of the chassis is formed by a nose-shaped nose, which is the crash element. Differences between formula cars and sports cars result from the dimensions and designs of the cockpit (single-seat, two-seat, open, closed) and the shape of the bodywork.

The main subsystems of racing vehicles are in detail:

- Cockpit: Accommodates the driver and protects him in the event of an accident.
- Frame (chassis): Houses the cockpit, absorbs all forces and connects other main assemblies together.
- Engine: Power source for vehicle and auxiliary systems. Gives the sport its name. Internal combustion engines and electric motors are used.
- Fuel system: Stores fuel and supplies the combustion engine with energy. In e-vehicles, the battery performs this function analogously.
- Drive train: Transfers engine torque to the wheels and converts engine speed and torque, or when recuperating, the circumferential force of the tires becomes a driving torque for an energy converter that fills the energy storage.

Fig. 1.15 Structure of a formula car. The centrally positioned cockpit hugs the driver as closely as possible in the area of the legs. The fuel tank is located between the driver and the engine. The gearbox is flanged to the engine, which also accommodates parts of the rear suspension. The heat exchangers are located in side boxes next to the cockpit. The nose of the car is attached to the cockpit and the front wing is attached to it. An undertray closes off the car at the bottom and generates part of the aerodynamic downforce

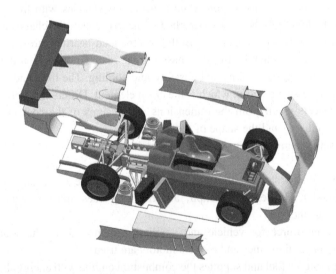

Fig. 1.16 Structure of a sports car. The bodywork parts of this two-seater sports car prototype are shown in a removed position, which clearly shows the close relationship to a formula car. The essential assemblies are in fact arranged in the same way

- Suspension: Guides and holds the wheels, responsible for function of the tires with the road.
- Steering: Enables maneuverability of the vehicle, providing one of the most crucial systems for the driver.
- Wheels and tyres: make contact with the road surface and are therefore one of the most important components.
- Brake system: Decelerates the vehicle. Can also – at least this is technically possible – be used for targeted stabilisation of the vehicle.
- Bodywork: Closes the vehicle to the outside, creates and transmits air forces; its shape and color dictate most of its appearance.
- Electrical and auxiliary systems: Ensures the flow of electrical energy and the ever-increasing flow of electronic data. It also includes hydraulic and pneumatic systems.

The *Bill of Materials* (BOM) for a complete vehicle project will still include the following components:

- Doors and flaps incl. Glazing
- Piping and tubing
- Lubricating oil system
- Cooling system
- Fire extinguishing system
- Wings completely front and rear
- Undertray
- Controls: steering wheel, foot pedal, external gearshift, display instruments
- Devices, gauges and tools for in-house production and assembly
- Data Acquisition System.

The following chapters explain and describe the fundamentals, design and construction of the complete vehicle system.[2]

References

1. http://www.fia.com/sport/Regulations
2. Andorka, C.-P., Kräling, F.: Formel 1, das Milliardenspiel. Copress, München (2002)
3. Newey, A.: How to Build a Car. HarperCollins Publishers, London (2017)
4. Braess, H.-H., Seiffert, U.: Vieweg Handbuch Kraftfahrzeugtechnik, 1st edn. Vieweg, Wiesbaden (2000)

---

[2] The development and tuning of racing cars is addressed in Vol. 5 Data Acquisition, Tuning and Development of the Racing Car Technology Manual.

5. Strambi, G.: Assembly technology for carbon fibre body structures. AutoTechnol. **6**(4), 56 ff (2006)
6. Birch, St.: Racing to Solve the Lubrication Challenges of Stop/Start-Equipped Engines. SAE-International. http://articles.sae.org/13903/. Accessed 18 Mar 2015
7. Indra, F.: Bringt der Motorsport Vorteile für die Serienentwicklung? Ist die Formel 1 auf dem richtigen Weg?. Die Evolutionäre Weiterentwicklung des Automobils, Heft II/94. Herausgeber: ÖVK, H.P. Lenz. Eurotax, Wien (1994)
8. Mezger, H.: Das 911-Triebwerk im Motorsport. In: Aichele, T. (ed.) Porsche 911, Der luftgekühlte Boxermotor. Sonderausgabe Edition Plus 2003. Erstausgabe. Heel AG, Schweiz (1997)
9. Völker, H.: Audi R10 TDI Power. Heel, Königswinter (2006)
10. Schedel, R.: McLaren Mercedes switching CAD systems on a live project. AutoTechnol. **1**(5) 74 f. Vieweg (2001)
11. Internetportal.: www.motorsport-guide.com. Accessed 30 May 2006
12. Piola, G.: Formel 1. Copress, München (2001)
13. O'Brien, J.: Return on investment, interview team Porsche. Professional Motorsport World Magazine, Heft März-Juni 2016 S. 26–30 (2016)
14. Katalog der Automobilrevue 2002. Büchler Grafino AG Bern (2002)

Vehicle Concept and Draft Design

2

The concept sets the course for the later detailed design. It is about the rough arrangement of the largest and heaviest parts and the basic characteristics of the car. Concept work should not be underestimated. Wrong decisions at the beginning of a project are often difficult to correct later. The devil, they say, is in the details, meaning that the concept is not so crucial. It must be added that the devil's ancestors were already in the concept.

2.1   Development *Process*

In motorsport, there is generally a vehicle from the previous racing season. The concept work for the following season therefore begins with an analysis of the previous model. Other influencing factors are the regulations, which are subject to constant change, and the timetable, or more precisely the date from which the new vehicle is to be available [4]. A new vehicle does not necessarily have to be available at the beginning of the new racing season. After the winter break, drivers need some racing practice again to test the limits of a car. For comparison purposes, it is therefore better to start with the familiar vehicle developed in the previous racing season and only test a modified concept later, when the drivers have reached the previous year's level [14].

For road vehicles, defining characteristics within the product description are defined according to the following order [1]:

- Vehicle class (size class, e.g. "compact class")
- Vehicle variants (e.g. notchback saloon 4-door, estate saloon 5-door)
- Aggregate assignment (motorization program, transmission range)
- Main vehicle dimensions
    - Exterior data (wheelbase, length, overhangs, width, height, track widths)
    - Interior data (length, width, height dimensions of seating systems, useful volumes)
- Technical description
    - Body design, variant concept
    - Engine versions and equipment (e.g. performance and country variants)
    - Transmission types (torque classes, automatic transmissions)
    - Chassis (axles, wheels and tyres, steering, control systems)
    - Technical equipment (e.g. air conditioning, electronic equipment, fuel system)
- Technical data
    - Weights, payloads, trailer loads
    - Driving performance
    - Consumption and exhaust gas target values.

The specifications are derived from the product description. In the next step, initial design representations are created to validate the selected dimensions. The basis for initial styling drafts is the so-called "Hard Point Package": A surface mountain is created from the required installation spaces for all necessary components and load-bearing structures as well as from the space requirements of the occupants, in which the bodywork is not yet taken into account.

2.2   Design Areas

The design ranges can be roughly subdivided for passenger cars [1]:

- Interior
- Front end
- Rear carriage
- Undertray

A new vehicle was generally developed from the inside out. In the course of concept development, interior studies with space and ergonomics investigations were at the beginning. However, the increasing role of design reversed the process. First the exterior dimensions and appearance are determined, then all the components have to be placed inside (packaging). This sequence is also common for racing cars. The reason, however, is the dominance of aerodynamics. Downforce-generating elements and the streamlined design of the bodywork dictate the outer shape alongside the regulations. Everything else must be subordinate to this.

The starting point for the dimensions of the interior is the position of the occupants in the front and rear seats.

These are placed on the seats according to favorable points of view, whereby for the front seats a seat adjustment with the extreme is to be provided. For this occupant position, the door dimensions are determined on the one hand, and on the other hand, from the eye points, the viewing angles to the front, side and rear.

When determining the fields of vision, it must be ensured that all roof posts have sufficient strength to achieve high body rigidity. Once the seating position is determined, steering wheel, instrument panel and foot lever mechanism must be fixed, taking into account the optimum direction of movement [2].

Somewhat delayed to the interior design are the elaborations in the area of the power unit (arrangements of engine, transmission, auxiliary units, front axle and steering train, consideration of safety features such as body structures, crash deformation zones).

The concept work in the undertray area focuses on transmission and powertrain, exhaust system, piping and body structure designs.

In the rear end, the focus is on layouts for body structures, rear axle, tank, exhaust system and luggage compartment optimisation. First variant investigations, e.g. for different rear versions or number of doors, are presented.

Dimension Definitions
The naming and definition of the most important dimensions of a vehicle are standardised in Europe by the ECIE (European Car Manufacturers Information Exchange Group), Figs. 2.1 and 2.2.

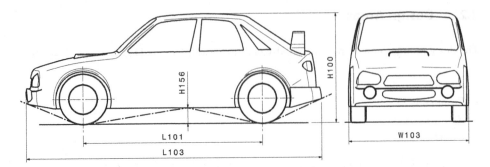

Fig. 2.1 ECIE exterior dimension definitions

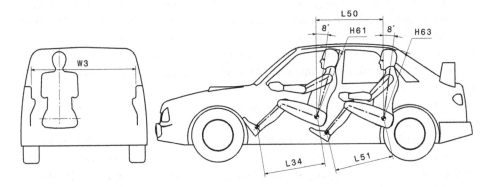

Fig. 2.2 ECIE interior dimension definition side view [1]

Table 2.1 provides some numerical values of selected dimensions of passenger cars for comparison.

2.3 Concept *Characteristic*

The following features generally shape the characteristics of a vehicle:

- Power unit position: front, rear, mid-engine, underfloor arrangement
- Drive concept: front, rear, all-wheel drive
- Unit installation: longitudinal, transverse
- Number of seats
- Comfort features: e.g. legroom
- Storage volumes.

Table 2.2 provides an overview of how the drive arrangement, i.e. the position of the engine and the orientation of the crankshaft in relation to the direction of travel, affects important characteristics of a vehicle.

Table 2.1 Comparison of dimensions of different vehicle classes, dimensions in mm [1]

Criterion	Vehicle class					
	Compact	At middle class	Middle class	Upper middle class	Upper class	Vans
Exterior dimensions						
Length (L103)[a]	3600–3800	3800–4400	4300–4700	4300–4700	4700–5100	4500–4800
Wheelbase (L101)	2350–2500	2400–2700	2500–2700	2500–2700	2700–3000	2700–3000
Width (W103)	1550–1650	1670–1740	1670–1770	1670–1770	1800–1900	1750–1900
Height (H100)	1350–1480	1330–1440	1360–1430	1360–1430	1400–1500	1650–1800
Ground clearance (H156)	130 … 150; SUV: Over 200					
Interior dimensions						
Front footwell (L34)	960–1080	970–1080	1000–1100	1000–1100	1000–1100	970–1080
Head space front (H61)	920–1000	940–1010	950–1010	950–1010	980–1020	1000–1050
Shoulder width front (W3)	1280–1360	1340–1440	1340–1460	1340–1460	1450–1500	1500–1650
Seat spacing front-rear (L50)	680–760	670–790	730–830	730–830	840–950	850–900
Footwell rear (L51)	730–920	760–880	750–920	750–920	900–1000	800–900
Rear head space (H63)	900–970	900–980	910–980	910–980	950–990	950–1000
Boot capacity	200–460	240–550	330–550	330–550	500–600	250–2500
Vehicle examples	VW Polo	VW Golf	Audi A4	BMW 5 series	Mercedes S	VW Sharan

[a]In (…) Dimension designation according to ECIE agreements

Table 2.2 Influence of the drive arrangement on the properties of a vehicle [40]

Motor location	Front, longitudinal			Front, crosswise		At the back		Middle	
Drive axle	Front	At the back	Both	Front	Both	At the back	Both	At the back	Both
Empty weight	+	0	−	++	−	+	−	+	−
Axle load distribution	+	++	++	+	++	+	++	++	++
Traction, dry	+	++	++	+	++	++	++	++	++
Traction, smooth	+	−	++	+	++	+	++	0	++
Traction, corner	++	0	++	++	++	+	++	+	++
Traction, gradient	0	+	++	0	++	+	++	+	++
Steering comfort	0	++	+	0	+	++	+	++	+
Straight ahead	++	+	++	++	++	−	0	0	+
Crosswind stability	++	+	++	++	+	−	0	−	0
Steering response	+	+	++	+	++	+	++	+	++
Braking behaviour	+	++	++	+	++	++	++	+	++
Interior use	+	+	0	++	+	−	−	−−	−−

Legend: ++ very good, + good, 0 medium, − poor, −− very poor

2.3.1 *Engine* Position and Drive Layout

The position of the engine and thus of the drive train has a significant influence on the layout of the installation space and the driving behaviour due to the axle load distribution and the position of the driven axle. Of the conceivable possibilities, however, only a few make sense, Fig. 2.3.[1]

Front Engine Layout
Features. The engine and transmission are installed in a block in front of the passenger compartment (longitudinally aligned (*north-south*) or transversely installed (*east-west*)). Water cooler and air-conditioning condensers are placed in front in the front area of the vehicle. Front-, rear- or all-wheel-drive versions. This engine arrangement is the most widespread in the passenger car market.

---

[1] Regarding the all-wheel drive, some basic considerations are made in thef Racing Car Technology Manual, Vol. 3 Powertrain.

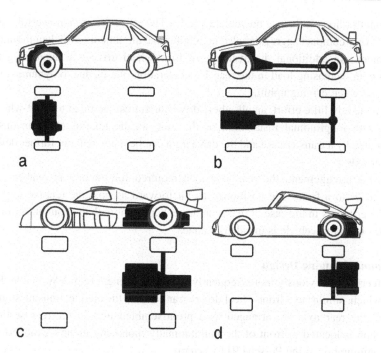

Fig. 2.3 Possible motor positions in the vehicle. (**a**) Front motor, transverse installation, front wheel drive, (**b**) Front longitudinal motor, rear drive (standard drive), (**c**) Middle motor, rear drive, (**d**) Rear motor, rear drive

Transverse installation is only used in front-wheel drive vehicles – in the case of rear-wheel drive, only longitudinal installation is used due to the simpler drive train and vibration-related advantages.

The transaxle principle (engine at the front, gearbox and drive at the rear) combines the advantages of front and rear engine.

Advantages. Compact design with short lines to all auxiliary units and to the coolers. The aggregate noises can be well partitioned off from the interior by the bulkhead. In the event of a frontal crash, early contact of the drive block with the bulkhead area leads to relief of the bodyshell structure from unit mass forces. Sufficient space is available for the exhaust system (especially silencers and catalytic converters) and fuel tank in the undertray and rear end areas.

When combined with front-wheel drive, the entire drive unit with front axle can be implemented as a compact pre-assembly unit and, in addition to a flat vehicle tunnel, enables a sufficiently high front axle load for good traction conditions. The advantage of rear-wheel drive over front-wheel drive lies in the increasing traction potential with increasing load in the rear, in acceleration phases or when driving uphill.

Disadvantages. In transverse installation, the engine size is limited to a maximum of 6 cylinders, and there are also severe restrictions in the gearbox size due to the limited overall length (direct influence on vehicle width). The effort for more powerful, but at the same time very compact aggregates also in longitudinal alignment (crash length) underlines this problem. An additional disadvantage of front-wheel drive is a decreasing traction potential with increasing load in the rear, in acceleration phases due to dynamic axle load shifting and when driving uphill.

With relatively little effort, an all-wheel-drive variant can be fitted to front-wheel-drive vehicles with longitudinal installation of the unit, as the longitudinal transmission is supplemented by a transfer case and the drive train can be made without further deflections to the rear axle.

With this arrangement, the rear axle load required for traction in rear-wheel drive vehicles leads to a front axle positioned as far forward as possible relative to the drive block. Nevertheless in the case of rear-axle drive, hardly more than about 50% of the rear-axle load (when the vehicle is unladen) realize.

Rear-Mounted Engine Design

Characteristics. Previously more frequently used arrangement (e.g. VW Beetle, Renault, Fiat) in which, similar to a front-wheel drive arrangement, the engine, transmission and, in this case, the rear axle are arranged as a pre-assembled unit in the rear section. The transmission is located in front of the longitudinally mounted engine. A modern vehicle in this configuration is the Porsche 911 Carrera.

Advantages. Due to the arrangement of the power unit behind the rear axle, very high rear axle load share (>60%), resulting in excellent traction characteristics, increasingly when accelerating and driving uphill, still very high when loaded in the vehicle interior or at the front. No heat load on the interior due to heat radiation from the power unit. Flat vehicle tunnel (no drive shafts or exhaust routing). With longitudinal installation of the power unit, all-wheel drive to the front axle can be easily implemented.

Disadvantages. The high rear axle load requires high-quality axle concepts (especially rear axle) to achieve good driving characteristics. Long lines result with water cooling with front-mounted radiators for the cooling itself as well as for heating and air conditioning. The body variability in the rear area is very limited by the space requirements of the power unit. In terms of space, competitive station wagons are not possible. Difficulty in designing an optimum exhaust system. Tendency to overdrive. Sensitivity to crosswinds.

Mid Engine Design

Features. Classic sports car configuration with engine arrangement in front of the rear axle. The alignment of the power unit is common both longitudinally (transmission behind engine) and transversely (analogous to front-wheel drive transverse installation). Due to the space requirements of the engine-transmission block, only a two-seat version makes sense. Monoposto racing cars (e.g. Formula 1, Formula 3, Formula Renault) are today exclusively designed with a mid-engine arrangement. The following design has become generally accepted for formula and production sports cars: The engine is co-supported by being bolted directly to the bulkhead behind the cockpit. The engine in turn carries the clutch housing to which the gearbox is bolted. The rear axle suspension is attached directly to the gearbox housing and sometimes also to the engine block.

Advantages. Due to the arrangement of the power unit in front of the rear axle, relatively high rear axle load share (>52%), resulting in very good traction characteristics, increasing when accelerating and driving uphill, remaining neutrally high when loaded. Vehicle concept with optimum driving dynamics potential due to balanced axle load distribution. The heat load of the interior by heat radiation of the aggregate is only small. A front luggage compartment is usually feasible. An additional luggage compartment in the rear is possible, the vehicle tunnel is flat (no drive shafts or exhaust gas routing).

Disadvantages. Front-mounted radiators with water cooling require long lines for cooling, heating and air conditioning. The body variability in the rear and interior area is strongly limited by the space requirement of the aggregate. Therefore, almost exclusively 2-seater vehicles are common (4-seaters lead to a very large wheelbase). The gearbox, which is placed behind the engine when installed longitudinally, requires long gearshift cables and precludes all-wheel drive, which is very complex even with a transverse installation of the power unit. Engine maintenance is also difficult.

A vehicle with this body is, for example, the Porsche Boxster.

Engine Compartment

The following considerations may be helpful in locating the engine. First, one will want good accessibility to high-maintenance areas. In addition, an inflow and outflow of cooling air is "vital" for the engine and peripheral parts and must therefore be ensured at all costs. In addition, it is important not to forget to plan for the location of the exhaust system with any necessary exhaust gas reactors. An engine as a supporting element facilitates engine removal and installation. However, the engine must be suitable for this requirement. In Formula 1 engines, for example, there is a loss of power of up to 50 kW during acceleration due to the deformation of the crankcase under the load [8].

Fuel Tank

The arrangement of the tank in passenger cars is determined by the necessary crash protection measures. A central position is preferred for racing vehicles. This means that the fill level has little influence on the driving behaviour. In general, the tank of a single-seater car is designed to be as short as possible without raising the centre of gravity when the tank is full. This leaves room in the length to move the engine and transmission for the desired axle load distribution. However, there are limits to this quest for a short tank. For a given fill volume at a given height, the tank becomes wider. The FIA regulations for Formula 1 cars, for example, limit the width to 800 mm.

2.3.2 Concept Comparison

Table 2.3 provides a rough comparison of different vehicle concepts according to selected criteria.

Table 2.3 Comparison of vehicle concepts [1]

Criterion	Front engine Front-wheel drive	Front engine Rear-wheel drive	Mid-engine Rear-wheel drive	Rear engine Rear-wheel drive
Traction capacity empty[a]	+	−	+	++
Traction capacity loaded[a]	−	+	+	+
Axle concept requirements	+	0	−	−
Interior size	++	++	−−	0
Trunk size	++	+	0	−
Rear body variability	++	++	−	−
Vehicle length requirement[b]	++	0	− (0)	0
Body structure loading in a frontal crash	++	++	−	−
Heat influence on interior	−	−	0	+
Noise influence on the interior	+	+	0	+
Suitability for all-wheel drive[c]	+ (++)	−	−	++
Total weight	++	0	+	+
Cable lengths	++	++	−	−
Production costs	++	+	+	+

Legend: ++ very good or very suitable, + good, 0 medium, − poor, −− very poor or unsuitable.
[a]For exact evaluation consideration of gradient, coefficient of friction, load conditions required
[b](0) for mid-engine and engine installation transverse instead of longitudinal
[c](++) for front engine and engine installation lengthwise instead of crosswise

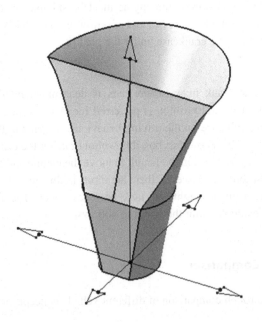

Driving *Behaviour and Driving Performance*
The following variables have a considerable influence on the driving behaviour [9] and should therefore be included in the concept considerations.

Mass

As a variable that determines inertia, mass has a direct influence on handling and performance. Rolling, acceleration and gradient resistance increase with increasing mass. The load on the tires due to higher lateral forces also increases. If the influence of tires remains constant, each kilogram of mass improves lap time by about 0.04 s for Formula 1 cars [11]. It is also calculated that an additional power of about 4–5 kW is required for each kg of additional weight [15].

Reducing the weight of the vehicle also makes sense if the minimum weight required by the regulations is not reached, because it is then possible to use ballast weights to distribute the mass of the vehicle in the direction of the optimum. Even movable masses are conceivable, which adapt the mass distribution for the respective driving situation (acceleration, cornering etc.). The Tyrell P34 (Formula 1, until 1977) had a movable fire extinguisher on board, the position of which could be influenced by the driver [17].

Low mass moments of inertia, i.e. the arrangement of all masses close to the centre of gravity, reduce the forces required to change the direction of the vehicle and thus improve its agility.

Furthermore, a distinction is made between sprung (wagon body, superstructure) and unsprung (wheel plus parts suspended from it). Physically, low unsprung masses mean small wheel load fluctuations and should therefore be aimed for. Due to the non-linear behaviour of the tyre rubber when building up and transmitting forces, a large wheel load fluctuation causes losses of lateral forces and longitudinal forces. On undulating road surfaces, the drivable lateral acceleration is thus smaller and the braking distance longer, whereby the effect is clearly more pronounced in the lateral direction [48]. Conversely, however, it must also be noted that this effect does not come into play on relatively even road surfaces because the wheel load changes caused by the road surface remain small. Numerous studies in the past [49] and present [50, 51] confirm the overestimated influence of the unsprung masses on level road surfaces. Additional masses were added to the front wheels and the drivers could not notice any difference in the grip level. The reason for this, apart from the low level of road excitation, is that tyres themselves also represent a spring (together with a damper). With off-road vehicles, which are exposed to large bumps, it is very important to keep the so-called unsprung masses small.

Centre of Gravity Position, Type of Drive, Wheelbase and Track Width

These variables have a decisive influence on driving stability.

Centre of Gravity Height

The centre of gravity height should be kept as low as possible. A low centre of gravity keeps the axle load shifting small during acceleration and braking and thus reduces, among

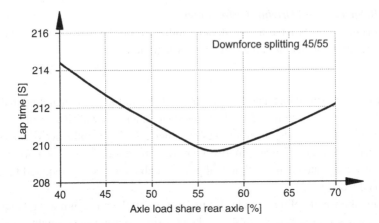

Fig. 2.4 Influence of the centre of gravity position on the lap time at Le Mans [22]. This simulation result shows how the axle load distribution of a rear-wheel drive vehicle with front 33/65–18 and rear 37/71–18 tires affects the lap time. With an aerodynamic balance from front to rear of 45/55%, the optimum results at 57% axle load share of the rear axle during the drive

other things, the effort required to distribute braking force front to rear. An extremely low centre of gravity also helps to keep the quality of the lateral force distribution high. A low centre of gravity also increases driving stability when braking in bends.[2] The centre of gravity height and also the centre of gravity distances to the axles generally change with the load.

Axle Load The centre of gravity distances and in connection with this the axle loads have an effect on the steering tendency. As the front axle load increases, the understeer tendency is encouraged. Since in most road vehicles the rear axle load increases more than the front axle load as the load increases, fully loaded is also the most critical loading condition in terms of driving stability in most cases.

Although Fig. 2.4 only shows the effect of shifting the centre of gravity in the longitudinal direction for a specific vehicle with given tyres on a specific track, the basic result is always the same: If the rear axle load is reduced from the optimum, the traction also decreases and the lap time deteriorates due to the lower acceleration. If you increase the axle load on the drive wheels, traction improves, but at the same time the tendency to oversteer increases due to the greater rear load. This goes so far that the lap time moves away from the best value again. So there is an optimum and this lies between the extremes of high front axle load and high rear axle load.

---

[2] See also Racing Car Technology Manual vol. 4 Chassis, Chap. 6 *Brake system.*

Table 2.4 Average axle load distribution of passenger cars, according to [31]

	Front wheel drive		Standard drive		Rear engine	
Loading	% in front	% rear	% in front	% rear	% in front	% rear
2 persons in front	60	40	50	50	42	58
4 persons	55	45	47	53	40	60

Fig. 2.5 Calculation of centre of gravity distances. The distance x of the total centre of gravity of two individual masses m_1 and m_2 results from the distances of the individual centres of gravity and the masses

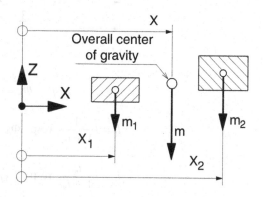

The average axle load distributions of passenger cars depending on the type of drive can be used as a reference value, Table 2.4.

The Formula 1 skirted cars with ground effect had about 45/55% front/rear distribution. The driver sat considerably further forward than now. After that, ratios of around 40/60 were chosen. With the wider front tyres from the supplier Michelin, the centre of gravity was shifted further forward again. Some teams drove with an axle load distribution of up to 46/54% [33].

The coordinates of the total centre of gravity of any number of individual masses m_i result from the distances of the individual centres of gravity from the origin of the coordinates, Fig. 2.5.

For example, in the *X direction:*

$$x = \frac{x_1 \cdot m_1 + x_2 \cdot m_2 + \ldots}{m_1 + m_2 + \ldots} \tag{2.1}$$

x, x_1, x_2	Distances of the centres of gravity in X-direction, mm
m, m_1, m_2	Masses, kg
	m total mass, so the following applies: $m = m_1 + m_2 + \ldots$

In Z-direction this equation applies analogously with the Z-distances of the centers of gravity.

In general, the axle loads follow from the moment equilibrium around the vehicle's centre of gravity V and it applies:

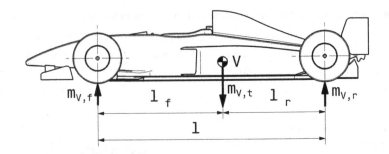

$$\frac{l_r}{l_f} = \frac{m_{V,f}}{m_{V,r}} = i_m \tag{2.2}$$

$$l_f = l\frac{1}{1+i_m} \quad \text{resp.} \quad \Phi_{A,r} = 100\%\frac{1}{1+i_m} \tag{2.3}$$

$$l_r = l\frac{i_m}{1+i_m} \quad \text{resp.} \quad \Phi_{A,f} = 100\%\frac{i_m}{1+i_m} \tag{2.4}$$

$$l = l_f + l_r \tag{2.5}$$

$$m_{V,t} = m_{V,f} + m_{V,r} \tag{2.6}$$

$l_f,\ l_r$	Centre of gravity distances front or rear, mm
$m_{V,f}$ or $m_{V,r}$	Axle loads front or rear, kg
i_m	Axle load ratio front/rear, –
$\Phi_{A,f}$ or $\Phi_{A,r}$	Percentage of front or rear axle load, %.
$m_{V,t}$	Total weight of the vehicle, kg

The static axle load distribution of a vehicle can be changed in various ways:

• Move an axle forward or backward.
• Shifting the overall centre of gravity, for example by changing the position of a ballast mass.
• Combination of these measures.

In addition, in the dynamic case, there are all the aerodynamic measures that influence the wheel forces in the vertical direction (downforce, lift), in addition to displacement effects of the axle loads due to inertia, more precisely due to the height of the overall centre of gravity of the vehicle.

Figure 2.6 illustrates the effect of various changes on the static axle load distribution on a Formula 1 car. It can be seen from this, among other things, that it is favourable if the

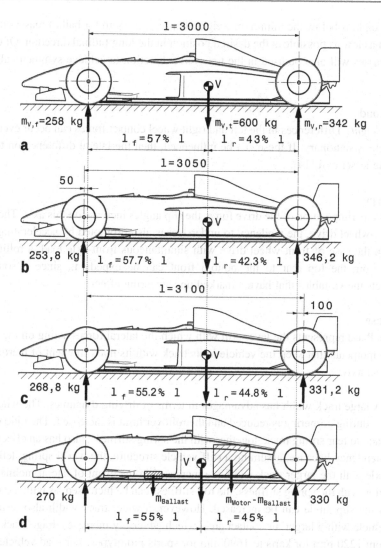

Fig. 2.6 Changes to the axle load distribution on a Formula 1 car. (**a**) **a Initial situation**. The total mass $m_{V,t}$ is 600 kg. The axle loads at the front $m_{V,f}$ and at the rear $m_{V,r}$ result from the position of the vehicle centre of gravity V. The wheelbase l is 3000 mm. The centre of gravity distances $l_f = 1710$ and $l_r = 1290$ mm provide an axle load ratio front/rear of $i_m = 0.754$; i.e. of 43/57% axle load distribution. (**b**) **Front axle** shifted forward by 50 mm. The wheelbase l changes accordingly to 3050 mm. (**c**) **Rear axle shifted** 100 mm to the rear. This can be achieved, for example, by a longer clutch housing or by an intermediate piece between the engine and the gearbox. (**d**) The **engine** has been lightened by 30 kg and the mass thus saved is arranged as ballast mass at the front (as low as possible). The distance to the original vehicle centre of gravity is chosen to be the same as it was before to the rear. This results in a new position of the total centre of gravity V'. In this case it moves to the front. The wheelbase remains unchanged at 3000 mm. The new centre of gravity distances l_f and l_r result in a new axle load distribution

vehicle weight is below the minimum weight. This leaves room for ballast masses that can be placed as low as possible at the desired position in the longitudinal direction. Of course, ballast masses will also be used in the transverse direction to achieve symmetrical wheel loads.

Wheel Load
(*corner weight*). Differences in the left and right wheel contact forces can occur even when the vehicle is stationary. If these can be influenced, then the lateral difference on the rear axle is the lesser evil [14].

Drive Type
With the superimposition of the drive force, the slip angles increase on this axle. Therefore, with front-wheel drive, the tendency to understeer is also promoted by accelerating in the corner. In the case of front-wheel drive with simultaneous front-axle load, a rolling axle inclined from the top rear to the bottom front can be expedient, since it somewhat counteracts the variables that have a marked understeering effect.

Wheelbase
The wheelbase represents the lever arm with which the lateral forces acting on the wheels form the moments that keep the vehicle in its track with its mass moment of inertia about the vertical axis.

Track A large track width has advantages in terms of driving dynamics. The wheel load difference during cornering is reduced and the rollover limit is increased. The ratio of front spring track to rear spring track, together with the spring stiffness, also has an effect on the steering tendency because, assuming a stiff vehicle structure, the wheel spring deflections on the axle with the larger track width are increased in the event of lateral inclination. In conjunction with the spring stiffness, the roll stiffness can be increased, as with a stabilizer, and thus the slip angle can be increased. However, a wider track width also results in a wider vehicle with a larger air contact area, which adversely increases drag. Track widths range from 1220 mm for karts to 1690 mm for sports prototypes. For road vehicles, track widths range from 1210 to 1600 mm [31].

The wheelbase to track width ratio is usually between 1.4 and 1.7:

$$\frac{\text{wheelbase}}{\text{track width}} = 1.4...1.7$$

The range of this ratio is from 1 for karts to 2.5 for historic racing cars [32]. The value 1.62 has the aesthetic advantage of the golden section [16].

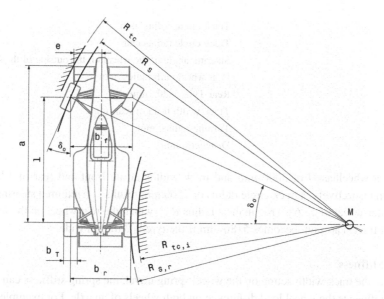

Fig. 2.7 Determining the turning circle. For low driving speeds, i.e. very small slip angles, the pole of the car M is on the rear axle. With increasing speed, the instantaneous pole moves towards the front axle

For most racing cars, the minimum wheelbase results from the requirement that the driver's feet are behind the front axle, see also Fig. 2.15. The values range from 1040 to 1220 (kart) to 3150 mm (Formula 1). For road vehicles, the wheelbases are between 2160 and 3040 mm [31].

Roughly, the following assessment applies. Long, slender vehicles have a small air attack area, exhibit great stability at high speeds and are insensitive to disturbances.

Short, wide cars, on the other hand, are slower in a straight line, behave more nervously, and are thus more agile on tight, twisty courses.

The wheelbase and track width also provide an initial estimate of the turning circle diameter as a function of the wheel steering angle. With this, a check can be made with known corner radii of race tracks whether the selected values are reasonable, Fig. 2.7.

The track radii result from the dimensions as follows:

$$\delta_o = \arcsin \frac{l}{R_S} \tag{2.7}$$

$$R_{tc,i} = \sqrt{R_S^2 - l^2} - 0.5(b_f + b_r + b_T) \tag{2.8}$$

$$R_{tc} = \sqrt{a^2 + \left(R_S \cos \delta_o + e - \frac{b_f}{2} \right)^2} \tag{2.9}$$

R_S	Track circle radius, mm
$R_{S,r}$	Track circle radius rear
δ_o	Steering angle of the wheel on the outside of the bend, °
$R_{tc,i}$	Rear wheel curb radius, mm
b_T	Rear Tyre width, mm
b_f, b_r	Track width front or rear, mm
R_{tc}	Turning radius, mm
a, e	Distances, mm

With a wheelbase l of 3000 mm and track widths at the front and rear of 1490 and 1540 mm respectively, a track circle radius of 7500 mm results in a required steering angle at the outer wheel of 23.6°. The smallest radius $R_{tc,i}$, which the rear wheel on the inside of the bend travels around, becomes 5186 mm if the tyre is 346 mm wide.

Spring Stiffness
Similar to the track width acting on the wheel spring travel, the spring stiffness can also be used to influence the wheel load difference on both wheels of an axle. For example, stiffer springs or, in the case of rigid axles, wider spring track, the use of a stabilizer or additional springs increase the wheel load difference and the slip angle on an axle. Softer acting springs or, for example, composite springs and compensating springs reduce the wheel load difference.

Driving Stability
Increasing the wheelbase promotes driving stability. In connection with the importance of rear axle tracking, it is important that the rear axle in particular is moved as far back as possible. Conversely, in all vehicles where there are still critical loading conditions in terms of driving stability, these occur when the load accumulates in the rear.

Moment of Inertia
With regard to the moment of inertia about the longitudinal axis of the vehicle: a small value means that dynamically smaller wheel load differences occur, the springs can be designed to be softer and the transverse inclination of the vehicle adapts more quickly to the transverse inclination of the road.

A difference in track widths, like all other measures that cause a difference in the rolling stiffness of the two axles, affects the slip angles.

Similarly, a low moment of inertia $J_{V,Z}$ about the vertical axis increases the agility of the vehicle. More precisely, the ratio l^2/i_z^2 should be large [41].[3] Thus, the mass of the car should be placed within the wheelbase l and as close as possible to the center of gravity of the vehicle.

---

[3] $i_z = \sqrt{J_{V,Z}/m_{V,t}}$ is the radius of inertia of the car about the z-axis. See also Racing Car Technology Manual Vol. 5 *Data Analysis, Tuning and Development*, Chap. 5 *Tuning*.

Elastokinematics of the Axles
This includes all variables that influence the position of the wheel plane in relation to the vehicle and the road. The camber, toe-in and caster changes are applied via the spring travel. Since the outer wheel compresses and the inner wheel decompresses when the vehicle rolls as a result of cornering, the roll control effect can be achieved via the toe-in and camber change. More elegant solutions are those using elastic deformation in the presence of lateral forces, as they do not interfere with straight-line driving on uneven road surfaces in the same way as wheel suspension kinematics. In racing vehicles, which are naturally operated close to the optimum of slip angles, elastokinematics proves to be ineffective or undesirable [41].

Table 2.5 compares some concept parameters.

A direct comparison of the individual aspects is not possible, because many dependencies are non-linear, the weightings of the individual influences are different and further, not shown (also mutual) dependencies must be considered.

Table 2.6 shows the basic influences of important parameters on the driving behaviour. The statements are based on the calculation results of a simple single-track vehicle model, but also apply to two-track vehicles.

For comparison and to classify the size ratios, some numerical values of executed vehicles, Tables 2.7 or 2.8 and Fig. 2.8.

l wheelbase, b_f track width front track, b_r track width *rear* track, l_f, l_r centre of gravity distances, V centre of gravity, L_t length over *all*, B_t width over all, H_t total height, h_V *centre of gravity height.*

Tilt Limit
The overturning limit of a rigid, unsprung vehicle is reached when the contact forces of the wheels on the inside of the corner become zero. Then the following applies, see also Fig. 2.9:

$$\frac{F_{V,Z,t}}{2} = \frac{F_{V,Y} \cdot h_V}{b} \quad \text{with} \quad F_{V,Y} = \frac{m_{V,t} \cdot v_V^2}{R} \qquad (2.10)$$

$F_{V,Z,t}$	Vehicle weight, N
$F_{V,Y}$	Lateral force at vehicle Centre of gravity, N
$m_{V,t}$	Vehicle mass, kg
v_V	Vehicle speed, m/s
R	Path radius, m

Table 2.5 Design of important passenger car concept parameters with regard to optimisation of active safety [2]

Property	Feature	Main dimensions				Mass distribution			Suspension			Body	
		Wheelbase	Track	Vehicle length	Vehicle width	Vehicle weight	Weight share front axle	Moment of inertia yaw axis	Position of drive axle	Tyre size	Springing (vertical, longitudinal, transverse)	Driver height above ground	Window area size
Driving and braking characteristics	Stability	Large	Large	–	(small)	Small	Large	Large	(FA)	Large	Hard	Low	–
	Dynamic steerability	(large)	Large	Small	–	Small	(large)	(small)	(HA)	Large	Hard	(low)	–
	Manoeuvrability	Small	Small	Small	(small)	(small)	Small	Small	(HA)	Small	–	–	–
	Traction	(FA: Large) (HA: Small)	–	–	–	–	FA: Large HA: Small	–	–	Large	–	–	–
	Braking behaviour	Large	Large	–	–	Small	Small	(large)	–	Large	Hard	(low)	–
	Behaviour in the event of wind disturbances	Large	(large)	–	Small	Large	Large	(large)	–	(large)	(hard)	Low	–
	Behaviour in the event of lane incidents	Large	Large	–	–	Large	–	(large)	–	Large	Soft	Low	–
Visual perceptual reliability		–	–	(small)	(small)	–	–	–	–	–	–	High	Large
Condition safety	Suspension behaviour	Large	Large	–	–	Large	–	–	–	(large)	Soft	Low	–
	Vibrations, noises	–	–	–	–	(large)	–	–	–	(large)	Soft	–	Small
	Interior heating (sun)	–	–	(small)	(small)	–	–	–	–	–	–	–	Small

Explanations: – clear statement not possible
( ) low significance
FA front-wheel drive
HA rear-wheel drive

Table 2.6 Basic influences of the vehicle design on the driving behaviour [6]

Influential feature		Impact		
		Coefficients of lateral stiffness with respect to the slip angle α		
		$c_{\alpha,f}$	$c_{\alpha,r}$	c_M
Wider tyres		+	+	
$l_f < l_r$ [a] Greater front axle load		+	−	
Harder torsion stabilizer in the front, harder body suspension in the front		−	+	
Low tire pressure		−	−	
Elasticities, kinematics of the wheel suspension depending on the design		±	±	
Body	More boxy, tail fins			−
	More drop-shaped, mostly also smaller c_w -value			+
Rear-wheel drive[b]			−	
Front wheel drive[b]		−		

Design target for (stable) understeering behaviour:
$(c_{\alpha,r} \, l_r - c_{\alpha,f} \, l_f) > 0$ and if possible $c_M < 0$.
Lateral forces acting on the front and rear tyres due to skewing:
$F_{W,\,Y,\,f} = c_{\alpha,\,f} \cdot \alpha_f \, F_{W,\,Y,\,r} = c_{\alpha,\,r} \cdot \alpha_r$.
[a]Since the influence of the axle load is incorporated non-linearly in $c_{\alpha,f}, c_{\alpha,r}$, the term $(c_{\alpha,r} \, l_r - c_{\alpha,f} \, l_f)$ nevertheless becomes larger if the centre of gravity is shifted forwards

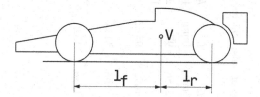

[b]The influence of the driving force has the effect of reducing the lateral force of the tyres while at the same time transmitting circumferential forces, especially on slippery road surfaces
Air moment acting as a yaw moment on the car: $M_L = c_M \cdot v_{rsl}^2 \cdot \tau_L$.
with: v_{rsl} resulting incident flow velocity (= wind speed − vehicle speed) τ_L Angle of airflow in relation to the direction of travel.

In order for a vehicle to actually tip, the corresponding lateral force must be applied by the tires. A coefficient of friction μ_{Kipp} is therefore applied at the tipping point:

$$\mu_{tilt} = \frac{b_f + b_r}{4h_V} \tag{2.11}$$

Table 2.7 Numerical values of concept parameters of some vehicles

Vehicle	L_t [mm]	H_t [mm]	l [mm]	l_f [mm]	l_r [mm]	h_v [mm]	b_f [mm]	b_r [mm]	l/b_m [–]	Axle load distribution front/rear	Turning circle $\varnothing$ [m]
Formula 1 [11]	4600	1100	3080	2156	924	260 [8]	1490	1405	2.13	30/70[a] to 45/55[b]	
Formula Renault [12]	4065	954	2645	1692.8	952.2	340	1471	1366	1.86	36/64	
Apollo super sports car	4460	1240	2700	1566	1134		1670	1598	1.65	42/58	
BMW 1 series	4227	1430	2660	1330	1330		1483	1497	1.79	50/50	10.7

Designations of the dimensions: see Fig. 2.8.

[a] $b_m = (b_f + b_r)/2$

[b] [21]

Table 2.8 Structural parameters of some road vehicles [10]

Vehicle	Drive	Payload	$m_{V,t}$ [kg]	Axle load f/r [%]	l_f [m]	l_r [m]	l [m]	h_V [m]	b_f [m]	b_r [m]	μ_{Kipp} [-]	$J_{V,z}$ [kgm²]	$J*$	$J_{V,y}$	$J_{V,x}$	$J**$
Small car	Front	2 persons	825	60/40	0.893	1.332	2.225	0.552	1.280	1.295	1.17	984	981	779	194	342
		4 pers. + 60 kg	1091	49/51	1.142	1.083		0.553			1.21	1275	1349	1021	236	452
Lower MK	Front	2 persons	1059	57/43	1.065	1.385	2.45	0.572	1.260	1.300	1.12	1652	1561	1270	240	434
		5 pers. + 60 kg	1315	48/52	1.270	1.180		0.595			1.08	2058	1971	1484	279	539
MK	Standard	2 persons	1171	55/45	1.134	1.371	2.505	0.537	1.298	1.275	1.20	1855	1821	1383	255	485
		5 pers. + 60 kg	1460	47/53	1.340	1.165		0.529			1.22	2240	2280	1635	300	604

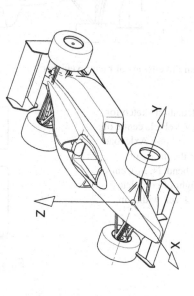

MK = Middle class

$J*$ is used for comparison with $J_{V,z}$: $J* = m_{V,f} \, l_f^2 + m_{V,r} \, l_r^2$. $J*$ corresponds to 2 masses exactly above the axes

$J**$ is used for comparison with $J_{V,x}$: $J** = m_{V,t} \, b^2/4$. $J**$ corresponds to 2 × 0.5·vehicle mass at distance 0.5·lane width

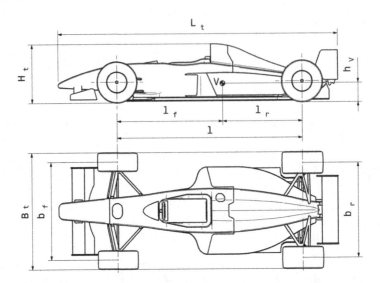

Fig. 2.8 Main dimensions of racing vehicles

Fig. 2.9 Calculation sketch for tipping limit. V Vehicle centre of gravity, $F_{V,Z,i}$ or $F_{V,Z,o}$ Axle load proportion on the inside or outside of the bend, N. h_V Centre of gravity height of complete vehicle, m. b Track width, m

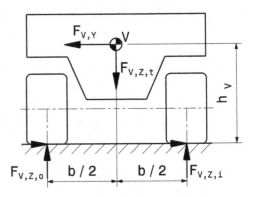

μ_{tilt}	Friction coefficient required for tilting, –
b_f or b_r	Track width front or rear, m

A sub-agency of the US Department of Transportation, NHTSA (National Highway Transportation Safety Administration), requires a value between 1.3 and 1.5 for passenger cars, for example. In most vehicles, however, it does not even get that far that it tips over because the tires have already reached their grip limit and have begun to slide. The coefficient of friction that can be transmitted by the tyres in this case is: $\mu_{W,Y} \leq v_V^2/(R \cdot g)$. From this follows the achievable maximum cornering speed without downforce to $v_{V,max} = \sqrt{\mu_{W,Y} \cdot R \cdot g}$. For more details see Sect. 5.3 Downforce.

On a level road surface, the practical tipping point is reached by elastic displacement of the wheel contact point with lower friction values. Aerodynamic downforce has a stabilizing effect, i.e. it shifts the tip-over limit to higher speeds because it increases the contact forces of the wheels without increasing the vehicle mass.

Table 2.9 provides a summary of the design options for influencing the driving behaviour of a vehicle.

Share of Individual Assemblies in Driving Performance
While it is difficult to determine the importance of individual assemblies in isolation from others (and completely disregard the driver), expert surveys of the contributions of individual assemblies to driving behavior or performance yield the following simple picture for racing vehicles [4]:

Tires.........40–50%
Chassis...30–40%
Motor........20%

The tyres as the connecting element between the vehicle and the road therefore have the predominant influence. The chassis, as the subsystem that determines the position of the tyres in relation to the road, which is important for power transmission, follows as the second most important variable. The influence of the engine is not as important as generally assumed. A high engine output, however, makes a high downforce possible in the first place. This is the real advantage of an "overpowered" vehicle. The possible top speed on the (relatively) short straights of common race tracks is achieved by such cars even with lower engine power.

Above this ranking, the aerodynamic aids (wings, undertray, wedge shape, . . .) must be seen as the most important components, because these multiply the effect of individual assemblies at high driving speeds. The frame or chassis system is responsible for the desired air forces. In such an overall view, which is more realistic for very fast vehicles, the distribution looks as follows [26]:

Chassis...50%
Tires.....35%
Engine.....15%

Typical concept decisions, which often force the weighing of conflicting demands, could best be made on the existing vehicle on real tracks. Such crucial questions could be: Should the center of gravity height be reduced or would it be better to raise the car and use the improved airflow to increase aerodynamic downforce? Should the track width be widened and the known disadvantages such as increased mass, increased air resistance as well as longer driving line in slalom be accepted?

Table 2.9 Design influences on the driving behaviour, according to [40]

System	Concept or design parameters
Complete vehicle	Wheelbase, track width, axle load distribution, Centre of gravity position, mass moments of inertia of the wagon body Drive concept and drive torque distribution (all-wheel drive) Aerodynamic properties (especially in the high-speed range)
Tyres	Dimensioning of wheel and Tyre Design of the tyres and possibly tread design Cornering stiffness
Axles	Axle concept and design Design and tuning of suspension, stabilisation (ARB), damping and additional springs Kinematics of the axles (longitudinal and lateral power steering) Longitudinal and lateral elasticities of front and rear axle Pitch kinematics of the axles (diagonal springing: Anti-squat, brake pitch compensation: Anti-dive) Roll kinematics of the axles (position of roll axis, roll support by suspension, additional springs, stabilization and damping) Distribution of roll support between front and rear axle
Wheel carrier	Caster angle, distance, offset, king-pin inclination, toe, etc. Wheel lift curves and change of wheel position during springing
Steering	Steering system concept and design, steering gear design Static and dynamic steering ratio Steering characteristics (steering moments, ratio) Elasticities, moments of inertia and damping in the steering train Interference lever arm, steering roll radius, arrangement of track rods (arrow) Design of kinematic steering return (caster, king-pin inclination) If applicable, the design and steering characteristics of a steering assistance system
Brakes	Brake concept, dimensioning and design Brake force distribution Brake pad characteristics
Powertrain	Arrangement of the aggregates and their mounting or fastening Elasticities and damping of the drive train Engine characteristics (torque curve, drag torque curve) Gear ratio Length and torsional stiffness of the drive shafts Locking characteristics of the differentials Characteristics of the drive control
Control systems	Launch control Traction control (traction control) Gearbox control Anti-lock braking system (ABS) Networking of control systems

In the case of further development of existing vehicles or vehicle types, comparative test drives are actually possible with the corresponding time and financial effort. In the literature, numerous principle tests are reported in order to clarify the question of the influence of the unsprung masses (see section on *mass*). A Formula 1 team investigated whether, in case of doubt, the height of the centre of gravity or the downforce should be given preference. Lead weights were attached to the roll bar and driven for so long that the degradation of the tyres was also included in the investigations. The series of measurements were carried out on three different tracks and produced the same result for all of them, namely that downforce plays the dominant role in lap time [52]. This explains why compromises are generally made in favour of aerodynamics in Formula 1 cars.

Simulations with realistic models of the vehicle, tires and track are an enormous help in this context. This makes it much easier and faster to carry out such influence analyses, even if no real prototype exists yet or the track may not be driven on (city circuits such as Monaco or Long Beach, tracks on public roads such as Le Mans).

A statistical view can also help to classify the importance of assemblies. Over a season, the operation of formula cars on permanent racetracks looks like this: 45–55% of the driving time the car is in a corner (characteristic: high yaw rates or yaw accelerations), 35–40% of the time the engine delivers full load torque (acceleration phases) and 10–15% of the time the brakes are in use to decelerate the vehicle [42]. During braking and cornering, the front tires are put under more stress than the tires on the rear axle. This is part of the reason why so many race cars are rear wheel drive. In this way, the tyres on both axles are subjected to a similar amount of stress, which ensures, among other things, an approximately constant driving behaviour during a race.

Although computer simulations only reflect reality to a limited extent, it is easier to consider influencing variables in isolation. For example, such studies of circuit races with sports car prototypes yield similar results. They show the greatest influence on lap time for tire grip, followed by vehicle mass. Engine power has less influence. Drag and downforce have the least influence, Fig. 2.10.

This diagram also invites reflection on the importance of the oft-cited engine power. For the improvement of the lap time, the power increase is the third most effective measure. Once aside from how hard a 10% increase in power is, doing so worsens fuel consumption by nearly 6%. This is worth detailed consideration, especially in endurance races where fuel stops can be race critical. The increased engine power also allows for a significant increase in top speed. Still, whatever compromises are necessary, you're going to choose the combination of improvements that gets the car to the finish line the fastest. After all, pure top speed is not awarded a prize in any event. In addition, it is important to remember that maximum engine power is related to an operating point. In fact, even a racing car is operated at partial load and in transitional ranges. Therefore, the drivability of the engine (throttle response, correlation accelerator pedal position – torque build-up, etc.) is an essential criterion in engine development and even becomes more important the more powerful the engine is or, more precisely, the more overloaded the tyres are. Torque delivery is another significant criterion when it comes to engine selection. A uniform

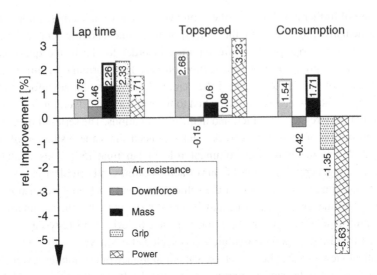

Fig. 2.10 Influence of a parameter improvement by 10% on lap time, top speed and fuel consumption [18]. This simulation was made for the 24 h of Le Mans for sports car prototypes. The lap time is generally the most important criterion for racing cars and there high grip and low mass are important. If the mass is reduced by 10%, the lap time is reduced by 2.26%. Engine power comes third in importance, but also causes a marked deterioration in fuel consumption

torque delivery that is gentle on the tyres keeps the tyre at a high grip level for longer – it does not degrade as quickly.

Simulations also show that as the engine power increases, its influence on the lap time improvement steadily decreases. Theoretically, there even comes a point at which no more lap time improvement is possible. This can be explained in practice: the tire potential is exhausted and maximum acceleration has been reached.

But this diagram also identifies the five essential parameters that determine the potential of a racing vehicle: Vehicle mass, engine power, tire grip, drag, and downforce. These follow from consideration of the physical equations of motion.[4]

Other simulations further show that the importance of individual variables depends on the course of the track. This manifests itself, among other things, in the fact that the setup of the same car is not the same for all tracks, even for the same weather conditions. To name two extremes, there are slow, winding tracks on the one hand and high-speed tracks on the other. For sprint races, on the other hand, the relative importance of individual assemblies is different from that on the circuit. Table 2.10 lists typical lap time improvements for individual measures for formula cars. This table can help with design decisions as to which solution should be given preference in the event of a conflict of objectives.

---

[4]For more details see Racing Car Technology Manual vol. *5 Data Analysis, Tuning and Development*, Sect. 6.2.2 *Simulation*.

Table 2.10 Lap time reduction in s of formula vehicles, according to [42]

	Vehicle type	a	b	c
	Total mass (excluding driver and fuel), kg	475	600	710
Measure	Engine power, kW	160	315	510
Weight reduction by 1		0.25	0.20	0.17
Reduction of Centre of gravity height by 1		0.02	0.02	0.03
1% improvement in tyre grip		0.30	0.32	0.35
Increase of downforce by 1		0.05	0.07	0.08
Reduction of air resistance by 1		0.13	0.12	0.10
Increase of engine power by 1		0.15	0.14	0.12

For rally cars, there are indeed major differences as far as the circumstances in the race are concerned – much more improvisation is required than on the circuit, there are no fixed maintenance points, on the contrary: the service team has to keep up with the cars, etc. – but hardly any as far as driving performance is concerned. The influences of the known criteria decrease according to the following series: Friction (grip), mass, engine power. Quite small is influence of air resistance and downforce [38]. Moreover, this statement is relatively independent of the distance. However, a greater influence of the driver on the driving time is found than on the circuit. For rally cars, the time gain/track [s/km] is used as a comparative value.

In road tests, the most neutral possible driving behaviour for sports cars proves to be optimal for slalom tests (30 m pylon spacing) [5]. The same statement also applies to circuit races with the highest speeds, such as the Indianapolis 500 miles [20]. This design allows the highest cornering speeds or the highest lateral accelerations. Driving at the limit has a disadvantage. In the end, the vehicle reacts oversteering or understeering depending on the driving speed (downforce distribution), tire mileage, stabilizer settings, etc. and is therefore almost unpredictable for the racing driver. A clear, consistent behaviour, such as understeering or oversteering, makes it easier to recognise the limit range and thus to hold the ideal line.

Extremely manoeuvrable vehicles, however, are characterised by low directional stability coupled with high steerability, in addition to the small mass inertias mentioned above. High (yaw) damping is required for controllability [41]. High yaw damping means large wheelbase together with high lateral stiffness of the tires. Yaw damping decreases with driving speed.[5]

If a vehicle is to be able to negotiate a corner at the highest possible speed, it is best if the lateral force potential of the front and rear axles is exploited simultaneously and equally.

---

[5] For more details see Manual of Race Car Technology vol. 5 *Data Analysis, Tuning and Development*, Chap. 5

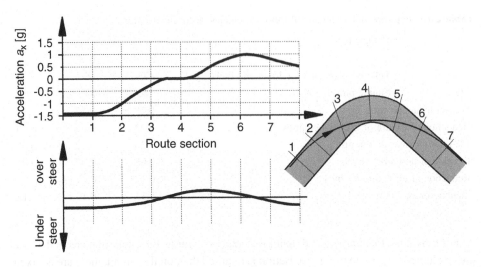

Fig. 2.11 Ideal self-steering behaviour of a racing vehicle when negotiating a corner, after [23]. The acceleration curve (top) and the self-steering behaviour (bottom) are plotted over the sections of a corner (right)

With a balanced weight distribution between the axles of about 50: 50%, this means that the slip angles of the tyres are approximately equal when the vehicle is not accelerating. In fact, the optimum may not be a certain constant self-steering behavior, but one that changes section by section. Figure 2.11 shows a conceivable course of a self-steering behavior during cornering as a function of longitudinal acceleration. The corner is driven through from stage 1 to 7. At the beginning of the corner, when braking, the car understeers, i.e. it is stable. At the apex of corner 4, the phase with the greatest lateral acceleration (and therefore no longitudinal acceleration due to the used-up tire force potential), the vehicle behaves neutrally and when accelerating out of the corner, a slight oversteer supports the yawing of the vehicle. Such a design would be unusable for series applications because it contains unstable driving conditions. At the same time, it also means that much higher demands are placed on the driver for such a competition vehicle than on the average driver.

Summarizing from the above considerations, some of which are theoretical, the following basic characteristics lead to high driving performance [3]:

- low vehicle weight
- high engine power or low power-to-weight ratio (e.g. in kg/kW)
- High transmission efficiency in the drive train
- high contact force of the drive wheels during acceleration
- wide tyres, especially for the drive wheels
- high friction tyre
- aerodynamic downforce aids.

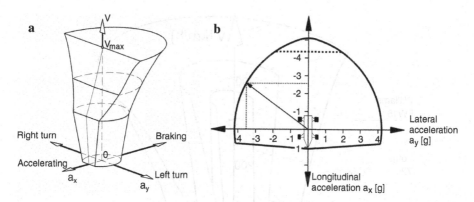

Fig. 2.12 Limits of drivability of a racing vehicle (g-g-v diagram [also: g-g diagram]), after [25]. The three-dimensional body (**a**) encloses the drivable area (a_x, a_y) as a function of the driving speed v. It can be seen that as the driving speed increases, the lateral accelerations a_y and the negative longitudinal accelerations a_x increase. The effect of the downforce becomes noticeable. The maximum speed v_{max} is reached when the drive acceleration becomes 0. (**b**) shows a horizontal section through this body at high driving speeds (g-g diagram). The dashed line results with a restriction of the braking power. In addition, a combined movement is entered: braking in a right-hand bend. So that the traction capacity of the tires is not exceeded, both the lateral acceleration and the braking deceleration must be reduced from the maximum values

The driving speed v has a significant influence on the driving performance due to the effect of aerodynamic aids, Fig. 2.12.

Figure 2.13 compares the performance of Formula 1 cars over several eras. The older cars did not yet have any downforce aids and tyres with much less grip. The most recent cars achieve their enormous driving performance primarily through the effect of aerodynamics. It can also be seen that aerodynamic aids only bring significant improvements from around 100 km/h upwards.

The maximum values currently achievable are -5.1 g during braking, 1.8 g during acceleration and over 4 g during cornering. Without aerodynamic downforce, hardly more than 1.4 g can be achieved in all directions, depending on the tyres and road surface, Table 2.11.

The influence of the engine power on the longitudinal acceleration of the vehicle can be roughly illustrated analytically. If we assume that the engine is operated at 50% of its rated power on average over time when a stepped transmission is used for maximum acceleration, the result is [24]:

$$t_a = \frac{v_{end}^2 \cdot m_{V,dr}}{12900 \cdot P_{M,\max}} \tag{2.12}$$

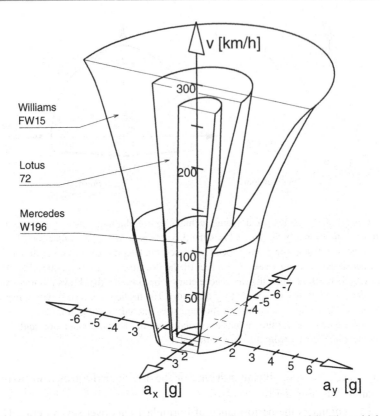

Fig. 2.13 Comparison of driving performance of Formula 1 cars, after [25]. The vehicle without downforce aids (Mercedes W196) shows as the only variable practically only decreasing acceleration with increasing driving speed. The Lotus 72 was the first vehicle with ground effect. The FW15 from Williams stands for a contemporary car. For the two vehicles with various aerodynamic aids (wings, undertray, ...) the driving performance increases noticeably from about 100 km/h onwards

Table 2.11 Driving performance of racing vehicles

Vehicle category	Longitudinal acceleration		Lateral acceleration
	Drive	Brakes	
Formula 1 [4]	Approx. 2 g	Approx. 5 g	Approx. 3.5 g
Formula 3	1.25 g	3.2 g	3.25 g
Sports car [5]			0.7 to 1 g
Sports Prototype [22]	1 g	3 g	2.5 g
Touring car, without ABS [7]		1.5 to 2 g	

t_a	Acceleration time, s
v_{end}	Final speed, km/h
$m_{V,dr}$	Vehicle mass with driver, kg
$P_{M,max}$	Nominal motor power, kW

Table 2.12 Power-to-weight ratio of some vehicle classes

Vehicle	Total mass, kg	Max. motor power, kW	Power to weight ratio, kg/kW	Time[a], s
Formula 1	620	560	1.1	1:42
Production sports car	650	345	1.88	1:55
Touring car	935	419	2.23	2:08
Formula Renault 2000	580	133	4.4	2:26
Passenger car middle class	1450	100	14.5	3:47

[a]The simulated running time refers to the "Großer Bergpreis von Österreich" (Rechbergrennen: distance 5050 m, average gradient 5.3%)

It can also be seen in the above relationship that the mass of the vehicle has the same (linear) influence on the acceleration time as the nominal engine power. The quotient of these two quantities (total mass/nominal power), the power-to-weight ratio, represents an essential characterization of racing vehicles. The lower the power-to-weight ratio, the greater the acceleration capacity. Table 2.12 compares this value of some vehicles. In addition, the simulated running time for a hill climb is entered. With the exception of the Formula 1 vehicle, the other vehicle classes took part in the race and the actual times deviate only insignificantly from the calculated result.[6] It can be seen that the power-to-weight ratio is a useful parameter for classifying vehicles.

2.4 Conceptual Design of Overall Vehicle *Layout of Overall Vehicle*

The ultimate goals of a racing vehicle development result from the considerations of the previous sections. These are physical considerations of extreme driving manoeuvres and these are independent of the regulations [19]:

- minimum weight
- high engine performance
- maximum rigidity of the frame and suspension parts
- low mass moments of inertia, especially around the vertical axis
- high aerodynamic downforce
- Extremely low centre of gravity
- stable handling, especially in transitional phases such as braking and acceleration.

---

[6]See also Racing Car Technology Manual Vol. 5 *Data Analysis, Tuning and Development*, Section 6.2.2 *Simulation*.

These points may seem obvious. The different success of executed racing vehicles ultimately results from the sum of their characteristics and the degree of approximation to these ideal characteristics.

Of essential importance for the basic concept is the balance of the vehicle and thus, in addition to the frame rigidity, the position of the centre of gravity. If this basic design is correct, the grip ratio between the axles is right and the first test drives begin with a noticeably coherent vehicle and a high level of grip. All further measures, as they are used for the setup, can serve practically only the fine tuning. Even the aerodynamic aids should not be used to "repair" an unsuitable balance, but should only be developed when the mechanical grip is suitable.

Further targets result from the planned use of the vehicle (long distance, sprint, rally, customer use). For example, the targeted service life can range from a few 100 m (dragster) to 6000 km (24 h of Le Mans), so it can vary greatly. Reliability within this period is just as much a goal as ease of repair. Repairs and maintenance work are partly done during the race and possibly by the car crew. In raid competitions a tyre change in the desert for driver and co-driver is nothing out of the ordinary. In such a case, the arrangement and fastening technique of high-maintenance parts is also a decisive criterion. In Formula 1, a tire change takes 3 s. Fitting another nose takes 11 s. For an engine change in the pits 45–60 min are scheduled, for a gearbox 30 min. In endurance races the quick exchange of systems like engine, gearbox etc. is even more crucial. Therefore a modular design of the vehicle is recommended. Here, the modules are functional groups and the number of interfaces between them is as small as possible, or the connections are arranged in such a way that they can all be separated and connected in one direction of movement.

The steps of the design activity result mainly from the influence of individual assemblies (Fig. 2.14) on the driving performance, as well as from the consideration that the driver can only achieve maximum performance in an ergonomically favourable posture, and of course the (unalterable) regulations. The following sequence of design can be derived from these

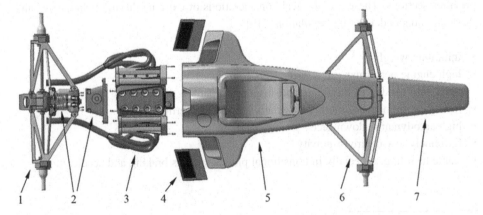

Fig. 2.14 Important assemblies of a racing vehicle. 1 Rear suspension, 2 Drive train: transmission, clutch, 3 Engine, 4 Cooling system, 5 Frame (chassis), 6 Front suspension, 7 Bow with crash element

considerations. The order of considerations is dictated partly by importance, partly by geometric logic. Of course, as always with a design process, only a rough sequence can be given. There will indeed be iterations, i.e. a step (or even several steps) backwards will be necessary again and again in order to incorporate findings that have arisen during the design of a building part into the preceding one.

1. Regulations
2. Driver position
3. Main mass distribution: engine, transmission, tank and heat exchanger position. Linked to this is the selection of an engine with transmission
4. Wheelbase, track width: Rough axle load distribution
5. Tyres
6. Wheels
7. Bodywork under aerodynamic aspects: Downforce, air resistance, cooling and combustion air ducting
8. Chassis geometry: roll center, instantaneous pole single wheel, pole distance single wheel, ...
9. Hubs
10. Brakes
11. Wheel carrier (knuckle, upright)
12. Body springs
13. Damper
14. Stabilizers
15. Steering
16. Drive train
17. Frame
18. Auxiliary systems: fuel system, electrics, piping, hosing, ...

The conceptual considerations therefore begin with an in-depth study of the regulations. It serves the safety and the competitive.

The regulations are very restrictive in terms of design and thus limit many areas of the vehicle, sometimes to a considerable extent. Nevertheless, it is precisely in these areas that the key to success often lies, because the design freedom required for competitive advantages can be created through clever interpretation of the regulations. Some exemplary geometric considerations for a single-seater are shown in Fig. 2.15.

The selection of individual crucial components is shaped by numerous considerations, which can be found in detail in the corresponding volumes of the handbook series.[7]

---

[7] Engine selection see Racing Car Technology Manual Vol. 3 *Powertrain*. Choice of tyres and wheels see Vol. 4 *Chassis*.

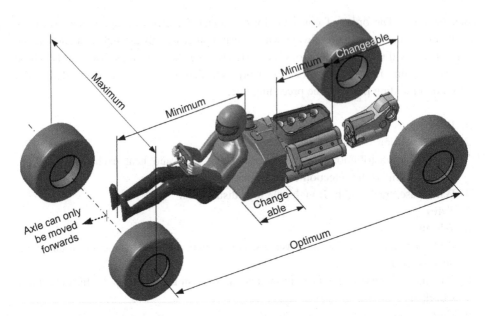

Fig. 2.15 Draft of the main dimensions of a single seater. The driver's feet must remain behind the front axle for safety reasons. The axle itself can therefore only be pushed forward in relation to the driver. The shape of the fuel tank hugs the driver's back and should be as short as possible so that the engine-transmission assembly can be shifted to achieve the desired axle load distribution. The engine should also be short for this reason. The gearbox is designed as slim as possible so that the air flow in the rear area of the vehicle is disturbed as little as possible

The further sequence can be different when designing based on an existing vehicle, especially when existing assemblies are adopted or when development priorities are set and identified weaknesses are to be eliminated. Further shifts in the order result from the importance of aerodynamics. In the case of vehicles that achieve lower speeds or where downforce aids are not permitted (such as Formula Ford), the bodywork will tend to be considered at the end, so to speak, as a cover for the design. If the frame is a CFRP shell that also includes parts of the exterior design from an aerodynamic point of view, this system will also be considered earlier than in the above list.

The art therefore lies in a clever arrangement of all parts and assemblies, the so-called packaging, which can be seen for a formula car in Fig. 2.16. It goes without saying that this can only be done with compromises. The question is always which function is considered more important than another. So there will hardly be one "best solution" for a problem. Furthermore, a solution that has proven itself in one racing class may prove unsuitable for other vehicles. Nevertheless, certain generally valid findings can be made.

Symmetrical Structure
A symmetrical design makes better use of the available space and eliminates the need to transfer cables from one side of the vehicle to the other [3]. The static wheel load

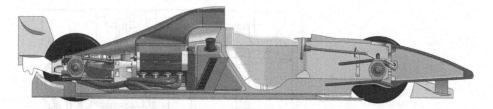

Fig. 2.16 Structure of a formula car. The available space is well used, with the aerodynamics dictating the outer shape (undertray, driver position, inflow wing). The fuel tank is centrally located behind the driver. The combustion air flows above the driver's helmet to the low-lying engine

distribution left to right is also more balanced this way. It will also result in a symmetrical outer contour, which avoids yaw moments caused by air forces. This is naturally easier to achieve with a monoposto. In sports and touring cars, attempts are made to move the driver's seat towards the middle of the vehicle if the regulations permit. For example, in the 1995 German Touring Car Championship (DTM for short) C-Class Mercedes, the gearbox and propshaft were offset and the driver sat closer to the centre of the car. This is a similar solution to that found decades earlier for the W154 (1938) [13].

Individual Focal Points
Heavy, spatially inseparable individual parts (motor, shafts, cables, etc.) should be arranged as low as possible and close to the driver.

Parts that are "immovable" (engine, gearbox, . . .) should be designed as light as possible. With ballast masses, the car can be trimmed to the required minimum mass and, above all, to the desired mass distribution. The area under the driver's legs is well suited for relatively heavy systems such as electrical parts (battery, cables, . . .).

Fuel Tank
The tank for the fuel system should be located as centrally as possible. This way it is protected and its unavoidable change in mass during the race has the least influence on the handling.

Small Total Moment of Inertia
The heavy masses should be arranged around the driver. In this way, the mass moments of inertia (especially around the vertical and longitudinal axes) remain low. In the event of conflict with the moment of inertia J_z about the vertical axis, however, preference should be given to the lower centre of gravity arrangement. J_z practically only comes into play in the corner entry and exit phase, i.e. when a yaw moment acts on the car due to deflection or steering [41]. Lap time simulations also confirm this. The influence of the centre of gravity height on the lap time is much greater [43].

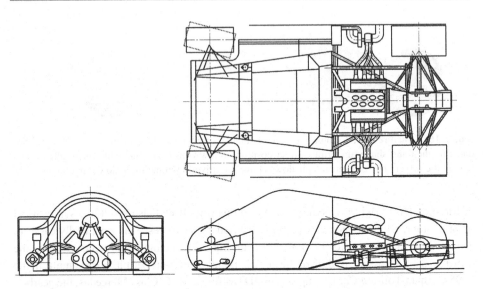

Fig. 2.17 Layout of a racing car (Sauber-Mercedes C9) [13]. A largely symmetrical engine, positioned as low as possible, is installed in front of the gearbox and thus close to the centre of the vehicle. The wheels represent the outermost vehicle contour. The two exhaust gas turbochargers are installed low down

Adjustability

In the case of adjustable systems (springs, dampers, stabilisers, wings, gear wheel sets, . . .), arrange the adjustment elements (knurled wheels, screws, bolts, valves, . . .) so that they can be reached without dismantling work (or at least under an easily removable cover) on the stationary vehicle. An exemplary overall design is illustrated in Fig. 2.17 with the arrangement of the most important parts.

Further steps of conceptual design:

- Estimation of the vehicle weight and the axle load distributions, derived from this, selection of the wheel and tyre dimensions (load capacity)
- Calculations of mileage
- Rough estimate of fuel consumption: derived from the required range Determination of fuel tank size requirements
- Determination of the required crash deformation lengths from the permissible acceleration values of the occupants.

Finally, on the concept, a realization that comes from studying significant concepts. Successful vehicles and engines are rarely distinguished by outstanding individual features, but appear almost disappointingly simple at first glance. On closer inspection, one discovers a clever combination of proven and well-known solutions. Conversely, overemphasizing a single criterion at the expense of others leads, as experience shows, to failure.

Conversely, a lot of useful information for the design of promising cars can be derived from the careful examination of unsuccessful vehicles.

2.5   General Principles of *Embodiment* Design

Certain constraints arise from the cost and deadline situation and the manufacturing capabilities of a company. Nevertheless, it is possible to name generally applicable design principles that are used in the design of vehicles and their components. The aim in each case is to meet the requirements within the economic, time and other project constraints. Not all of the following principles can be applied at the same time. One principle may be decisive for a task, others even contradictory. Which principles are used depends on the requirements and general conditions. For example, a spring should behave elastically, i.e. flexibly over a large distance, while those parts of the chassis that transmit forces to it should be as stiff as possible, i.e. show the least possible deformation.

Simplicity
Technical structures are simple when they are clear. A solution appears simpler if it can be realised with fewer components or parts, because, among other things, less machining and assembly work, fewer wear points and less maintenance work can be expected. This is also true if these parts are geometrically simple and their arrangement is not complex. As few parts as possible with a simple design are therefore generally to be aimed for. As a rule, however, a compromise must be made: The fulfilment of the function requires a minimum of components or parts that cannot be omitted.

Symmetrical shapes prove to be favourable in the sense mentioned. During manufacture, under load and under the influence of temperature, they lead to manageable deformations.

A simple design inevitably leads to easy maintenance, inspection and repair.

Simple designs generally perform the same functions with fewer components than more complicated ones. A comparison between similar systems with different numbers of components can be made on the basis of their reliability. Ultimately, the decisive factor is system reliability, i.e. with what probability does the vehicle reach the finish line or, conversely, with what probability does the failure of an important part lead to failure in the race (apart from the many other possible influences, starting with the driver via the weather etc., which make racing interesting even in the computer age).

For example, if 50 intact components are vital for a race car (spark plugs, injectors, tires, drive shafts, fuel pump, lines, etc.) to see the finish line, the probability of finishing the race depends on the reliability of the individual parts. In this numerical example, the individual parts should all have a 99.99% reliability over a race distance, i.e. 0.01% or 1 part in 10,000

Table 2.13 ABC analysis, according to [34]

A-parts	B-parts	C-Parts
Risky	Risky	Low-risk
Calculable service life	Lifetime not calculable	Lifetime not calculable
Shafts	Seals	Retaining rings
Gears	. . . .	Screw plugs
Bearings		. . . .
. . .		

fails. Then the probability of the car crossing the finish line is $0.9999^{50} = 0.995$ or 99.5%. That is, conversely, the probability of failure is 0.5%, i.e. 1 car in 200 fails.

If this vehicle had only 10 critical components, the calculation would look like this:

$0.9999^{10} = 0.999$ or 99.9%. So the probability of failure drops to 0.1% or only 1 car in 1000 fails. [28]

A modern Formula One car is made up of about 80,000 parts, but obviously not every one of them is vital . . .

Reliability

For all the performance of a racing vehicle, the famous phrase still applies: To finish first, you first have to get to the finish line.[8] The reliability of the most important parts is therefore also decisive for victory or defeat. But what are the most important parts of a racing car, when it only consists of the most essential? Important parts are definitely those whose failure leads to immediate failure in the race. These include wheel-carrying chassis parts and power-carrying parts of the engine and drivetrain. Such a list of all vital parts (also called Critical Items List, CIL) helps to focus the development on the crucial areas. This means that at the beginning there will be an analysis of which parts or systems are particularly important for the reliability of the car and which are less so. A subdivision into three classes has proven to be useful. This so-called ABC analysis is shown in Table 2.13.

In the case of B-parts, one has to rely on empirical values and test results, because their service life cannot be calculated in the same way as for A-parts. C-parts are reliability-neutral and are therefore no longer taken into account in further consideration.

Further analysis methods are the FMEA (Failure Mode and Effects Analysis) and the FTA (Fault Tree Analysis).

An FMEA is a largely formalized method for the systematic recording of possible defects and for estimating the associated risks (effects). Using a form, potential defects are

---

[8] "In order to finish first, you must finish first."

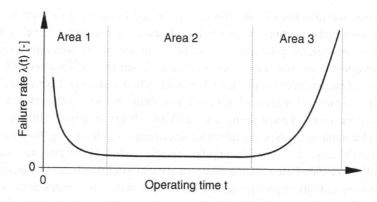

Fig. 2.18 Areas of failure behaviour (bathtub curve). The failure rate indicates the probability with which a part fails during the operating time $t + dt$. It is indirectly proportional to the survival probability. The bathtub curve not only maps the behaviour of individual components, but is also observed for entire systems

listed with their consequences and causes and the probability of the occurrence of a defect, its consequences and its detectability are estimated. This is followed by a ranking according to the level of risk of individual errors. Furthermore, the planned test measures (actual state) and recommended or ultimately taken remedial measures are compiled.

In FTA, the functional structure of a system is considered and the individual functions are assumed to be "not fulfilled" in sequence. The effects on the overall system are then considered and the functional structure is subsequently rearranged or expanded so that individual malfunctions do not lead to total failure.

The typical failure behaviour of components over their service life is characterised by three areas, Fig. 2.18.

Early failures (area 1): These are the effects of design errors (wrong material, design unfavourable due to notches, design for too low service load, deformation too great...) and production errors (adhesive connection not executed according to instructions, weld seam porous, screw connection loose,...).

Random failures (area 2): These failures can occur at any time during operation, the failure rate is constant (although low due to targeted measures). Causes of failure can include misuse or operating errors, poor maintenance, foreign parts (contamination) during assembly or operation. Electronic components usually fail unpredictably and spontaneously [46].

Wear and fatigue failures (area 3): Towards the end of the (planned) service life, failures due to age and wear increasingly occur. Material fatigue, creep, embrittlement of seals, reduction of the load-bearing cross-section due to corrosion or wear, pitting of gears, etc. are typical causes. Depending on the area, the following measures can be taken to reduce the failure rate and thus increase reliability. For area 1: Incoming goods inspection, preliminary tests, tight tolerances and complete production control, supervised testing of components before racing, non-destructive material testing. For area 2: avoidance of overloads (see also principle of safe existence), adherence to maintenance intervals, regular checks of wear rates, continuous monitoring of operating equipment (level, condition, volume flow, pressure, temperature, . . .). As a rule of thumb, purely electronic components such as transistors or integrated circuits contribute 10%, sensors and actuators 30%, but the connections of the components to each other and to the outside world contribute 60% to failures. If avoidance measures are not sufficient, fault tolerance measures (e.g., multi-channel, self-monitoring) must be used to mask the effects of a fault [46]. For area 3: Suitable calculation methods; comprehensive determination of loads, e.g. from data of the previous year's races; creep tests with cyclic dynamic loading of components, assemblies and complete vehicles based on the recorded data of completed races – especially in endurance competitions (hydro-pulse test bench and shaker for components, dynamic test benches for engine and transmission, 7-stamp test bench for complete vehicles).

Another approach to filtering out the significant parts is to look at the reasons for failure of cars in races. It turns out that it is usually the rotating hot parts that cause trouble. These include the engine, gearbox, driveshafts, brakes, wheel bearings and tyres. Of course, teams know this and pay special attention to these parts from design to manufacturing to maintenance. Nevertheless, these parts remain at the top of the list of reasons for failure [25].

With the help of Boolean algebra, a simple probabilistic consideration of an overall system can also be carried out. The following simplifying assumptions are made: The individual components of the system have only two states: Intact or Failed. The elements operate independently of each other, i.e. the failure of one element has no effect on the other elements. Let the survival probability of the elements be known through experimentation. The reliability $R(t)$, i.e. the probability that the component will work until time t, is exactly the complement of the probability that the component will fail until then:

$$R(t) = 1 - F(t) \tag{2.13}$$

$R(t)$	Survival probability of a component until time t
$F(t)$	Probability of failure of a component at time t

In order for this reliability consideration to be possible, the following steps must be carried out, especially for complex systems [53]:

- System analysis
- Determination of component reliabilities
- Determination of system reliability.

System analysis transforms the overall system into a mathematical model [54]. recommends starting with a breakdown into functional units for adaptronic systems:

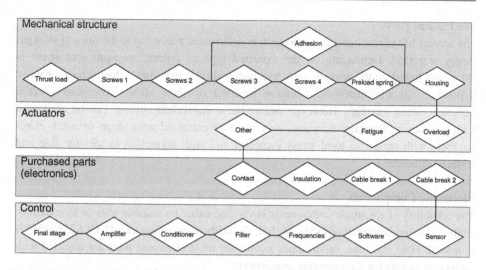

Fig. 2.19 Example of a reliability structure, after [54]. The system has largely serial structures of the elements. Only bonding and screwing act in parallel

Fig. 2.20 Example of a reliability structure reduced to the essentials, after [54]. This example belongs to the structure from Fig. 2.19

- Mechanical design
- Electronics
- Regulation
- Actuators.

The system is divided into functional units and further subdivided according to the failure modes known from experience, Fig. 2.19. In any case, the aim is to represent the overall system by a functional diagram. A component can also appear in several places in a reliability diagram because it does not reflect the design structure but the functional structure of a technical object.

Subsequently, the subsystems can be decomposed into individual components. The functional structure can be simplified to reduce the mathematical effort [54]: Components with no or little influence on reliability are disregarded. Different failure modes of a component can be combined into a single component failure. Figure 2.20 demonstrates this using the example of the structure of Fig. 2.19.

The procedure for determining the individual component reliabilities depends on the assignment to the outline mentioned above.

Mechanical Parts

The service life of mechanical components is determined according to the rules of strength theory, e.g [55]. Depending on the expected type of failure, an equivalent stress is calculated based on the normal stress hypothesis (brittle materials), shape change energy hypothesis (ductile materials, vibration loading) or shear stress hypothesis (ductile materials, static loading). However, before these equivalent stresses can be compared with a material characteristic value, they must be corrected with shape or notch effect numbers with regard to local stress increases and with correction factors for size and surface influence.

Electronic Components

The reliability of electronic components is verified either by suitable tests or by means of analytical estimation methods [54]. In [54], the method according to IEC TR 62380 [56] is proposed. Here, the basic failure rates known for most electronic parts are adjusted with correction factors for the operating conditions.

Regulation

For software, standards have been developed in the automotive industry for specific areas, such as Automotive SPICE[9] (Software Process Improvement and Capability Determination) for evaluating the software process capability of suppliers and AUTOSAR[10] (AUTomotive Open System ARchitecture) an open and standardized software architecture for control units (ECU). However, general methods for determining reliability do not yet exist.

Actuators

One source for determining the service life is manufacturer information, which is generally based on very detailed reliability tests.

The function structure serves as the basis for calculating the overall reliability with Boolean system theory. Generally, to perform a function, a signal, substance or energy will flow from the input to the output of the overall system. If there is an upright connection from input to output, the system is intact. Basically, there are a few arrangement possibilities of components, Fig. 2.21. The most common arrangement in mechanical engineering is the series arrangement (Fig. 2.21a). It represents the minimum amount of parts and is therefore cheap and easy. However, if only one element in this chain fails, the entire system fails. For the reliability R_t of the entire system applies:

---

[9] http://www.automotivespice.com/fileadmin/software-download/Automotive_SPICE_PAM_30.pdf
[10] https://www.autosar.org/

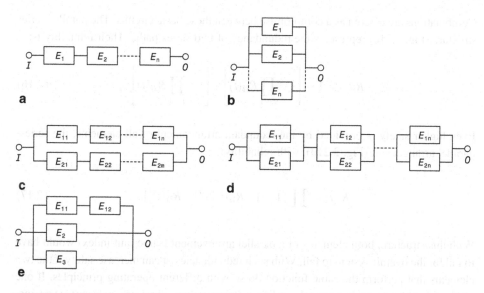

Fig. 2.21 Basic arrangements in reliability schematics. (**a**) Logical series arrangement, (**b**) Logical parallel arrangement, (**c**) Logical parallel-series arrangement, (**d**) Logical series-parallel arrangement, (**e**) Logical mixed arrangement (example). *I* Input, *O* Output, E_n Element *n*

$$R_t(t) = R_1(t) \cdot R_2(t) \cdot \ldots \cdot R_n(t) = \prod_{i=1}^{n} R_i(t) \qquad (2.14)$$

R_t	Reliability (survival probability) of the overall system. $R_t \leq 1$
R_n	Reliability of the device E_n. $R_n \leq 1$

Thus, the overall reliability is at most equal to the reliability of the worst element, confirming the old adage that a chain is only as strong as its weakest link.

If elements are arranged in parallel (Fig. 2.21b), the reliability is increased. If one element fails, the elements connected in parallel take over the function and the entire system remains intact. Total failure only occurs when all n elements have failed. However, the structural complexity of such systems is greater, which is why they are primarily used in safety-critical systems (dual-circuit brake system, dual position sensors in dual-clutch transmissions, redundant angle sensors in e-gas systems, double-walled fuel lines, …). The reliability of the parallel arrangement is calculated to:

$$R_t(t) = 1 - [1 - R_1(t)] \cdot [1 - R_2(t)] \cdot \ldots \cdot [1 - R_n(t)] = 1 - \prod_{i=1}^{n} [1 - R_i(t)] \qquad (2.15)$$

Combinations are offered as a compromise between these basic circuits. The parallel-series structure (Fig. 2.21c) represents the redundancy of two series paths. Their reliability is:

$$R_t(t) = 1 - \left[1 - \prod_{i=1}^{n} R_{1i}(t)\right] \cdot \left[1 - \prod_{j=1}^{m} R_{2j}(t)\right] \qquad (2.16)$$

Even higher safety is provided by the redundant arrangement of all elements, the series-parallel arrangement (Fig. 2.21d), with reliability:

$$R_t(t) = \prod_{i=1}^{n} \left\{1 - [1 - R_{1i}(t)] \cdot [1 - R_{2i}(t)]\right\} \qquad (2.17)$$

With this structure, both elements of a parallel arrangement (same unit index) would have to fail for the overall system to fail. With such redundancies, it can increase safety if the two elements that perform the same function do so with different operating principles. If one element fails due to environmental conditions (temperature, vibrations, pressure, corrosion, etc.), the other element does not necessarily suffer a loss of function as a result. If, on the other hand, both elements are of the same design, the probability is high that both will fail after the same runtime in the event of a principle-related failure. In practice, mixed structures (e.g. Figure 2.21e) consisting of the above arrangements will occur.

More in-depth considerations on this topic can be found, for example, in [27, 47, 53].

In a series circuit, every component is therefore important. In the field of electronics, programmable power distribution systems help to increase reliability. Here, a control unit monitors the currents occurring in the individual circuits and switches off the current briefly in the event of defined overcurrents. However, unlike a fuse, which interrupts the circuit forever, the circuit is subsequently reenergized and continues to be monitored. If the cause was not a dynamic effect, but e.g. a mechanical problem (blocked pump, clogged pipe,...) and thus the current remains permanently too high, the control unit switches off the circuit completely or it can switch the system to a substitute system according to a predefined algorithm and thus maintain the function.

The failure probability $F(t)$ of components can be represented mathematically with a Weibull distribution – assuming corresponding test data or empirical knowledge. In [54], this was done for an active interface, which is installed between the head bearing of the strut and the strut dome of a McPherson front suspension and keeps road excitations away from the car body by active length change. The Weibull mesh of the overall system of this interface is based on considerations of the reliability structure Fig. 2.20 and is shown in Fig. 2.22. The time history of the failure probability of the overall system (green) initially follows the failure behavior of the controller (magenta). The controller and amplifier (blue) show constant failure rates typical for electronic systems. Subsequently, the failure behavior of the actuator (dashed red) dictates the reliability of the overall system. This only changes at very long life, when the first failures of the preload spring (light blue) become

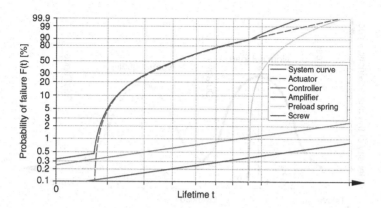

Fig. 2.22 Example of a Weibull mesh, after [54]. The reliability structure of the overall system is shown in Fig. 2.20

likely. The failure progression of the screw cannot be seen at all in this illustration. The service life axis would have to be extended for this. The screws have no significant influence on the behavior of the overall system. In conclusion, it can be seen that in this system the actuator has the greatest influence on reliability. In order to increase the reliability, it is necessary to start with the actuator. Subsequently, the electronics could also help to increase the service life of the actuator, for example by preventing overloads through software measures.

Under the above keyword "simplicity", among other things, the number of components was considered as a criterion for fail-safety. This method (*parts count method*) has its justification and is often used for electronic products. In the case of mechanical structures, however, another consideration is of greater significance. In this case, reliability is strongly influenced by the interaction between resistance and stress [39]. In other words, the greater the stress on a component, the greater the risk that it will fail. For example, the reliability of a planetary gearbox is increased by adding planetary gears, although the number of parts (gears, bearings) has increased even more. However, dividing the power among several gears reduces their stress and thus has a positive effect on service life. In comparison, a stepped gear set with two gears requires fewer parts, but can be less reliable than a planetary gear set with the same power due to the higher stress on the gears.

Reliability is therefore also a question of component safety. This can be achieved by means of three basic safety principles, whereby the designer must decide on one in order to fulfil a function safely:

1. Principle of safe existence (safe-life behaviour)
2. Principle of limited failure (fail-safe behaviour)
3. Principle of redundant arrangement (redundancy).

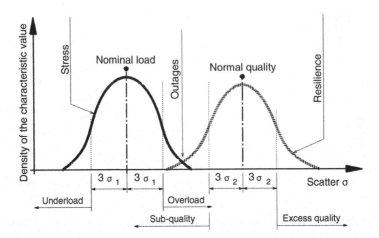

Fig. 2.23 Scattering of stress and resistance, after [34]. Neither the quality of the material nor the level of the load always have the same value. On the contrary, both variables fluctuate around an average value. If a component with low quality suffers an overload, this part will fail

Safe Life

The principle of safe existence assumes that all components and their interrelationship are such that during the intended period of use all probable or even possible occurrences will be survived without failure or malfunction.

To this end, it must be borne in mind that not only the load but also the resistance of the component to it cannot be reduced to a constant value. On the contrary, there are unavoidable scattering of the values of load and material properties, which have an effect on the component behaviour. Different loads result from different operating conditions. The material characteristic value fluctuates due to quality and manufacturing influences (work hardening, casting: Cooling behaviour of different wall thicknesses, tolerances, . . .). Figure 2.23 shows this relationship schematically.

Safe life is thus ensured by:

- appropriate clarification of the acting loads and environmental conditions, such as expected forces, duration, type of environment, etc. (avoids wear and fatigue failures).
- sufficiently safe design based on proven hypotheses and calculation methods (e.g. FEM, see appendix)
- numerous and thorough checks of the manufacturing and assembly process (avoids early failures)
- Component or system investigation to determine the durability under partly increased load conditions (load height and/or number of load cycles) and the respective environmental influences.(avoids wear and fatigue failures)

- Defining the scope of application outside the range of possible failure circumstances (avoids random failures)
- optimized geometry (no notches, favorable force flow)
- high quality materials with guaranteed specifications.

Characteristic of this principle is that safety lies only in the exact knowledge of all influences with regard to quality and quantity or in the knowledge of the failure-free range. This principle requires either relevant experience or a considerable amount of preliminary investigations, line tests and continuous monitoring of the material and component condition, i.e. time and money.

In Formula 1, for example, quality control goes so far that around 5000 individual parts, 1000 of which are different, are inspected and individually marked (e.g. with bar codes) during engine production before they are assembled into a racing engine in an average of 80 working hours [37].

During the ride, neuralgic points are continuously monitored by sensors: temperatures and pressures can be used to draw conclusions about the condition of systems and, if necessary, to react by reducing power before a total failure occurs.

Reliability is – as so often – also a question of budget and discipline. Regular maintenance with consistent replacement of parts when they have fulfilled their time quota, regardless of whether they still look usable or not, undoubtedly increases reliability.

In order to ensure a corresponding component safety, its stress must remain below a certain limit value. Materials research provides the designer with material limit values for the individual elementary types of stress (tension, compression, bending, shear and torsion) on a test bar, i.e. generally <u>not</u> on the component itself, beyond which fracture occurs. The permissible stress σ_{zul} thus follows from a measured material limit value σ_n weakened by a form influence as well as a nominal safety S:

$$\sigma_{zul} = \frac{\sigma_n \cdot k_O \cdot k_G}{S} \qquad (2.18)$$

σ_{zul}	Permissible stress, N/mm^2
σ_n	Material characteristic value, N/mm^2
k_O	Surface influence (production, machining), –
k_G	Size influence (stress gradient, etc.), –
S	Safety factor, –

Rough surfaces contain numerous notches and thus have a strength-reducing effect under cyclic loading, Fig. 2.24. This effect increases with increasing material strength. The higher the strength of a material, the more important it is to have a smooth surface under dynamic loading. The size of a component also has an effect. Strictly speaking, the values determined in the so-called tensile test only apply to the test bar with a diameter of 10 mm.

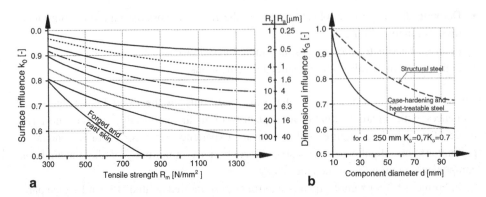

Fig. 2.24 Influences on the strength of steel components, after [34]. (**a**) Influence of surface finish R_z average roughness depth, R_a arithmetic mean roughness value (DIN EN ISO 4287, 4288). (**b**) Influence of component size

If the component has larger dimensions, its strength decreases. This influence is naturally even more pronounced in heat-treated steels. One can also see from this an advantage of thin-walled cast components. Apart from the fact that they are lighter and do not have shrinkage cavities, they also have higher strength values than comparable thick-walled constructions.

Limited Failure

The principle of limited failure allows for malfunction and/or rupture during the period of use without serious consequences. In this case

- maintain a function or capability, albeit limited, that avoids a dangerous condition
- the limited function is taken over by the failing part or another part and performed until the part can be replaced
- the error or failure becomes apparent.

In this case, the warning is essentially accompanied by a restriction of a main function through: Increasing running unsteadiness, leaking, decrease in performance, impaired movement, etc. without immediately causing a hazard. Warning systems are also conceivable that notify the driver of the onset of failure.

The principle of limited failure presupposes knowledge of the damage sequence and a design solution that assumes or maintains the limited function in the event of failure.

For example, a bolt that is inserted into a connection from above will still hold chassis parts together even if the nut has fallen off due to vibrations or similar. The greater play that occurs signals the fault to the driver. If the bolt is installed from below, it falls out due to its own weight and the connection is completely loose.

Redundancy

The principle of redundant arrangement is a means of increasing both the safety and the reliability of systems. Redundancy (=abundance) in the technical sense means multiple arrangement of parts or systems. Resilient safety tanks, for example, must be housed in a sealed container. The walls that hold the fuel are therefore redundant. Redundancy leads to an increase in safety as long as the possibly failing system element does not cause a hazard on its own and the further system element arranged either in parallel or in series can take over the full or at least limited function.

The arrangement of several fuel pumps, multi-strand cable pulls as well as multi-circuit brake systems are examples of active redundancy: All components actively participate in the task. In the event of a partial failure, there is a corresponding reduction in energy or performance.

Several sensors can also have the same task. For example, two sensors are used to detect the gear stage. The additional weight is more than compensated for, as there is a risk of engine damage due to overrevving in the event of failure.

If reserve units – usually of the same type and size – are provided, which are switched on in the event of failure of the active units, e.g. standby pumps, this is referred to as passive redundancy, the activation of which requires a switching operation.

If a multiple arrangement is the same according to its function but different according to its operating principle, this is called principle redundancy. The frequently required double closing spring of the butterfly valve may serve as an example. One spring can be a tension spring acting via a lever, the second a spiral spring acting directly on the shaft.

Safety-increasing units can be arranged in parallel (e.g. spare oil pumps, . . .) or also in series (e.g. filter systems). In many cases, however, such simple circuits are not sufficient, but circuits with crosswise linking are required, e.g. to ensure continuity despite failure of several components.

However, the redundant arrangement cannot replace the principle of safe existence or limited failure: The redundant arrangement of fuel pumps has no safety-enhancing effect if the pump itself tends to overheat, thereby endangering the entire car.

Safety enhancement is only given if the redundant elements satisfy one of the above principles of safe existence or limited failure.

Notches

Stress concentrations occur at notches, which prove to be a weak point especially under dynamic loading. Notches cannot always be removed, but the component environment can be designed to reduce the life-reducing effect of notches.[11]

For fail-safe bolted connections that are highly and especially dynamically stressed, notch effects and the force flow in the stressed parts must be taken into account, Fig. 2.25.

---

[11] How the notch effect of shafts and axles can be reduced by appropriate design is described in the Racing Car Technology Manual, Vol. 3 *Powertrain*, Section 5.3.1 *Power Transmission*.

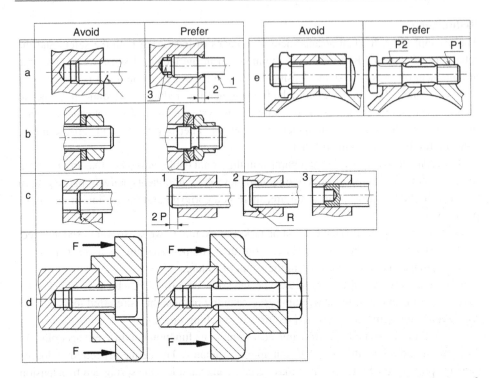

Fig. 2.25 Design of highly stressed bolted joints, after [36]. Explanations see text. (**a**): In the case of stud bolts with the usual jamming of the thread run-out in the blind hole, there is a risk of fatigue fracture (arrow). This can be reduced by the measures shown: Flexurally soft expansion shank (1), blind hole thread overhangs screw thread (2) and tensioning of the screw via lug (3). (**b**): In the case of particularly notch-sensitive screw materials (e.g. titanium), the standard nut support on a flat washer is not sufficient. Fractures often occur in the area of the contact surface between nut and washer. The remedy is a design with two spherical washers and a nut that is designed so that the screw thread ends inside the nut. (**c**): In blind hole screw connections, the highest stress occurs at the end of the screw thread. Fatigue fractures can occur in this area (arrow). The stress curve in the nut thread can be compensated for by various measures. The bolt overhangs the nut thread (1) by approx. 2 times the pitch P. This can also be done within the part by countersinking the thread with a fillet (2) or by drilling out the bolt (3). (**d**): The durability of bolted joints can be increased by increasing the elastic compliance of the bolt (longer bolt by appropriate design of the components or with a sleeve and slender shank) and by shifting the point of application of the operating force F towards the parting line. (**e**): An elegant method to achieve a fixed preload in the smallest space is a differential thread. The two threads have the same pitch direction but different pitches (P1 > P2). The smaller the difference, the greater the preload forces that can be achieved with the same tightening torque. The force flow in this case is also much more favourable than with the conventional screw connection. The clamping forces are introduced at both ends of the screw over a longer range via the thread

The best remedy against loosening of bolted connections are high-strength pretensioned bolts. In off-road applications (raid, ice speedway, ...), some bolted joints loosen with regular pitch. As a remedy, some teams resort to drastic means and compress the nut (gluing is generally not sufficient). The loosening of such a connection is done with an impact wrench. However, the thread is then unusable and the bolt and nut must be replaced.

Because a bolt elongates under the pretensioning force, but the nut thread is pressed, opposite elongations occur. As a result, in principle only the last threads in engagement bear. If you want to distribute the pre-tensioning force more evenly over the length of the screw, you have to use a tension-compression nut, which also elongates to some extent. This requires more space and is heavier than the initial solution. Formula 1 therefore uses screws with a thread pitch that is adapted to the deformation, i.e. variable. Although this is technically elegant, it is complex and expensive.

There is also the case where a notch is intentionally provided in a component, namely at a predetermined breaking point. This point is created by applying a cross-sectional crack or constriction and its effect can be additionally increased by heat treatment. Local embrittlement occurs, which leads to fracture at the intended overload. Such locations are found, for example, on wishbone linkages of single-seaters. In the event of an accident, the control arms break away and do not destroy the expensive frame or monocoque. In general, care is taken to ensure that in a continuous load path the cheapest component, which causes the least damage to the overall vehicle, always breaks first.

In conclusion, 12 basic requirements for developing a reliable product, according to [53]:

• draw up precise specifications
• as few individual parts as possible
• abandon risk elements
• easy exchangeability of wear parts
• do not exhaust the last safety reserves during design
• carry out early component tests
• simulate the practical use of the product during development on the computer
• clarify the dynamic behaviour before practical use by means of vibration calculations
• for series products: Carry out extensive test bench and field testing with practical stress collectives. In the case of racing vehicles, use appropriately recorded data from past competitions.
• strictest quality controls at the suppliers and in our own production
• plan random sample checks in (series) assembly: constantly ensure conformity with specified quality characteristics
• evaluate warranty and goodwill statistics.

Lightweight Design
The weight of a design generally depends on the material, the design method, the design and the connection effort. To save weight, one will use materials that have high strength with low density. More details are given in the next section. The integral design method is a weight-saving method. In extreme cases, no connections are required at all. Among other things, connections have the disadvantage that overlaps and additional connecting elements

are required, which stand in the way of slimming down. The design, i.e. the dimensioning of components, can be characterized by safety factors determined by regulations, or it can be so exhausted for special loads that the component can only withstand them for a certain time. Lightweight design is characterised by optimisation of the structure and optimisation also always means specialisation, i.e. limitation to a specific area of application with a precisely defined period of use.

Lightweight design therefore means, above all, designing a component precisely for its expected load(s) (fatigue strength calculation). For this purpose, on the one hand the load must be known exactly and on the other hand the response of a material to this load, i.e. the material behaviour, must be known exactly. Both are generally only possible within a scatter range. Loads are superimposed by impacts etc. and materials are subject to the usual quality fluctuations of a production (Fig. 2.23). The ideal would be a component in which all areas are equally (and highly) stressed. This would mean that no material was "wasted". Another step to keep the mass of components low results from the design type. Durable parts are naturally much larger and heavier than time-resistant parts. An operationally stable design includes [44]:

- Consideration of the phenomenon of material fatigue,
- Creation of design concepts for reliable structures,
- Fracture-mechanical description of crack propagation under fatigue loading and experimental determination of crack propagation parameters,
- Lifetime prediction and selection of inspection intervals,
- Crack-resistant design (e.g. installation of crack traps).

Unlike in many areas of technology, where dynamically loaded parts are designed to be fatigue-resistant, in vehicle design it is necessary to keep parts light. One possibility for this is offered by the dynamic material behaviour, Fig. 2.26. In the case of alternating loading, (fatigue) fracture occurs after a certain number of load cycles. In this case, the stress that can be borne decreases with increasing number of load cycles (oscillations, cycles,...) and, for many materials, after a certain number of load cycles (for steel $10 \cdot 10^6$), it reaches a value that hardly changes, the fatigue strength. If the alternating stress remains below the fatigue strength, no fracture will occur even at the highest number of load cycles. If we now know the number of load cycles that a component under consideration is to endure (e.g. number of crankshaft revolutions during the targeted engine service life), the part can be designed for a specific higher stress (fatigue strength), i.e. the material cross-sections can be made smaller for the same load.

The shorter the desired service life of a component, the more lightweight it can be designed. However, the development effort for such parts is considerably complex. Load spectra (see appendix) must be recorded, material batches must be tested regularly, and proven conditions must be observed during manufacture and assembly. In the case of parts subjected to vibration stress, for example, an incorrect grinding direction (scoring transverse to tensile stress) leads to a significantly lower number of load cycles up to fracture.

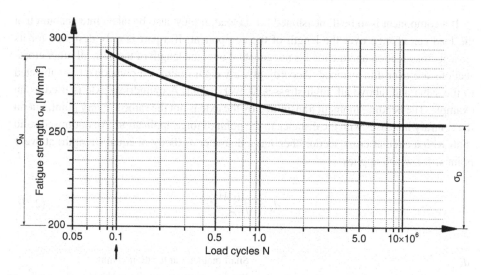

Fig. 2.26 Principle strength diagram of materials. Limit stress line (Wöhler line) of steel. σ_D fatigue strength, σ_N creep strength for N load cycles. A part made of this material is fatigue-resistant if its stress remains below σ_D. If, for example, a service life of 10^5 (arrow) load cycles is sufficient, the stress can be higher, i.e. at σ_N

For the team, this means that all parts must be individually marked (letter/number combination on adhesive labels or automatic identification systems – auto-ID: barcode, data matrix code, chip cards or RFID technology)[12] and their use (operating hours, kilometres, …) must be precisely recorded. Of course, odometers or hour meters must be on board for this purpose. Manual input of the relevant data is time-consuming and error-prone. Auto-ID systems are technically more complex, but their logistical accuracy makes them worthwhile, especially for a large fleet. With RFID systems, it is no longer even necessary to have visual contact with the parts. Before the parts reach the end of their service life, they must be replaced in good time (*lifing*, runtime control) if they are not to reach the end of their service life just during the race. The areas of a vehicle with the components that have to be replaced most frequently are the chassis, the engine compartment including drive shafts, the pedal system, the rear end, the undertray or the torque tube in the case of the transaxle concept [57].

In general, the range between the exchange of parts is very different. There are parts that are changed after every race (e.g. tyres), and then there are GT3 cars and endurance prototypes that drive 6000 km without a planned change of parts. For Formula 3 cars it is recommended to change the following parts after 25,000 km (or 2 years): steering shaft, steering gear including tie rods, brake pedal, wheel bearings, wishbones including bearings, drive shafts, wings and the rear wing suspension.

---

[12] Radio Frequency Identification.

If a component is to be dimensioned for its load, it must also be taken into account that the load can change over the length of the component. If a part is only designed for its critical cross-section and the load is not constant over the length of the part, all other cross-sections are over-dimensioned. Accordingly, a weight-optimized component is subjected to the same and, above all, high stresses at every point. A cylindrical axle is used as an example. The bending moment transmitted by the axle is not the same at every point, but on the contrary increases with the distance to the point of force application. An optimised solid shaft with a circular cross-section then has a diameter d_x (body of equal strength) at every point x, for which applies:

$$d_x \approx \sqrt[3]{\frac{10 M_{x,b}}{\sigma_{b,zul}}} \tag{2.19}$$

d_x	Shaft diameter at location x, mm
$M_{x,b}$	Bending moment at point x, Nmm
$\sigma_{b,zul}$	Permissible bending stress of the material, N/mm^2

In terms of production technology, casting, forging and sintering are suitable for complex components with cross-sectional transitions and different wall thicknesses. However, all of these are more suitable for large quantities. Racing parts are therefore more likely to be produced from the solid by machining. This is faster and more economical for small quantities. For small batch sizes and somewhat larger planning horizons, sand casting is also an option. Large-area parts are laminated (built up in layers) in a sandwich design of fibre-reinforced plastic around a foam or honeycomb core. Additive manufacturing (AM additive manufacturing, rapid prototyping, generative processes, 3D printing) is ideal for short development cycles and tight delivery times. Plastics of all kinds (ABS, PA, PPS, PEEK, PVA, PC, PLA, TPE,...), core sand and recently also metals (Al, steel, Ti, CoCrMo, NiCr) can be built up directly from a 3D CAD model layer by layer to form a casting or laminating mould or a component.

Even though the computational and experimental effort required for extreme lightweight design quickly becomes large (Fig. 2.27), simple rules and strategies can be specified with which material and thus component mass can be saved even without FEM analyses (computer-aided numerical stress calculation carried out with computers, see Appendix). material and thus component mass can be saved. Although series-production vehicles should also have a low mass, cost-optimized lightweight design is pursued for economic reasons. In the case of racing vehicles, other goals are in the foreground with significantly lower quantities, which is why weight reduction is pushed much further (extreme lightweight design).

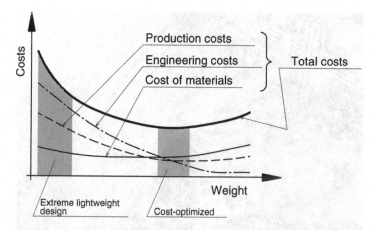

Fig. 2.27 Relationship between weight and cost of structures. The total costs (composed of manufacturing, engineering and material costs) depend directly on the weight of the design. Engineering costs increase with decreasing weight, as do manufacturing costs. In addition, the material costs increase with lighter materials as well

Some of these principles are:

- Direct load application
- Realization of high surface moments of inertia
- Integral design
- Active area parallelism
- Speed increase
- Overload limitation
- Improved cooling for thermally stressed designs
- Reduction of notch effect due to favourable force or stress flow
- Use of high-strength materials

Principle of Direct Load Path

If a force or moment is to be transmitted from one point to another with as little deformation as possible, then the direct and shortest load transmission path is the most expedient: Only a few zones are loaded and the load transmission paths, whose cross-sections must be designed accordingly, are minimised in terms of material expenditure (weight, volume) and resulting deformation. This is especially true if the task can only be solved under tensile or compressive loading. In contrast to bending and torsion, these types of stress result in lower deformations. Fig. 2.28 shows how a bending beam can be replaced by tension and compression bars through appropriate design.[13]

---

[13] For a bell crank, this is done in the Racing Car Technology Manual Vol. 4 Chassis, Chap. 4, Fig. 4.65.

Fig. 2.28 Design of a turbocharger suspension consisting of tension and compression struts. (Renault Formula 1 engine from 1984, V6 1.5 l displacement). This design, consisting of three struts, is much lighter than a boom arm that is stressed to bend. The struts are attached to the cylinder head at one end and meet with the other end at a point where a bracket supports the exhaust gas turbocharger via traction

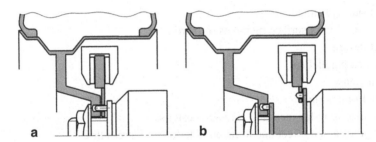

Fig. 2.29 Diagram of the load line for a disc brake. (**a**) Wheel hub with one flange, (**b**) Wheel hub with two flanges. In case (**a**), a direct load conduction takes place between the disc and the wheel. In case (**b**) a section of the wheel hub is also stressed

Another example is provided by the brake system. The braking torque is transmitted from the brake disc to the brake shoe via bolts in the wheel, i.e. there is no detour via flanges, wheel hub, etc. This would only increase the deformation in the system and the wheel hub and other parts would have to be dimensioned more strongly, i.e. more heavily, Fig. 2.29.

For stiff, light designs, tensile and compressive loads should be preferred. Whereby compression is disadvantaged by possible instability with slender component shape, namely by the failure form buckling (see appendix) or buckling. Figure 2.30 shows an

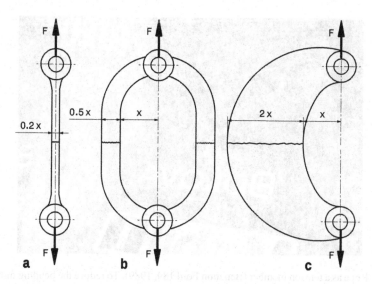

Fig. 2.30 Design of components transmitting tensile forces, according to [29]. (**a**) pure tension member, (**b**) ring-shaped member, additional bending moment. (**c**) sickle-shaped member, one-sided bending moment. The same tensile force F leads to significantly larger cross-sections (width $0.5x$ or $2x$ compared to $0.2x$ for the pure tension member) and thus to significantly heavier components, despite the same material strength under additional bending stress

example of the obvious influence of additional bending stresses in components on their mass.

Closed and symmetrical profiles or designs prove to be more favourable than open or asymmetrical cross-sections.

This principle of only allowing tension/compression stresses leads to truss designs. Their disadvantage is that a large construction volume is required for large loads.

Double wishbone axles also basically consist of struts that transmit tension and compression.

The ideal element in the sense of this principle is the rope. In fact, it is used in many places when it is necessary to support or stiffen parts with a low mass, Fig. 2.31.

Realization of High Surface Moments of Inertia
In order to achieve the greatest possible bending stiffness and strength for a given cross-section (and therefore mass), the load-bearing surfaces of a cross-section must be arranged as far apart as possible (cf. also Sect. 6.2.1 Tubular Space Frames). This is achieved by using tubular cross-sections instead of solid sections, sandwich structures (Sect. 6.2.3), three-dimensional load-bearing structures, corrugations in plates, ribbing and fine structuring of structures, shell formation in plates (see also Sect. 6.2.2), pre-curvature of structures against the main load direction, etc.

Fig. 2.31 Rope as a tension member (Benetton Ford 189, 1989). To reduce the bending moment due to the weight and downforce of the front wing, it is tensioned with a rope towards the upper side of the nose

Integral Design

This refers to the combination of several individual parts into one workpiece. Typical examples are cast constructions instead of welded constructions, extruded profiles instead of joined standard profiles, forged flanges instead of joined flanges. In this way, joints are eliminated and tight tolerances become possible, which are not feasible when constructing a structure from several individual parts due to the accumulation of individual tolerances.

Preferred manufacturing processes are: Casting, especially investment casting and injection molding, sintering, sheet metal forming, deep drawing, forging, erosive or electrolytic removal, lamination of fiber mats and machining from the solid.

Conversely, differential design, i.e. the division of a component into several parts with different tasks, also has its advantages: Each part can be stressed to its limit and the choice of material can be concentrated on the partial task without compromise. The individual parts can be manufactured in parallel, which can be particularly important for time-critical components. Furthermore, typical replacement parts can be limited to individual wear points and the entire component does not have to be disposed of in the event of maintenance.

Composite design involves combining favorable material properties in one component by building a part from several materials that are inseparably bonded together. An example of this is CFRP control arms with glued-in metal joint eyes.

Active Area Parallelism

A parallel connection of active surfaces results in a power division, whereby the entire system can be designed smaller with the same power. An example of this is provided by planetary gears, which can also be found in differentials. The torque to be transmitted is divided among several planetary gears, so that each individual gear theoretically only has to transmit a fraction of it. In fact, the gears must be designed for a somewhat larger torque because manufacturing tolerances mean that the input torque is not evenly distributed among all the planetary gears. Another example can be seen in the all-wheel drive of a vehicle. The power to be transmitted is distributed to all wheels in a certain ratio. These are therefore further away from the slip limit and can transmit a greater lateral force for the same total power (cornering).

Speed Increase

If the speed of a power transmission system is increased while the power remains constant, the torque to be transmitted is reduced (power = torque times speed). It therefore makes sense to flange the gearbox directly to the engine. Due to the high input speed, the input shaft only has to transmit a relatively small torque and is correspondingly small, as is the entire gearbox. The increase in torque required for the vehicle drive occurs as late as possible, in this case in the rear-axle gearbox, but the power flow to the drive wheels is divided between two side shafts (parallel connection of active surfaces).

Overload Limitation

The mass of components results from the dimensioning. If components are designed for a load that occurs extremely rarely, they are overdimensioned for all other loads, i.e. they are too heavy. As a result, neighbouring components (bearings, housings, flanges,...) must also be dimensioned larger, which makes the entire system heavier. Remedy is offered by a reduction of impact effects, e.g. by softer drive or installation of elastic intermediate elements. Another possibility is to fix the highest external load exactly by a defined limitation. Such a possibility is offered by slipping clutches, fluid couplings, predetermined breaking points, pressure relief valves, isolating switches, etc.

Innovations

Like everywhere else where you want to stand out from the competition, racing is also, or rather especially, characterized by the eternal search for tricks, improvements and innovations with which you can literally leave the competition behind. In doing so, the potential of new solutions must be weighed against the time required for the innovation to mature. All too often, you find that the (immature) brand new is beaten by the mature old – at least initially. Here, "teething troubles" and lack of experience of the users (drivers, race engineers and mechanics) at the beginning of the development play a major role. And not infrequently, it is not the person who introduced a new system, but the person who adopted and further developed it, who reaps the rewards of the idea. So every innovation does not have to be synonymous with winning the first race, rather a lean period will be the result during which the new system has to be developed. In that context, reliability is an issue. By introducing a new system, a new concept or a new solution, you may be abandoning the tried and tested and risking failure. On the other hand, you can't think too conservatively either, or one day you'll be driving behind. Any team planning a major innovation would be well advised to develop in parallel. That is, the new car is built with proven solutions and, independently, part of the team tackles the development of the new solution. Only when the flawless functioning of the new solution has been assured will it be used in the vehicle. Of course, this requires an appropriate team size and a coordinated budget.

Maturity curves of numerous technical systems describe a typical course of innovation: Initially, they work purely mechanically and with fixed, compromised settings, followed by electrically influenced solutions with partial adjustability, and the final stage is represented by electronic systems that are also adaptable and capable of learning. In the search for new solutions, it is advantageous to keep the technical ideal in mind, even if it seems superficially unattainable. In many cases, the linguistic formulation of the actual problem proves to be a big step in the right direction.

Innovations often result from a change of material or manufacturing process. A change in the operating principle can be just as effective (gas spring instead of metal spring, liquid damping instead of friction damping,...).

2.6 Materials

The designer must know about materials, as they have a lasting influence on the design. For example, shaping depends on the manufacturing process, which in turn depends on the material. Naturally, the properties of materials vary, such as strength, hardness, brittle, non-corrosion resistant, tribologically unfavourable, conductive, creep tendency, hot forming resistant, weldable, ageing, chemical resistance, and so on. Furthermore, the joining technique cannot be selected without knowledge of material properties.

Therefore, a brief overview of the construction materials follows first, followed by a comparison, and finally advice on selection is given.

2.6.1 Overview of Common *Materials*

Construction materials are divided into four groups, which can be further subdivided:

- Metals
  - Ferrous metals
  - Non-ferrous metals
- Powder and sintered materials
- Non-metallic substances
  - Plastics
  - Wood
- Composites

Metals

Ferrous Metals

Steel. Steels are still among the most important materials in vehicle design. This is also reflected in a favorable price per kilogram. Even in a modern passenger car, steel accounts for more than half of all materials used by mass. There are numerous different grades whose properties can be specifically modified by alloying and heat treatment. Steels are generally easy to forge and weld. Shafts, gears, springs, bolts and exhaust systems, among other things, are made of steel.

Cast steel. Is a form of steel characterized by the manufacturing process (casting followed by annealing) with practically the same properties as steel, i.e. forgeable, weldable and alloyable.

Cast iron. Cast iron has a high content of carbon, most of which is present as graphite (lamellar, spherical or cubic) in the structure. Its outstanding features are above all its compressive strength, favourable running properties and its damping capacity. Many engine blocks are therefore still made from this material today. So-called austenitic cast iron is more suitable for lightweight designs because its strength values are almost twice as high as those of unalloyed cast iron grades.

Non Ferrous Metals

Aluminium. Aluminium and especially its alloys are an important lightweight design material. Similar to steel, a wide range of properties can be achieved by alloying. There are *cast* and *wrought alloys,* which are characterised by low density and high strength. Generally, wrought alloys are stronger and tougher. However, this can be compensated for by special casting processes (vacural casting, squeeze casting, thixo forming).

Sintered aluminium and foamed aluminium are interesting variations for lightweight design. Aluminium parts have a wide range of applications and can be found in engine and gearbox design as well as in the frame and chassis area: housings, covers, levers, pistons, screws, wheel carriers, brake calipers and brackets.

Magnesium. Magnesium alloys are characterized by extremely low density with useful strength. They are easy to cast (extremely good in die casting) and easy to machine. Wrought alloys are also available. However, a number of disadvantages stand in the way of their obvious widespread use. One of these negative points is its easy flammability: for this reason, its use is prohibited by some regulations in certain places, such as the cockpit. In addition, the components are very notch-sensitive and, depending on the alloy composition, also susceptible to corrosion. The low elongation at break means that magnesium parts are sensitive to impact and shock. Nevertheless, some alloys prove to be more fatigue-resistant than higher-strength aluminium alloys when subjected to vibrating stresses. Magnesium alloys are therefore recommended for gearbox housings and other parts that are subjected to dynamic loads over long periods of time. Wheels are also made from this metal.

Titanium. Titanium has some properties that make it interesting for racing cars. At about half the density of steel, it sometimes outperforms even high-strength steels in strength. It exhibits low thermal expansion and is very corrosion resistant. Titanium and its more commonly used alloys are not without their drawbacks. Once apart from the extremely high costs, titanium proves to be a poor running partner. Thus, collar bearing shells or coatings are required on connecting rod eyes to prevent wear on the steel crank webs. The notch sensitivity requires particularly careful design of cross-section transitions and force introduction (especially screw connections). Nevertheless, titanium has a high fatigue strength, provided that the surface quality is high. The value for fatigue strength in relation to tensile strength is about 0.7, which is significantly higher than for other materials [27]. In the case of very highly stressed components, however, failure can also occur without any noticeable previous deformation. Some titanium alloys have different processing properties. Some are readily weldable, while others exhibit high high-temperature strength. Machining is difficult with all alloys. Well-known parts made of titanium alloys are coil springs, screws, connecting rods, mushroom valves, wheel suspension parts, wheel hubs, shafts, housings and exhaust systems.

Sintered Materials (*Powder-Metal Material*)
They are manufactured using powder metallurgical processes. The density of the material can thus be varied within wide limits from porous to dense. This process can also be used to bring metals with the highest melting points (tungsten, molybdenum,...) into a desired shape. Sintered parts are mostly small filigree parts, such as gear wheels, chain wheels, encoder wheels, etc. or – in porous form – also bearing and filter inserts.

Nonmetallic Materials
Plastics. There is a wide range of plastics, which are further divided into thermoplastics and thermosets. Resins are important for load-bearing components, namely as a matrix material for fibre-plastic composites.

Composites
A composite material consists of at least two components which are present next to each other, i.e. which are not dissolved in each other. Such composites are produced, among other things, by sintering, special casting processes or impregnation of porous semi-finished products. By cleverly combining individual materials, it is possible to produce composites that combine the positive properties of the components while masking their negative properties.

Fibre composites
Composites consist of a combination of several material types (name!). A "fibre-reinforced plastic", for example, consists of plastic as the base material (the so-called matrix), in which fibres are embedded for reinforcement. By combining the fiber with the base material, it is thus possible to create composite materials with properties specifically tailored to an area of application. These materials exhibit the best properties in the direction of the fibres. Therefore, materials with combined fiber directions are also produced (bi- and

Table 2.14 Comparison of the properties of fibres

	Glass fiber	Carbon fiber	Synthetic fibre
Advantages	Inexpensive production High resistance to chemicals	High strengths High stiffness in tension and shear High potential for increase/ improvement Cost-effective production	Extremely high tensile strength High tensile stiffness Non-brittle Good thermal resistance High glass transition temperature High dimensional stability
Disadvantages	Low stiffness	Brittle Anisotropic thermal expansion	Moisture absorption (hygroscopic)

multidirectional fibers) or typical components are represented as shell bodies by a laminate of several unidirectional layers. Monocoques of monoposti and driver cells of Le Mans prototypes are classic representatives of such parts.

The most commonly used base materials are thermosets (resins) and thermoplastics. However, light metals and ceramics can also be reinforced in this way. Glass, carbon and synthetic fibres (aramid, Kevlar) can be used as fibres. Compared to solid materials, the fibres have considerably higher tensile strengths and are flexible despite their high hardness. The advantages and disadvantages of common fibres are summarised in Table 2.14.

Such fiber composites have good specific properties, but also a more complex material behavior, which complicates the calculation and component design.

CFRP

(*carbon-fibre reinforced plastic*). In this case, carbon fibres are embedded in a resin matrix. They have a high stiffness and tensile strength at low weight. It is therefore not surprising that about 60% of Formula 1 cars are made of CFRP. These include the monocoque, nose including crash element, bodywork parts, wings, parts of the wheel suspension, covers as well as housings on the engine, parts of the transmission, clutch discs and air ducts for brakes. But not only the static strength is outstanding, also the fatigue strength related to the tensile strength is higher than some steels or aluminum alloys.

Until now, this composite material has simply been too expensive to purchase and process for mass-produced vehicles. Recently, however, developments in this direction have been increasing. After all, lightweight design is a key issue when it comes to reducing fuel consumption and thus CO_2 emissions in vehicles. In addition to their low fatigue tendency, CFRP components also have good crash properties and high corrosion resistance.

MMC

(*metal matrix composites*). Just as the matrix of CFRP is made of plastic, it can also be metallic. In order to save weight, light metal (aluminium, magnesium alloys) is preferred. The fibres or particles are made of steel, carbon or ceramics. MMC are more creep resistant than unreinforced light metal alloys and their strength decreases only slightly with an increase in temperature. Known parts made of MMC with an aluminum matrix are pistons, crankcases and wheel carriers. Pistons can also be constructed with a magnesium matrix.

2.6.2 Material Comparison

Absolute values are unsuitable for direct comparison of different materials in question. Instead, relative values must be used. This starts with the density, i.e. the mass of the material in relation to the volume it occupies. However, density alone is not sufficient as a comparative value for the choice of material; instead, the required mass and – depending on the external load – the strength of the material (permissible stress; permissible deformation, ...) must be put into perspective.

The specific stiffness $E/(g\ \rho)$ compares the longitudinal deformation behaviour (E = Young's modulus, see appendix) in relation to the weight. The tear length $R_m/(g\ \rho)$ clearly represents the length at which a suspended bar would tear under its own weight (R_m = tensile strength, see appendix). The buckling stiffness of bars in relation to their weight is quantified by the term $\sqrt{E}/(g \cdot \rho)$. Table 2.15 provides an overview of important property values of common material groups.

Beryllium not only exhibits extremely high stiffness, but also provides this with a low specific weight. Its tear length is correspondingly large. However, unidirectional AFK is

Table 2.15 Comparison of materials, according to [27]

Material	ρ [kg/dm^3]	E [N/mm^2]	R_m [N/mm^2]	$E/(g\ \rho)$ [km]	$R_m/(g\ \rho)$ [km]	$\sqrt{E}/(g \cdot \rho)$ [m$^2/\sqrt{N}$]
Steel	7.85	210,000	500	2675	6.37	5.95
Al alloy	2.70	70,000	350	2593	12.95	9.99
Mg alloy	1.74	40,000	330	2299	18.96	11.12
Ti alloy	4.50	102,000	900	2267	20.00	7.23
PA 6 (dry)	1.15	2500	80	217	6.96	4.43
GRP-UD[a] (50%)	1.95	40,000	800	2051	41.03	1.95
CFRP-UD[b] (50%)	1.40	250,000	1000	17,857	71.41	36.41
AFK-UD[c] (50%)	1.35	65,000	1500	4815	111.11	19.25
Beryllium	1.85	245,000	400	13,243	21.62	27.27
Wood	0.50	12,000	100	2400	20	22.33

[a]Glass fibre reinforced plastics unidirectional
[b]Carbon fiber reinforced plastics
[c]Aramid fibre reinforced plastics

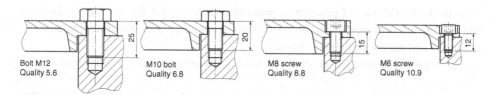

Bolt M12   M10 bolt   M8 screw   M6 screw
Quality 5.6   Quality 6.8   Quality 8.8   Quality 10.9

Fig. 2.32 Influence of material strength on weight, after [35]. The external load is the same for all four connections. However, the bolts have different strengths and therefore the dimensions of the bolts must be different. The resulting stress (tension) is the same for all of them. Mass can therefore also be saved with higher-strength material

unsurpassed in this respect. In general, all fibre composites are ahead in this table in terms of tear length. Wood can also be interesting as a structural material for lightweight design – it has the same tear length as titanium alloys. When it comes to longitudinal stiffness at low weight, unidirectional CFRP is the first choice. It has the highest specific stiffness. For weight-saving compression bars, magnesium alloys can be an interesting alternative. They have a higher specific buckling stiffness than, for example, steel and titanium and aluminum alloys, although they have a much smaller modulus of elasticity than these.

The effect of higher material strength is clearly illustrated in Fig. 2.32. The higher strength bolt can be made smaller for the same load. This means that the area around the bolt can be made correspondingly smaller, which in turn makes the entire structure lighter. Hexagon socket screws (DIN 912) are particularly favourable in this context.

2.6.3   Material Selection

For the selection of the material, economic and technical aspects are considered. Which ones predominate depends on the task and the budget. One criterion must be ensured in any case: The strength must be sufficient to fulfil the function. Furthermore, the choice of material and manufacturing process is not always possible independently of each other. Certain materials can only be given the desired shape using certain manufacturing processes (castings, sintered parts, . . .) and, conversely, some manufacturing processes restrict the choice of material (welded parts, erosion processes, drawn parts, . . .). Forming processes have the advantage to weld material defects (flaws, pores, . . .) and to enable a force-flux-appropriate fibre flow. This makes them suitable for safety-relevant parts that are subject to continuous vibration stress. However, the tooling costs for small quantities are too high and the time required for any design changes are disproportionately greater in comparison to welded or milled parts Parts for such applications that are machined from the solid undergo numerous production phases: Among other things, they must be stress-relieved, shot-peened, polished and individually inspected.

In the highest classes of motorsport, materials are often used that are not standardized and are difficult or impossible to obtain on the open market. Such materials are naturally expensive and have long delivery times.

The following criteria can generally be used for material evaluation and selection:

1. To ensure function: strength (time, temperature, ...), hardness, modulus of elasticity, elongation at break; corrosion; thermal expansion, thermal conduction; electrical properties, tribological properties; haptics (see appendix); damping properties, acoustic properties.
2. Weight of the finished component
3. Manufacturing properties: Castable, forgeable, weldable, deep-drawable, etc.
4. Cost of materials

For larger quantities, the following criteria are added:

5. Investments required for processing and testing
6. Running costs in production
7. Recycling possibility

Some considerations on individual criteria listed above: It is not sufficient to aim for higher strength alone. Toughness, i.e. plastic deformability, enables stress peaks to be relieved in the case of unevenly distributed stresses and is one of the most significant safety factors that a material can offer. In general, the toughness of materials decreases with higher strength. Thus, care must be taken to maintain a minimum toughness to ensure the benefits of plastic deformability. Dangerous cases are those in which the material becomes brittle with time or for other reasons (e.g. radiation, corrosion, temperature or due to surface protection) and thus loses the ability to deform plastically when overstressed. This behaviour is particularly true for plastics.

The weight of the finished component depends on its load and thus on the type of load. In addition to design measures (see Sect. 2.5 General design principles – lightweight design), the decisive criterion here is the choice of materials. Table 2.16 provides a summary of the selection criteria for parts with the lowest mass from common semi-finished products under different loads.

For series designs, costs are high on the list of requirements. Here, an economical approach to material selection will be preferred. The most cost-effective material that can just bear the load will be used. If one wants to save weight by using lighter material, the costs will increase because, in general, material prices increase with decreasing density.

The machinability of the material is essential for the production costs. This is particularly important for large components and the production of large quantities. The easier a material is to machine, the more favourable it is for production. Materials are generally easier to machine the lower their strength (hardness). Grey cast iron behaves roughly like medium strength steel. Copper alloys, plastics, but especially light metals are generally

Table 2.16 Specific material properties for stiffness and strength-oriented parts at minimum mass, according to [45]

Semi-finished product	Load	Specified dimension	Changeable Dimension	Orientation according to	
				Stiffness	Strength
Round bar	Tension	Length	Diameter	$\dfrac{E}{\rho}$	$\dfrac{R_{ref}}{\rho}$
Rectangular profile tube	Bending	Length and width	Profile height	$\dfrac{E^{1/3}}{\rho}$	$\dfrac{R_{ref}^{1/2}}{\rho}$
Round rod or shaft	Torsion	Length	Diameter	$\dfrac{G^{1/2}}{\rho}$	$\dfrac{R_{ref}^{2/3}}{\rho}$
Flat plate	Bending	Length and width	Board thickness	$\dfrac{E^{1/3}}{\rho}$	$\dfrac{R_{ref}^{1/2}}{\rho}$
Round bar	Compression	Length	Diameter	$\dfrac{E^{1/2}}{\rho}$	$\dfrac{R_{ref}}{\rho}$

Explanation: E modulus of elasticity, G shear modulus, ρ density of the material, R_{ref} relevant strength value (tensile strength, shear strength, compressive strength)

easier to machine than steel. High-strength, austenitic steels or special steel castings (stainless and/or heat-resistant) are difficult to machine.

In addition, the choice of material, especially for "exotic" materials, should already be made in the concept phase of a project. It is not uncommon that the desired material cannot be used because it is not available at short notice.

2.7 Costs

This section could be invalid if the following statement is true: Racing cars are designed strictly from a purely technical point of view – apart from the regulations – and therefore cost is not an issue. However, it is true that costs are always an issue. The only question is with what importance they are included in the considerations when designing a vehicle. In the case of series-produced vehicles, costs, especially manufacturing costs (and here the cost of materials), are at the top of the list. Here, optimisation of the manufacturing and assembly processes pays off in the truest sense of the word. Things are different for racing cars, of which often only a few individual pieces are built. Nevertheless, even a Formula 1 team has a (finite) budget with which it must make do and which must be divided into individual areas according to various criteria, Fig. 2.28. When designing, therefore, all designers must reach for the ceiling, even though it naturally does not hang equally high – or better – low for everyone. Cost, then, is an issue and every designer will do well to consider what is worth to him and how much. The findings from Sect. 2.3.2 *Concept comparison*, share of individual assemblies in driving performance (which may seem purely theoretical to some) can be helpful in these considerations.

Figure 2.33 shows (on a double logarithmic scale!) that Formula 1 is outstanding in terms of team size and budget and is unsuitable for comparative purposes when considering

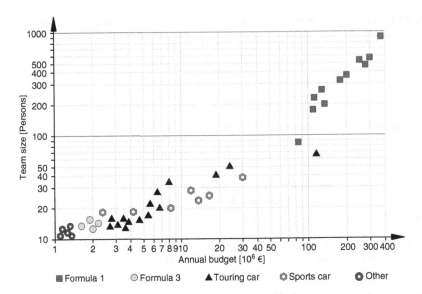

Fig. 2.33 Overview of budgets and team sizes in motorsport, after [30]. For five typical motorsport categories, annual budgets are compared with employee numbers. By far the largest teams and the highest budgets are found in Formula 1 (note the double logarithmic scale!). At the other end of the field, micro teams with 10–20 members manage a season at a high level with a fraction of the expense

costs because it represents a world of its own. Private teams, which participate in the long distance race in Le Mans, partly do without weight-saving materials, because with a regulation-conditioned minimum weight of the vehicle, the additional costs are not to be justified. Thus, steel conrods are used in the engine instead of titanium conrods. Even the valves in the cylinder head are made of steel and not titanium. Of course, titanium valves would have advantages for the entire valve train due to their lower mass – a possible weaker valve spring creates less friction at the camshaft and requires lower opening forces, etc. – but the bottom line is that the advantages are just not so great that the expensive material pays off.

Teams also don't like it when their drivers drive excessively hard at the limit. That only puts a strain on the budget. Likewise, teams try to work as efficiently as possible during the testing phase, for example by planning individual test points in advance. Development methods that can reduce testing on the track are now also worth considering for Formula 1 teams. After all, the bottom line is that a test kilometer costs around €650.

Material Costs

Although the absolute prices of the materials are subject to the usual market fluctuations, the price ratios remain fairly constant over the years and behave roughly as in [27]:

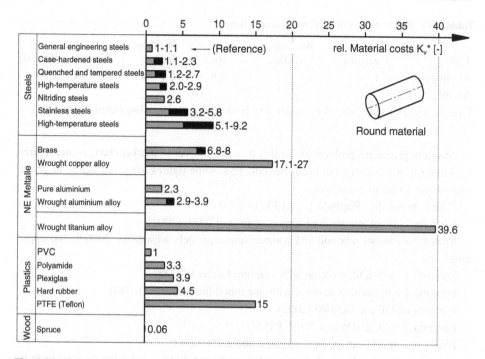

Fig. 2.34 Relative costs of selected materials, after [35]. Round structural steel material is used as a reference for the relative material costs. Scatter ranges result for materials with different grades. The costs are gross costs related to the volume. For example, 1 dm³ titanium wrought alloy costs about forty times that of mild steel of the same volume

Steel: Aluminium: GRP: AFRP: CFRP = 1: 5: 10: 100: 500.

With decreasing specific weight, the materials therefore become more expensive.

Relative material costs result in comparative values which hardly change their relation to each other over longer periods of time – even in the case of price fluctuations. Some examples of relative costs according to VDI 2225 are given in Fig. 2.34. Here, the costs per volume are related to a common material (structural steel round material).

When comparing material costs, however, it is important to compare the quantity needed for the same load. Higher-strength (and therefore generally more expensive) materials can actually be more cost-effective on balance, because less material is needed for the same application, see Fig. 2.32.

If sheets are to be joined together, as may occur in frame design, wingss or bodywork parts, a joining technique must be selected. In addition to functional (joints, overlaps, . . .) and strength considerations, costs can also play a role, Table 2.17.

Table 2.17 Relative costs of sheet metal connections [35]

Link	Spot welding	Bonding (Araldite)	Rivets	Welding (arc)	Screws	Brazing
Relative cost figure	1	1.7	2.6–3.5	2.9–4.4	3.6–4.4	3.7–6.9

3 mm steel or aluminium sheets, batch size 200, production costs excluding material costs

Absolute prices are problematic in that they are subject to market changes and figures therefore quickly become outdated. Nevertheless, some figures are given below to give a sense of the order of magnitude.

Carbon brake disc Formula 1 ... 1300 to 4000 €.

Formula 1 wheel, magnesium forged: approx. 1500 € (2004).

Monotube damper rebound and compression separately adjustable: 2000 €, "no upper limit".

Formula 1 carbon fibre clutch with titanium basket €13,000.

Formula 1 longitudinal gearbox with integrated final drive €110,000.

Formula BMW car €45,000 (2002).

Formula 3 vehicle (Dallara 2007) €88,000.

Production sports car (without engine): about 100,000 €.

If you may not be able to deduce much from it, one thing is immediately clear from it: nothing works in racing without sponsors!

References

1. Braess, H.-H., Seiffert, U.: Vieweg Handbuch Kraftfahrzeugtechnik, 1st edn. Vieweg, Wiesbaden (2000)
2. Fuhrmann, E.: Kraftfahrzeugkonstruktion. Skriptum zur Vorlesung, TU Wien (1983)
3. Hack, I.: Formel 1 Motoren, Leistung am Limit, 2nd edn. Motorbuchverlag, Stuttgart (1997)
4. Tremayne, D.: Formel 1, Technik unter der Lupe. Motorbuch Verlag, Stuttgart (2001)
5. Berkefeld, V., Dworzak, U.: Sportliches Fahren auch in der Zukunft? ÖVK-Bericht (2002)
6. Desoyer, K.: Skriptum zur Vorlesung Fahrzeugdynamik. TU Wien, Wien (1984)
7. Voigt, T.: Tourenwagen Story '98. Sport, Hamburg (1998)
8. Schöggl, P.: Aktuelle Trends und Methoden in der Rennfahrzeugentwicklung (22.01.2003). Vortrag im Rahmen der ÖVK Vortragsreihe. Graz (2003)
9. Bosch: Kraftfahrzeugtechnisches Taschenbuch, 22nd edn. Vieweg+Teubner, Wiesbaden (2001)
10. Henker, E.: Fahrwerktechnik. Vieweg, Wiesbaden (1993)
11. Piola, G.: Formula 1. Technical Analysis 2005/2006. Giorgio Nada Editore, Mailand (2006)
12. N.N.: Formula Renault 2000 Manual, Renault Sport Promotion Sportive (2001)
13. Ludvigsen, K.: Mercedes Benz Renn- und Sportwagen, 1st edn. Motorbuch Verlag, Stuttgart (1999)
14. McBeath, S.: Competition Car Preparation, 1st edn. Haynes, Sparkford (1999)
15. Stulle, O.: Motor pur, Die Entwicklungsgeschichte der Formel 1 Motoren. Manuskript
16. Staniforth, A.: Race and Rallycar Source Book, 4th edn. Haynes Publishing, Sparkford (2001)

17. Incandela, S.: The Anatomy & Development of the Formula One Racing Car from 1975, 2nd edn. Haynes, Sparkford (1984)
18. Ulrich, W.: Audi- der Sieger von Le Mans. Vortrag der ÖVK Vortragsreihe, Mai, Wien (2004)
19. Hölscher, M.: Carrera GT. Der neue Hochleistungssportwagen von Porsche. Vortrag der ÖVK Vortragsreihe, Okt, Wien (2003)
20. Bamsey, I.: The Sound of Speed. In: Race Tech Okt./Nov. 2004, vol. 56, p. 32. Racecar Graphic Ltd., London (2004)
21. Bamsey, I.: Raising the Bar, in Race Tech Okt./Nov. 2004 56, 22 ff. Racecar Graphic Ltd., London (2004)
22. Appel, W.: Development of the chassis for the R8. In: AutoTechnology, vol. 3, pp. 56–59. Wiesbaden Vieweg Verlag (2003)
23. Smith, C.: Tune to Win. Aero Publishers, Fallbrook (1978)
24. Soltic, P., Guzzella, L.: Verbrauchsvergleich verschiedener Verbrennungsmotorkonzepte für Leichtfahrzeuge. In MTZ Heft 7/8(2001), 590 f. Vieweg, Wiesbaden (2001), 59596
25. Wright, P.: Formula 1 Technology, 1st edn. SAE, Warrendale (2001)
26. Tremayne, D.: The Science of Formula 1 Design, 1st edn. Haynes Publishing, Sparkford (2004)
27. Klein, B.: Leichtbau-Konstruktion, 5th edn. Vieweg, Wiesbaden (2001)
28. Dubben, H.-H., Beck-Bornholdt, H.-P.: Der Hund, der Eier legt. Erkennen von Fehlinformationen durch Querdenken. Rowohlt Taschenbuch, Hamburg (2006)
29. Hintzen, H., et al.: Konstruieren und Gestalten, 3rd edn. Vieweg, Braunschweig (1989)
30. Müller, J.: Synergieeffekte und Mehrwert durch Automotive-Entwicklungsdienstleistung im Rennsport. Vortrag auf der RaceTech, München (2006)
31. Reimpell, J., Betzler, J.: Fahrwerktechnik: Grundlagen, 4th edn. Vogel, Würzburg (2000)
32. Staniforth, A.: Competition Car Suspension, 3rd edn. Haynes, Sparkford (1999)
33. Formula One Technology 2006/2007: A Race Engine Technology Special Report. High Power Media, Somerset (2007)
34. Lechner, G., Naunheimer, H.: Fahrzeuggetriebe, 1st edn. Springer, Berlin (1994)
35. Ehrlenspiel, K., Kiewert, A., Lindemann, U.: Kostengünstig Entwickeln und Konstruieren. Kostenmanagement bei der integrierten Produktentwicklung, 3rd edn. Springer, Berlin (2000)
36. Muhs, D., et al.: Roloff/Matek Maschinenelemente, 17th edn. Vieweg, Wiesbaden (2005)
37. Van Basshuysen, S. (ed.): Lexikon Motorentechnik, 1st edn. Vieweg/GWV-Fachverlage, Wiesbaden (2004)
38. Mühlmeier, M.: Virtual Design of a World Rally car. Vortrag auf der Race.Tech, München (2006)
39. Gandy et al.: Berücksichtigung von Ungenauigkeiten bei Zuverlässigkeitsanalysen in frühen Entwicklungsphasen. Konstruktion. 6. Springer, Berlin (2006)
40. Heißing, B., Ersoy, M.: Fahrwerkhandbuch, 2nd edn. Vieweg+Teubner, Wiesbaden (2008)
41. Milliken, W., Milliken, D.: Race Car Vehicle Dynamics, 1st edn. SAE Inc, Warrendale (1995)
42. Toso, A.: Race cars. In: Mastinu, G., Ploechl, M. (eds.) Road and Off-Road Vehicle System Dynamics Handbook, pp. 497–515. CRC Press, Boca Raton (2014)
43. Rechberger, L.: Konzeptentwicklung vom idealen Formula-Student-Fahrzeug basierend auf einer Rundenzeitsimulation. FH Joanneum, Diplomarbeit Graz (2014)
44. Schumacher, A.: Optimierung mechanischer Strukturen: Grundlagen und industrielle Anwendungen, 2nd edn. Springer, Berlin (2013)
45. Mallick, P.K.: Fiber Reinforced Composites. Materials, Manufacturing, and Design, 3rd edn. CRC Press, Boca Raton (2008)
46. Bauer, H., et al.: Bosch, Kraftfahrtechnisches Taschenbuch, 25th edn. Springer Fachmedien, Wiesbaden (2003)
47. Naunheimer, H., Bertsche, B., Lechner, G.: Fahrzeuggetriebe, 2nd edn. Springer, Berlin (2007)

48. Rompe, K., Heißing, B.: Objektive Testverfahren für die Fahreigenschaften von Kraftfahrzeugen. Quer- und Längsdynamik, 1st edn. TÜV Rheinland, Köln (1984)

49. Bennett, N.: Inspired to Design. F1 Cars, Indycars & Racing Tyres: the Autobiography of Nigel Bennett. Veloce Publishing, Poundbury (2013)

50. Bickerstaffe, S.: Going Full Circle. In Automotive Engineer, Heft Dez., 37 (2013)

51. van Schalwyk, D.J., Kamper, M.J.: Effect of Hub Motor Mass on Stability and Comfort of Electric Vehicles. University of Stellenbosch, South Africa (2013)

52. Newey, A.: How to Build a Car. HarperCollins Publishers, London (2017)

53. Bertsche, B., Lechner, G.: Zuverlässigkeit im Fahrzeug- und Maschinenbau, 3rd edn. Springer, Berlin (2004)

54. Flaschenträger, D., et al.: Zuverlässigkeitsbewertung eines adaptronischen Systems zur aktiven Vibrationsminderung. Konstruktion **1/2**, 52–66. VDI Fachmedien, Düsseldorf (2011)

55. Niemann, G., Winter, H., Höhn, B.-R.: Maschinenelemente, Band 1 Konstruktion und Berechnung von Verbindungen, Lagern, Wellen, 3rd edn. Springer, Berlin (2001)

56. N. N: International Electronic Council: Reliability Data Handbook, 1. Aufl. IEC, Geneva (2004)

57. Reuter, B. (ed.): Motorsport-Management. Grundlagen – Prozesse – Visionen. Springer Gabler, Berlin (2018)

Safety

3

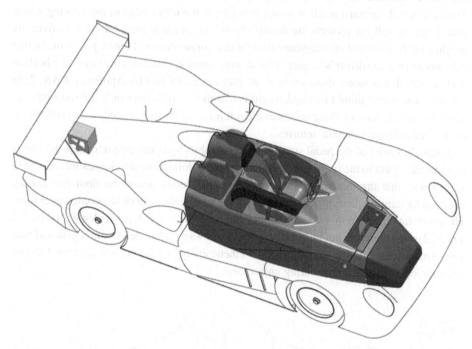

Racing vehicles are used to try to drive through a certain route as quickly as possible. The limits are set primarily by the engine power, the grip of the tires and the skill of the driver. The often quoted driving at the limit is a tightrope walk, where the limit marks are not or not always visible. Driving as fast as possible can therefore also mean exceeding the limits.

© The Author(s), under exclusive license to Springer Fachmedien Wiesbaden GmbH, part of Springer Nature 2023
M. Trzesniowski, *Complete vehicle*,
https://doi.org/10.1007/978-3-658-39667-1_3

Because this often happens at high speed, the protection of the driver is of great importance. In this context, the following quotation indicates the direction of development:

Speed does not kill, but a sudden lack of it does. (Henry Labouchere)

It is not the speed that is fatal, but its sudden loss.

3.1 Vehicle Construction

With an open cockpit, special measures must be taken to protect the driver. The structure of such vehicles is to be considered here as an example for all two-track vehicles, Fig. 3.1. In principle, this structure is also found on other competition vehicles.

The driver is placed in the vehicle so that his feet are behind the front axle. The transverse bulkheads of the main structure are formed by three bulkheads or roll bars. The central roll bar must reach at least the height of the upper edge of the steering wheel rim. A second roll bar protects the driver's head area and is located behind the driver. Its height shall be such that an imaginary straight line connecting the highest points of the two bars passes over the driver's helmet in the driving position (hands on the steering wheel) in such a way that a safety distance (e.g. 80 mm according to FIA Appendix J Art. 258) remains. The idea behind this requirement is that the car rolls over on a flat road surface. Then the vehicle rests on these points of the roll bars and the driver needs a survival space, even if the rollover structure deforms.

Located in front of the pedal system is a shock-absorbing area that transmits the forces of a frontal impact to the frame and, through its own deformation, decelerates the vehicle in such a way that the maximum acceleration values remain below the limit that can be tolerated by humans. During acceptance tests, the acceleration curve is recorded over time. The acceleration may only exceed a certain value for a few ms. A crash element (Figs. 3.2 and 3.22) must therefore not be too stiff so that it allows a certain deformation path and thus a deformation time during the impact. The kinetic energy of the vehicle is converted into deformation energy and heat during an accident.

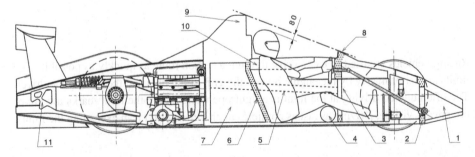

Fig. 3.1 Vehicle body diagram. 1 front protection area, 2 front bulkhead front, 3 middle bulkhead, 4 fire extinguisher, 5 side impact protection, 6 firewall, 7 fuel tank deformable structures, 8 front roll over hoop, 9 rear roll over hoop, 10 driver's shoulder protection area, 11 rear impact element

Fig. 3.2 Example of a front protection zone on a monoposto (Formula BMW). This crash element is made of CFRP laminate and is bolted to the front bulkhead only in the area of the outer bulkheads. The nose fairing, which carries the front wing, is put over it

To protect the driver in the event of a side impact, lateral deformation elements (side impact protection) are provided.

The fuel tank and engine area is separated from the driver's area on all vehicles by a flameproof bulkhead. If the fuel tank is installed in the vehicle interior without a bulkhead facing the driver, the tank must be housed in its own fireproof containment.

Further enclosing containers, e.g. made of GRP, are prescribed if the battery is housed in the vehicle interior. These must be electrically insulating and collect any liquids leaking from the battery.

The rear end of the vehicle has a rear impact element. This is attached to the rear end of the gearbox housing or to the frame, Fig. 3.3.

Survival Cell

Many regulations now stipulate a minimum area to ensure the driver's survival in the event of an accident. Figure 3.4 shows an example of the external and internal dimensions of such a cell. It is used for single-seaters and can therefore be used as a basis for the design of a single- or two-seater vehicle when determining the main dimensions.

Such cells are also provided for two-seater vehicles. The driver and passenger then each sit in their own "monocoque", which is not the load-bearing structure of the vehicle, but on the contrary is anchored in the vehicle like a seat.

3.2 Switches

It should be possible to operate the fire extinguisher from outside, e.g. by a marshal. The handle for the release cable or the switch must be marked with a symbol (see Figs. 3.5 and 3.7).

Fig. 3.3 Rear impact element of a Monoposto (Dallara Formula 3). The element is bolted to the rear end of the gearbox housing. The element is made of CFRP laminate in sandwich construction and is pot-shaped hollow. It has a relatively large overall wall thickness of about 20 mm. This is necessary so that the element can sufficiently decelerate a rear impact despite its small length

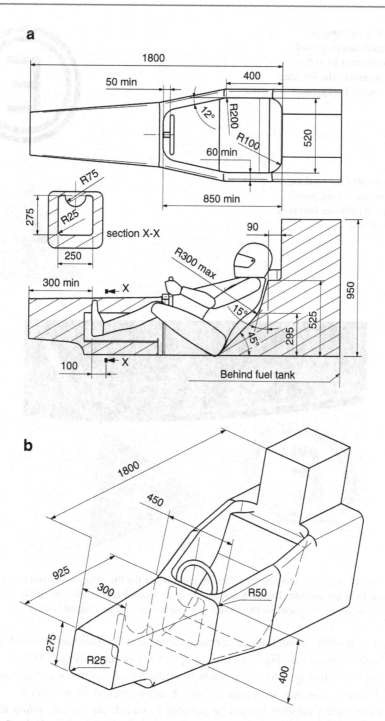

Fig. 3.4 Survival cell for a Formula 1 car. (**a**) Indoor area, (**b**) Outdoor area. The cell must start at least 300 in front of the feet (or the pedals in the unactuated state) and include the fuel tank. The minimum area between the driver's back and the tank is specified. Minimum dimensions for the entrance opening are also specified. The minimum dimensions of the inner area are set from 100 mm behind the sole of the foot (Sect. X-X)

Fig. 3.5 Fire extinguisher
release identification symbol.
This marking must be at least
10 cm in diameter. The lettering
is red on a white background

Fig. 3.6 Symbol for marking
the main switch for the power
supply. A red lightning bolt on a
blue background

Fig. 3.7 Safety switch on a touring car. The two controls for the fire extinguisher and power cut can be operated from the outside. The switches are therefore located on the left side of the front of the vehicle in front of the windscreen. The identifying symbols are located on the bonnet

It must be possible to interrupt the circuit from the outside. This master switch must be marked with a symbol (see Fig. 3.6). It must open without sparks (Fig. 3.7).

The main switch (Fig. 3.8) is accessible from the outside so that marshals can easily reach it. It is also clear how to operate the switch and where the off position is.

On circuit racing vehicles it must be possible to switch off the fuel pump from the driver's seat, e.g. by means of a toggle switch.

Fig. 3.8 Arrangement of the main switch. The actuating direction and the off-position of this main switch are indicated in an exemplary manner

3.3 Protective Devices

Rollover Structure

Roll cages as the main component of a rollover device are an essential part of the protective devices for production sports cars, touring cars, GT vehicles, rally cars, etc. (Fig. 3.9). The main components, Fig. 3.10, are laid down in the FIA regulations [1], which form the basis for many national regulations. Roll cages that do not comply with these design regulations can also be used, but must pass a static load test, Sect. 3.5 Tests. Alternatively, this test can be replaced by an FEM analysis[1] from an accredited institute.

The main hoop is located behind the front seats and spans the entire passenger compartment in one piece. The bending radius of the centre line must be at least three times the tube diameter. Connected to this, either two lateral stirrups or a continuous front stirrup form the further structure. Diagonal braces additionally stiffen this structure. The flank protection is provided by diagonal braces in the door openings. In the direction of travel, the main hoop is held by the rear supports. In the side view, the entire rollover device must be located between the points of the chassis that absorb the wheel forces, i.e. between the attachments of the body springs and shock absorbers. Additional struts may be incorporated as reinforcement. These may also be removably attached, for example by bolting.

When positioning the roof diagonal(s), it must be taken into account in some classes (e.g. GT3) that the driver's seat may only be moved within narrow limits and that a roof hatch (cf. also Fig. 3.27) is fitted above the driver to facilitate the removal of the helmet when recovering an injured driver.

---

[1] See attachment.

Fig. 3.9 Racing accident. Shortly before, the vehicle had touched a stack of tyres in a slight right-hand bend, continued on two wheels and finally rolled over to the left into the ditch, causing it to roll over several more times. The driver was uninjured and was able to get out of the car himself

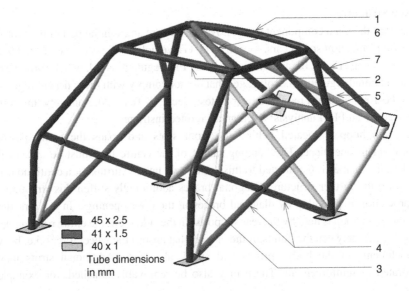

Fig. 3.10 Main elements of a roll cage, according to FIA Appendix J Art. 253 [1]. **1** main hoop, **2** front hoop, **3** lateral hoop, **4** door cross-struts, **5** diagonal struts, **6** roof diagonal member, **7** backstay

Figure 3.11 shows the possibilities of constructing roll cages from stirrups. These stirrups must all be made from one piece without joints. All pipes of the cage must not conduct any liquids.

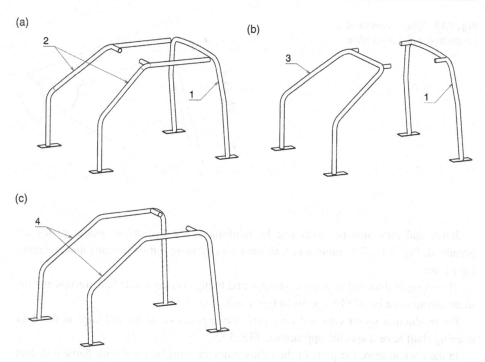

Fig. 3.11 Roll cage assembly options. (**a**) main bar (1) with two side half bars (2), (**b**) main bar (1) and front bar (3), (**c**) two side bars (4) with connection behind front seats

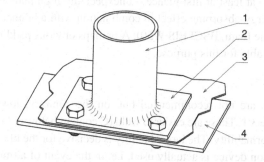

Fig. 3.12 Attachment of roll cages to the frame. **1** Bracket or strut end, **2** Mounting foot. At least 3 mm thick and no thinner than the tube to which the plate is welded, **3** Sheet metal of the chassis or bodywork, **4** Reinforcing plate. At least 3 mm thick and at least 120 cm^2 surface. For the feet of the rear support, 60 cm^2 is sufficient. This plate can also be connected directly to the mounting foot. In this case, however, it must be welded to the body

The roll cage must be attached to the frame or chassis by means of mounting feet. Each bracket end and support must have a foot, i.e. each cage must be provided with at least six feet. These feet must be bolted to the frame via reinforcing plates with at least three M8 bolts with a quality 8.8, Fig. 3.12.

Fig. 3.13 Reinforcement of a
connection with gusset plate

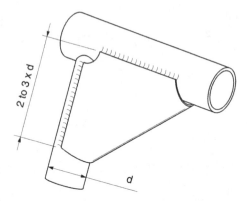

Joints and crossings of struts can be reinforced. Short struts or gusset plates are permitted, Fig. 3.13. The minimum wall thickness of these reinforcements must be more than 1 mm.

The fixing of detachable struts to stirrups and fixings between side half stirrups and the main stirrup must be of FIA approved types only, Fig. 3.14.

For production sports cars and rally cars, the appearance of the roll cage in the door opening shall have a specific appearance, Fig. 3.15.

In the cockpit area, the parts of the rollover device must be fitted with flame-retardant padding (FIA Standard 8857-2001 Type A or B, SFI Specification 45.1) that could come into contact with the driver or passenger, Fig. 3.16. What is remarkable about the permissible padding is its – at least at first glance – unexpectedly high hardness. They also only have the desired energy-absorbing effect in combination with a helmet (e.g. Confor CF42 or CF45 polyurethane foam, BSCI EIS W50). A perhaps obvious padding with a soft foam is completely unsuitable for this purpose.

Material
High-strength steels are not recommended but, on the contrary, low-alloy, low-carbon steels, see also Table 3.1. These are easier to weld (C content below 0.3 mass %) and, above all, have a large deformability. The large ductility is decisive for the life-saving effect when the rollover protection device is actually used, i.e. in the event of a (multiple) rollover.

Rollover Structures
For two-seater production sports cars with open or closed cockpits, two rollover structures are required by the FIA, Figs. 3.17 and 3.19. The front and rear parts of the main structure must be a certain horizontal distance apart and symmetrical to the longitudinal plane of the vehicle. The driver's helmet in the driving position must be a safe distance from an imaginary connection across the two rollover elements. In addition, there shall be a second rollover structure behind the driver to protect the driver in the event of failure of the main

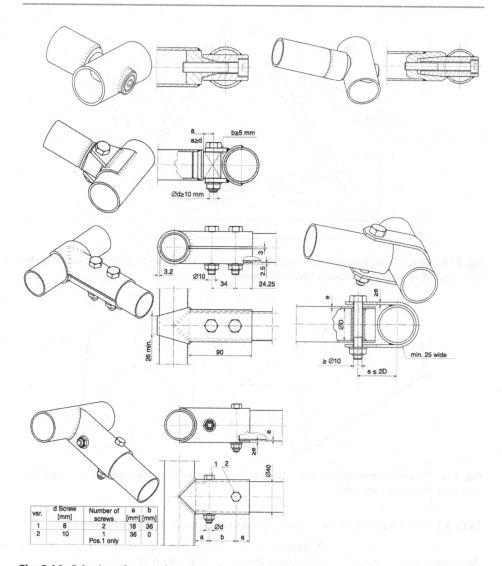

Fig. 3.14 Selection of connections recognised by the FIA

var.	d Screw [mm]	Number of screws	a [mm]	b [mm]
1	8	2	18	36
2	10	1 Pos.1 only	36	0

structure. It must project above the helmet when viewed from the front and have a minimum diameter of 280 mm.

The rear rollover device can also be used to recover the vehicle after an accident, Fig. 3.18. However, for this the manufacturer must give his written consent to the race control (Fig. 3.19).

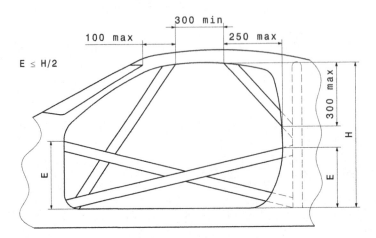

Fig. 3.15 Door opening with roll cage, prescribed dimensions [1]. Parts of the roll cage must comply with certain dimensions in relation to the door opening

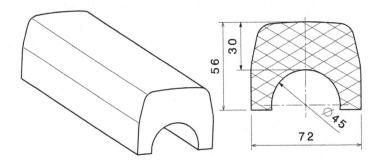

Fig. 3.16 Example of protective rollcage padding. Shown are the view and typical cross section of a casing for 45 mm diameter pipes

Table 3.1 Requirements for roll-over protection structure tubes according to FIA Annex J, Art. 253

Minimum quality	R_m, N/mm^2	Minimum dimensions, mm	Application site
Seamless, cold-formed Unalloyed carbon steel	350	45 × 2.5 or 50 × 2.0	Main hoop, side hoops and their rear connections
		38 × 2.5 or 40 × 2.0	Lateral half-bars, struts

AFP Halo (*Additional Frontal Protection* Halo)

The open cockpits of Formula 1, 2 and E cars have to have a ring-shaped cover since the 2018 season, Fig. 3.20 (FIA Standard 8869-2018: AFP – single-seater additional frontal protection). The mandatory strength tests reflect the idea behind it. The halo must withstand the impact of a wheel (approx. 20 kg mass) hitting it from the front at approx. 225 km/h.

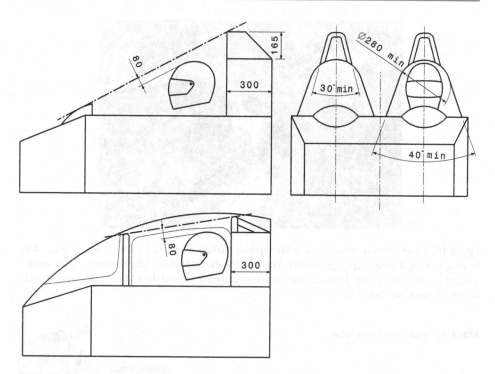

Fig. 3.17 Rollover structures on production sports cars, according to FIA Annex J Art. 258A. The rollover structures for open and closed vehicles are mandatory for the protection of the driver. Front and rear structure are connected with an imaginary line. The driver's helmet must have a safety distance from this line

Fig. 3.18 Roll bar on a monoposto (Formula 3). The bracket is mounted behind the driver. The engine receives its air through an airbox mounted on the side of the engine. Therefore, this bracket does not include the air intake of the engine. A solution that is otherwise used on many single-seaters

Fig. 3.19 Front rollover structure on a Monoposto (Dallara Formula 3). Above the steering shaft bearing is a small nose-shaped protrusion. This absorbs the front contact force with the road surface when the vehicle rolls over. The imaginary line between the nose and the rear roll bar runs above the driver's helmet (see Fig. 3.1)

Fig. 3.20 Halo on a formula car

The halo must also withstand the impact of a wheel (approx. 20 kg mass) hitting it from above. A force of 116 kN is exerted from above and 46 kN from the front on the triangular connecting node of the hoop. The resulting force is therefore 125 kN. In another quasi-static test, the halo is loaded horizontally at an angle from the front (combination from the side and to the rear) also with 125 kN. The structure is made of titanium alloy (TiAl6V4 (DIN) heat treated or Ti6Al4V Grade 5(ASTM)) and must be well matched with the chassis structure because of the high test forces. The entire halo itself weighs 7 kg and increases the vehicle mass by up to 20 kg due to the additional monocoque stiffeners.

For the open cockpits of the IndyCar Series, transparent covers similar to the cockpits of fighter planes have been in development for several years. However, for the 2019 season, only trapezoidal deflector elements (approximately 80 mm high, 20 mm thick on average) were provided in the middle of the vehicle at the front edge of the cockpit, which are essentially similar to the add-on element of Fig. 3.19. The main purpose is to deflect whirled-up vehicle debris away from the driver. From the 2020 season onwards, IndCars will run with a halo to which a windscreen is also bolted.

Crash Element (Impact Attenuator)

In addition to a stiff driver cell, the vehicle needs energy absorbing elements on the relevant sides (front, rear, side), Fig. 3.22. These must begin to deform under the action of a certain force and, ideally, their length continues to decrease while they oppose the change in length with a constant force. This results in a greater deceleration period and the acceleration acting on the driver remains at a tolerable constant value. The area under the curve of a force-deformation displacement diagram is a measure of the energy absorbed by the crash element. Stiff structures that produce too great an initial force on impact are unsuitable, as are overly soft ones whose maximum deformation path is used up too quickly. Suitable structures are aluminium honeycomb structures,[2] certain foams and cylindrical hollow bodies made of fibre composites. For frontal impacts, the energy absorption elements can be "hidden" under the outer skin or the vehicle nose itself forms the crash structure, Fig. 3.21. The latter is ideal for formula vehicles. The weight advantage speaks in favour of the latter, because no additional elements are required. The development effort for the nose structure, which can be considerably more expensive due to numerous crash tests, speaks against this (Fig. 3.22).

While metals absorb energy through plastic deformation, this occurs in composites through fracturing and splitting into small parts. In addition, there is always the friction between the crash structure and the counterpart. The deformation path (= deceleration path), which is decisive for the basic design, results directly from the initial speed and the permissible average acceleration:

$$s_{eff} = \frac{v_{test}^2}{2a_{m,\lim}}$$

(3.1)

s_{eff} Effective deformation path, m
v_{test} Impact velocity during test, m/s
$a_{m,\lim}$ Permitted mean acceleration, m/s^2

---

[2] See, e.g., https://www.plascore.com/de/ein-angebot-an-produkten-von-weltrang/energieabsorber/plascore-crushlite/

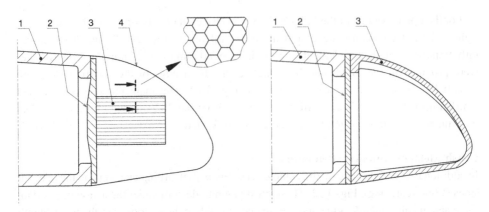

Fig. 3.21 Types of crash elements. Shown is the cross-section through the nose area of a Fomel wagon. (**a**) Al honeycombs as crash element, (**b**) vehicle nose as crash element. 1 Monocoque/frame, 2 Carrier plate (a) or protective plate (b), 3 Crash element, 4 Bodywork. In the case of design (**b**), the protective plate (2) (*anti-intrusion plate*) can be omitted (provided that the regulations permit this) if the driver's footwell is far enough away from the area where the nose is located or the penetration of fragments is prevented in another way in the event of a crash

Fig. 3.22 Crash element at the front of a production sports car. The crash element is made of CFRP laminate and is attached to the front end of the frame. The frame is a steel tubular space frame planked with aluminium plate

The executed length of the crash element is chosen slightly larger depending on the structure, e.g. $1.45s_{eff}$ for Al honeycombs.

The cross-sectional dimensions of the crash element result from the strength of the selected structure and the average deceleration force:

$$A_{cr} = \frac{m_t \cdot a_m}{\sigma_{cr} \cdot f} \tag{3.2}$$

A_{cr} Cross-sectional area of the crash element, mm^2
m_t Total mass of the vehicle or mass required by the regulations for crash test, kg
a_m (selected) mean deceleration, m/s^2
σ_{cr} Compressive strength of the crash structure, N/mm^2
f Factor that takes into account the strength-increasing influence of the deformation rate, -. $f = 1 - 1.2$

During the impact, the kinetic energy of the vehicle is converted into deformation energy and heat in the crash element. During acceptance tests, the force-displacement curve is recorded and the process filmed with a high-speed camera (around 1000 frames/s). Decisive criteria – in addition to the maximum deformation path – include the maximum and average deceleration, see Sect. 3.5.

The actual average deceleration a_m follows from the recorded course of acceleration.

$$a_m = \frac{1}{t_t} {}_{t_2}\!\!\int_{t_1} a \, dt \tag{3.3}$$

t_t Deceleration duration, s. $t_t = t_2 - t_1$
t_1, t_2 Start or end time of the deceleration, see Fig. 3.23
a Measured deceleration, m/s^2

Limits set by the FIA for tests on Formula 1 frontal crash elements are $_{amax,lim} = 10\,g$ for the first 150 mm of deformation and $a_{m,lim} = 40\,g$ for the whole test.

The suitability of composite materials for crash elements is usually investigated in preliminary tests in drop towers, e.g. with cylindrical or conical specimens. Only then are the more complex crash tests carried out with the selected laminate structures. Thanks

Fig. 3.23 Acceleration curve during a pendulum impact test (schematic). The values of interest for technical acceptance are max. deceleration a_{max} and mean deceleration a_m. The limit values of these quantities permitted by the regulations are marked with the index lim. In the case shown, the crash element would have passed the test

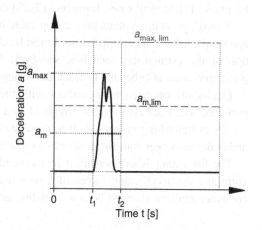

Fig. 3.24 Switch for electrically triggered fire extinguishing system. The switch is located in a production sports car on the side cockpit wall

to major advances in hardware and software, crash tests can also be simulated meaningfully on the computer [2]. However, a real test is (still) required for final approval by the FIA.

Fire Extinguisher

For rally vehicles, both a built-in extinguishing system and a hand-held fire extinguisher must be carried. For circuit races, slaloms and hill climbs, one of the two is sufficient.

The number and size of fire extinguishers depends on the individual regulations and the extinguishing agent. Only certain extinguishing agents can be used, namely AFFF, FX G-TEC, Viro 3 and powder.

According to FIA regulations, two extinguishers with 2.5 and 5 kg filling are carried in Formula 1 [3]. In most cases, however, a 2.25 l extinguisher will satisfy the regulations [4].

All extinguishing systems must be operable both from the driver's position (Fig. 3.24) and from outside. The external control must be close to the main switch. For the identification of the extinguisher switches, see Sect. 3.2. In the case of two-seater vehicles, passengers must also be able to reach the extinguisher easily.

Electrically operated extinguishers with their own battery and separate wiring are preferred. There are also systems triggered by a cable.

The extinguishing systems shall be capable of fighting fires in the engine compartment and in the passenger compartment or two separate extinguishing systems may be installed.

The fire extinguisher itself must be mounted in the vehicle in such a way that it can withstand the acceleration forces of a race run. Specifically, the FIA requires that the container attachments must be able to withstand a deceleration of 25 *g*. Furthermore, the

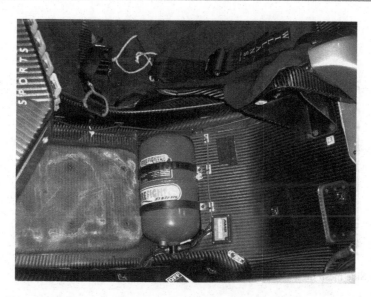

Fig. 3.25 Fire extinguisher in the cockpit of a racing vehicle. This fire extinguisher is placed to the left of the driver's seat, which is possible on a two-seat sports prototype. The container is fixed with two metal straps

fixings must be made of metal and be equipped with a quick release system. Two metal straps are thus the minimum requirement.

Fire extinguishers must be located in the cockpit. In single-seaters they are usually located below the driver's knees or in the nose of the car, in two-seater sports prototypes at the place of the "passenger" (Fig. 3.25).

Life Bottle
In many racing vehicles, a rescue system is installed that supplies the driver with breathing air via a fireproof hose line to the helmet in the event of an emergency. The air tank can supply the driver with air for about 30 seconds.

This tank is usually located below the driver's knees or in the bow of the car in single-seaters.

Safety Belts
Seat belts are prescribed by the individual regulations. Due to the extreme driving conditions, however, they are also required for "normal driving". For more details on the design and fastening, see Sect. 4.8 *Restraint Systems*.

Steering Wheel Quick Release
An easily removable steering wheel allows the driver of a single-seater to quickly exit the narrow cockpit. But quick-release locks are also mandatory for sports prototypes and other

Fig. 3.26 Side mesh on a touring car

vehicles with wide passenger compartments. The quick release is integrated in the steering wheel hub.[3]

Side Net (Window Net)
Side nets are used on touring cars in the area next to the driver towards the outside of the car in the head and shoulder area, Fig. 3.26. They can thus be mounted on the driver's door or directly on the roll cage. Door mounting has the advantage of allowing the driver to get in and out easily, but once the door is open, protection is no longer provided [5]. There are also side net mounts that are equipped with a quick release system, allowing them to be released at the touch of a button.

The net shall have width × height dimensions of 400 × 405 mm (4-door cockpit) or 525 × 467 mm (2-door vehicle).

Roof Hatch
On touring cars with the low-slung seat shells and side struts, the roof is fitted with a removable escape hatch above the driver's seat, Fig. 3.27. In an emergency, the opening is used to facilitate first aid and rescue of the driver by the extrication team.

---

[3] A description of this can be found in the Racing Car Technology Manual, Vol. 4 *Chassis*, Sect. 5.3 *Steering Shaft*.

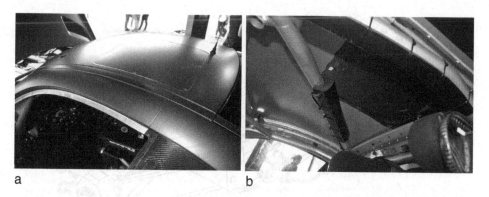

a   b

Fig. 3.27 Escape hatch in the roof of a touring car (Mercedes-AMG GT3). (**a**) View from outside, (**b**) View from inside

Fig. 3.28 Rear light on a monoposto (Ferrari F1). The rear light is mounted in the plane of symmetry of the vehicle above the diffuser outlet

Taillight

A red 15 W tail light at the rear of the vehicle is switched on by the driver in bad weather and in the pit lane, Fig. 3.28.

Tether Ropes

The wishbones of the wheel suspension have predetermined breaking points at the articulation points to the vehicle. If a wheel collides with an obstacle, the control arms deform and break away. To prevent the exposed control arm ends from breaking through the cockpit wall and injuring the driver's legs, they are connected to a longitudinal strut behind the predetermined breaking point.

On Formula 1 and Formula 3 cars, the wheel carriers are additionally secured to the frame with ropes so that they cannot hit the driver in the event of an accident, Fig. 3.29. The

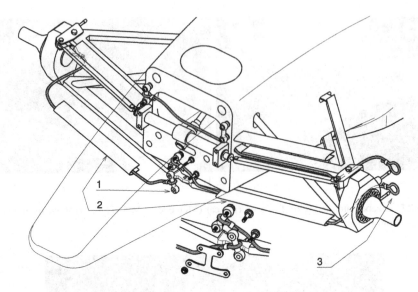

Fig. 3.29 Retaining cables for a front axle (Dallara F306). The picture shows a general view and a detailed enlargement. Not all ropes are shown for reasons of clarity

example shows a front axle. Each wheel carrier is secured with two retaining ropes. They are fastened via the screw bosses for the brake calliper (3) and via their own retaining brackets (1) on the chassis side. The ropes are routed crosswise, i.e. the ropes of the left wheel carrier are bolted to the right bow side. The cables are routed along the wishbones and held under covers (2).

The ropes in Formula 1 must be at least 8 mm in diameter and must be able to withstand a minimum tensile force of 50 kN.

Safety Flaps on NASCAR Vehicles (*Roof Flaps*)
An arguably unique safety feature is found on the North American cars of the Stock Car Winston Cup. The near-series external shape of the vehicles and the high speeds often led to a dangerous phenomenon in this racing series. If the cars had a spin and continued to roll backwards or nearly backwards, the now "reversed" inflow created a lift that could be so strong from 260 km/h that the car lifted off the track despite its 1590 kg mass. The remedial measures (see Fig. 3.30) consist of 12.5 mm high lateral sheet metal strips, which are attached to both sides of the roof run and, firstly, provide a certain stabilisation when driving straight ahead and, secondly, above all, allow the flow to be detached in the case of a large oblique flow. In addition, two 510 × 205 mm flaps are arranged in the roof area in such a way that they open due to the dynamic pressure when the flow comes from behind or diagonally to the right (the vehicles always drive around to the left in the oval) and act as an air brake. One flap is mounted exactly perpendicular to the direction of travel, the other perpendicular to 140° vehicle yaw angle.

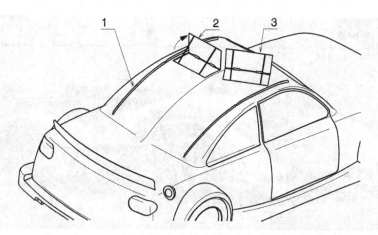

Fig. 3.30 Safety flaps on NASCAR cars. 1 Roof rails, 2 Flap against direction of travel, 3 Flap 140° against direction of travel. The flaps are shown in the open state. During normal travel they are closed and barely protrude the roof contour

Breakaway Valves
Breakaway valves on lines to and from the tank prevent fuel from leaking if these lines break.

Vertical Air Guide Device
Since 2011, a vertical aerodynamic device above the engine cover has been mandatory for LM P1 and P2 (Le Mans Prototype) category vehicles as part of the safety package. Such a tail fin (*shark's fin*) stabilises the vehicle at high speeds, if the center of pressure (centre of air attack) is behind the vehicle's centre of gravity and thus the air ensures a reverse momentum if the sideslip angle becomes too large, see also Sect. 5.7 Fig. 5.92. Figure 3.31 shows the design of such a tail fin on an LM P1 vehicle.

Wheelhouse Ventilation (Air Extraction Holes)
To avoid overpressure in the wheel arch of the front wheels, LMP vehicles must have an opening on the upper or inner side of the wheel arch, Fig. 3.32. Overpressure that can build up in the front area of a rotating (enclosed) tyre, in combination with an unfavourable undertray position (air loss rear tyre, pitching vibration, . . .), can lead to the vehicle lifting off.

Tyre Pressure Monitoring System (TPMS)
In such a system, tyre temperature and pressure are permanently monitored on a wheel-by-wheel basis for safety reasons. Apart from the fact that incorrect tyre pressure has a negative effect on tyre slippage, underinflation increases the temperature of the tyre during

Fig. 3.31 Audi R18 TDI at Le Mans [Audi Motorsport 2011]. The winning car of the famous 2011 Le Mans endurance classic features a tail fin that extends from the air intake above the cockpit to the rear wing

Fig. 3.32 Wheel house ventilation (Porsche 919 Hybrid). You can see the rectangular opening on the wheel arch of the left front wheel of the winner of the 24-h of Le Mans 2015 and 2016

operation to such an extent that a tyre blowout can be the result.[4] The sudden loss of lateral control and contact force causes the vehicle to become unstable. The car breaks away, sits up and – if a rear tyre is destroyed – gets underinflated (lift!), which can lead to serious

---

[4] See also Race Car Technology Manual Vol. *4 Chassis*, Sect. 1.2.3 *Wheels and Tyres: Rolling Resistance.*

accidents, especially at high speeds. Some teams therefore install such systems and monitor the condition of the tyres in practice as well as in the race via telemetry. If a loss of pressure is detected, the car is immediately brought into the pits. In the USA, tire pressure monitoring systems are required by law for passenger cars, and the EU recently introduced a similar regulation for new vehicles.

Marshalling Message Display

In some racing series, there must be a small screen in the cockpit within the driver's field of vision that provides information about current decisions by the race control. In this way, the driver can immediately recognise yellow flags, safety car phases, etc. even before he passes a marshal. In endurance races, this also reduces the risk of flag signals being overlooked due to limited visibility at night or in the rain.

Accident Recorder (ADR)

A device (ADR – *Accident Data Recorder*) under the driver's seat of Formula 1 cars continuously records data from two acceleration sensors mounted on the driver's survival cell. The memory capacity is sufficient for 2 minutes. In the event of an accident, the last 2 minutes before the event are thus recorded. The evaluation of this data is used, among other things, for improvements to safety equipment.

3.4 Fasteners

For structural bolting and other highly stressed connections, racing primarily uses bolts conforming to the following American standards: AN (Army/Navy), MS (Military Standard) or NAS (National Aerospace Standard). For the remaining detachable connections, ISO quality 8.8 bolts are used – as is also common in the automotive industry.

So-called K-nuts are often used as locknuts. According to aviation standard LN9338 with metric standard thread or LN65409 with fine pitch thread and according to MS21042 with UNJF thread (inch thread with fine pitch). These nuts are supplied with a pre-compression in the thread area, which leads to an increase in the loosening torque. However, it also damages the bolt thread, which is why these nuts should only be used for one-time tightening. Connections that require frequent loosening (adjustment work, maintenance covers, gearbox housing covers, . . .) are better provided with stud bolts and/or self-locking nuts with nylon ring (DIN EN ISO 7040, 7042, 10 511). K-Nuts are available in versions that can be used up to 235 °C and those that can withstand temperatures up to 425 °C. Figure 3.33 shows a K-nut, the most important dimensions can be taken from Table 3.2.

The material for K-Nuts with an operating range of up to 235 °C is 1.7220.3 (35CrMo4), quenched and tempered to 1250–1600 N/mm^2, and that for up to 425 °C is 1.4944.4 with a tensile strength of $R_m = 960$ N/mm^2 (at room temperature). The latter nuts are additionally silver-plated to increase the surface resistance.

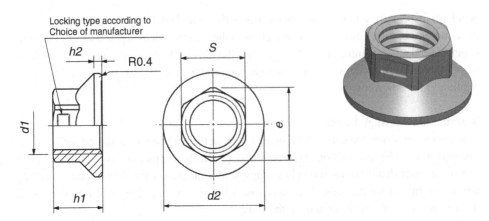

Fig. 3.33 K-nut. Locknut with collar according to aviation standard. Numerical values of the dimensions see Table 3.2

Table 3.2 Main dimensions of selected K-nuts according to Fig. 3.33

Dimension	Thread $d1$	$d2$ mm	$h1$ mm	$h2$ mm	Width across flats s mm	e mm
	M4	8	4	0.6	5	5.51
	M5	9.2	5	0.6	6	6.64
	M6	10.5	5.4	0.9	7	7.74
	M7	12.6	6.2	1	8	8.87
	M8	13.6	7	1.1	10	11.05
	M10	16.6	8.5	1.3	12	13.25

Screw Lock

The automatic loosening of dynamically loaded screw connections can become critical to the safety of structural connections and chassis components. For the design of such connections see Sect. 2.2 *Design Areas*, 2.5 *General Principles of Embodiment Design*.

In many racing classes, chassis bolts must be positively secured, e.g. by castellated nut and cotter pin or safety wire.

In wire locking, screw heads, which must have a transverse hole, are connected to each other or to another component in such a way that the screws cannot untwist, Fig. 3.34.

The usual wire for this application is made of stainless steel with a thickness of 0.8 mm. However, 0.5 and 1 mm are also used. The hole in the screw head has a diameter b around 2 mm (up to M6 thread b is 1.2 mm, from M8 b is 1.8 mm).

Self-locking nuts with a plastic insert (elastic stop nut with nylon collar) are not temperature-resistant and therefore cannot be installed in the vicinity of hot parts (brake, exhaust system, engine, heat exchanger, …). The maximum operating temperature is 120 °C.

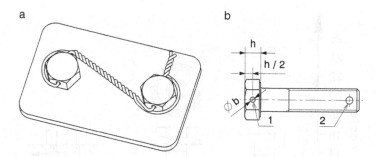

Fig. 3.34 Wire securing of screws. (**a**) The wire connects the two screw heads in such a way that loosening of the screws (with right-hand thread) is prevented. (**b**) Hexagon screw with holes for safety wire (1) and split pin (2)

In multi-shear bolted joints, oscillating transverse movements quickly cause the nut to loosen as soon as the transverse force exceeds the frictional force caused by the bolt pretensioning force, resulting in a relative movement of the clamped sheets [6]. For such cases, ratchet bolts or nuts are suitable for non-hardened surfaces. Investigations have shown significant advantages in the durability of the bolted joint for this type of bolt, Fig. 3.35.

General recommendations for bolted joints to avoid loss of preload or loosening, especially under dynamic loading:

Provide high-strength bolts and high pretensioning force, taking into account the contact pressure under the bolt head – if necessary, place a rigid washer underneath. Use threads with fine pitches. Minimize the number of parting lines and make the joint surfaces as smooth and even as possible. Design screws with high flexibility, the clamping length l_k should be $>5d$ (d is the screw shaft diameter). A more elastic screw also has the advantage that it keeps the additional screw force smaller, Fig. 3.36.

In the case of high-strength materials ($R_m > 1000$ N/mm^2), it must be noted that their deformability decreases with increasing strength and their sensitivity to stress corrosion cracking increases. In particular, the presence of hydrogen seriously degrades the performance of these materials (hydrogen degradation, formerly hydrogen embrittlement).

The clamping length not only depends on the dimensions of the braced components, but can also be increased advantageously by design, e.g. by means of a compression-resistant spacer sleeve. Figure 3.37 shows connections with increasing durability.

At critical bolted joints, teams also mark the bolt head and nut with a highly visible paint stroke so that any loosening of the parts can be easily detected during a visual inspection. Torque checking of critical bolted joints should also be on the pre-race checklist before the car is cleared for the start.

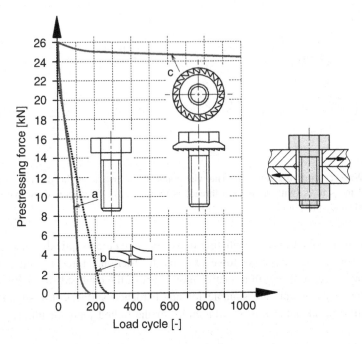

Fig. 3.35 Loss of prestressing force under oscillating shear force loading of bolted joints (junk test). (**a**) single bolt, (**b**) Screw with spring washer, (**c**) serrated bolt. The simple bolt (a) initially loses even less preload force with increasing number of load cycles than the connection with a spring washer (b). Only after further load changes does a certain locking effect of the spring washer become noticeable. Nevertheless, both connections are unusable for this load. Only the ratchet screw (c) exhibits safe behaviour

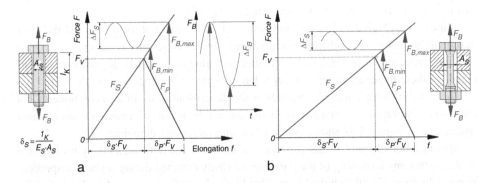

Fig. 3.36 Stressing diagrams for bolts with dynamic axial working load. (**a**) Screw with low compliance δ_S, (**b**) Screw with high compliance δ_S. The external operating load F_B varies by the amount ΔF_B and causes a larger force deflection ΔF_S for the stiffer screw (a). With the more elastic screw (b), the force deflection is smaller and the connection is more durable. F_V preload force, A_S *shank* area of the bolt, l_k grip length, E_S modulus of elasticity of the bolt, F_S bolt force, F_P compression force of the plates, F_B external working load

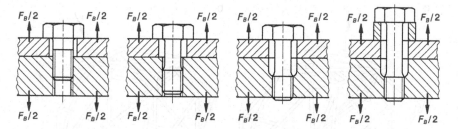

Fig. 3.37 Bolted joints with increasing fatigue life, after [7]. The loose rotation safety increases from left to right in the designs shown. F_B External operating load

3.5 Tests

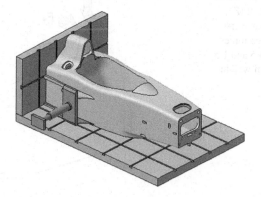

The individual motorsport authorities prescribe different tests for assemblies and components. These tests are of interest to the designer in that they provide size values for loads and deformations and these components are only released if they pass the test. If no criteria are specified for a vehicle project, such values can be used as a realistic basis for dimensioning. In general, in the absence of specifications for the design of protective devices, a deceleration of 10 g can be assumed as the minimum size.

Roll cages for touring cars and similar vehicles must endure a static load test for their release [1], which consists of two partial tests. The basic set-up is shown in Fig. 3.38. The main bar is loaded with a vertical force F_Z and the front bar with an oblique acting force F, which correspond to a multiple of the dead weight incl. two persons of 75 kg:

$$F_Z = 7.5 \cdot (m_V + 150) \cdot g$$

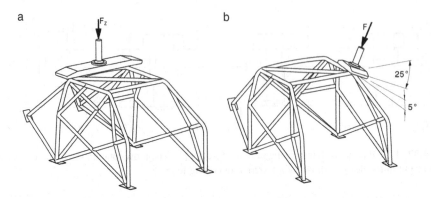

Fig. 3.38 FIA load test for roll cages. (**a**) Test of the main yoke, (**b**) Test of the front yoke. The angles of the longitudinal and transverse axes of the stem to the horizontal are shown

Fig. 3.39 Roll bar test. The bracket is mounted on the frame and loaded in the vertical (Z), longitudinal (X) and transverse (Y) directions with three forces corresponding to 7.5, 5.5 and 1.5 times the vehicle's own weight including the driver

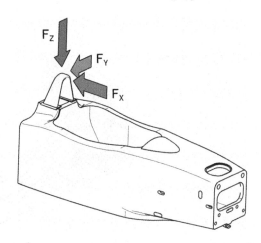

$$F = 3., 5 \cdot (m_V + 150) \cdot g$$

F_Z, F Test forces according to Fig. 3.38, N
m_V Tare weight of the vehicle, kg
g Acceleration due to gravity. $g = 9.81$ m/s^2

The roll-over protection structure as a whole shall not exhibit any breakage or any specified plastic deformation in the direction of the force. The maximum values of plastic deformation are set at 50 mm for the main yoke and 100 mm for the front yoke.

According to FIA standards, for example, the following tests are carried out on the vehicle body and bow. The rear rollover structure on sports prototypes and monoposti is statically loaded with a spatial force using a flat punch of 200 mm diameter. Here, the three force components correspond to $F_X = 1.5\ G$, $F_Y = 5.5\ G$ and $F_Z = 7.5\ G$, where G is the vehicle weight including the driver at 75 kg, Fig. 3.39. Here, the deformation in the

Fig. 3.40 Testing the chassis. The vehicle hull is subjected to various tests with transverse forces between 10 and 20 kN

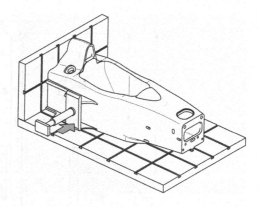

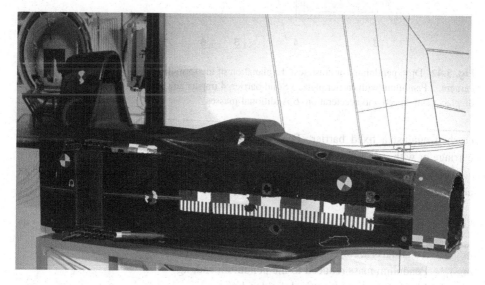

Fig. 3.41 Chassis of a monoposto after the crash test (Ferrari F1)

direction of force must not exceed 50 mm and any failure of the supporting structure must not exceed 100 mm measured in the vertical direction from the apex of the roll bar. A further test of chassis sections is shown in Fig. 3.40. In addition, complete monocoques are also run on a sled against solid barriers, Fig. 3.41.

These acceptance tests were only introduced after several tragic accidents and increase the passive safety of the vehicles enormously. They are also reflected in the design of a monocoque. Whereas a typical Formula 1 monocoque weighed about 35 kg in 1994 (i.e. in the time without tests), today's monocoque weighs about 65 kg.

In a crash test for an energy absorbing element, see Sect. 3.3 Protective devices, the test object (crash element) together with a certain mass is accelerated to the desired speed and

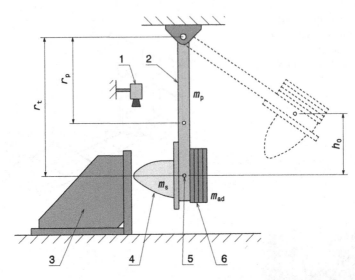

Fig. 3.42 Drop pendulum for crash test. Explanation of the formula symbols see text. 1 High-speed camera, 2 Pendulum with carrier plate, 3 Solid barrier, 4 impact attenuator (test object), 5 Sensors for displacement, speed and acceleration, 6 Additional masses

crashed against a fixed barrier. The acceleration is effected by a horizontally guided carriage or with a drop pendulum, Fig. 3.42.

The total mass m_t, which is decisive for the test and whose kinetic energy must be absorbed by the crash element, is calculated from the pendulum structure as follows

$$m_t = m_{red,p} + m_s + m_{ad} \tag{3.4}$$

m_t — Total mass according to regulations or vehicle mass, kg
$m_{red,p}$ — Pendulum mass reduced to the pendulum end, kg
m_s — Mass of the test object incl. holder, kg
m_{ad} — Additional mass to achieve the desired mass, kg

The reduced pendulum mass follows from the consideration of the kinetic energy to:

$$m_{red,p} = m_p \left(\frac{r_p}{r_t} \right)^2 \tag{3.5}$$

m_p — Mass of the pendulum, kg
r_p — Centre of gravity distance of the pendulum, m. See Fig. 3.42
r_t — Distance between pivot point and effective axis of the test object (measuring point), m

Fig. 3.43 Checking the nose
cone. The nose attachment is
subjected to a lateral load of
20 kN. A frontal impact test with
nose mounted to fuselage is
performed at 10 m/s

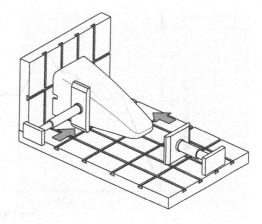

The required drop height of the pendulum is determined by the desired impact velocity of
the test object:

$$h_0 = \frac{v_{test}^2}{2g}$$
(3.6)

h_0 Drop height of the pendulum, m. See Fig. 3.42
v_{test} Impact velocity during test, m/s
g Gravitational acceleration, m/s^2

Thus the kinetic energy absorbed by the crash element increases:

$$E_{kin} = \frac{1}{2} m_t v_{test}^2$$
(3.7)

E_{kin} Kinetic energy, J

The regulations usually specify a minimum value for the kinetic energy that must be
achieved by the test specimen.

In the test arrangement shown in Fig. 3.43, a reversed arrangement to Fig. 3.42 was
chosen. The crash element is stationary and an impactor acts on it. The corresponding
evaluation can be seen in Fig. 3.44.

Impactor mass	560 kg
Impact speed	10.56 m/s
Average acceleration	10.7 g
Max. acceleration	15.26 g
Max. deformation	448 mm

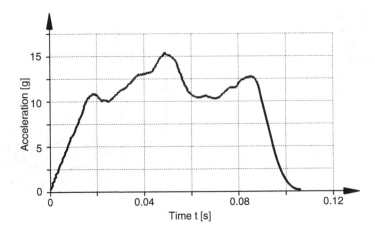

Fig. 3.44 Evaluation of a crash test according to arrangement as in Fig. 3.43 [8]. The nose is mounted on the chassis. An impactor acts frontally on the nose cone

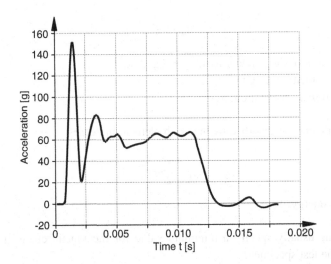

Fig. 3.45 Impact test on a steering wheel of a racing car (Formula Renault 2000) [8]. The 8 kg impact weight is moved at 7 m/s onto the steering wheel. The acceleration may exceed 80 *g* for a maximum of 3 ms. In the test shown, this period is 1.12 ms. The steering shaft has two flex joints in Z-arrangement. (The steering shaft is illustrated in the Racing Car Technology Manual Vol. 4 Chassis, Chap. 5 (Fig. 5.34)). This allows the steering wheel to deflect axially

The steering wheel and its mounting are also subjected to a test, Fig. 3.45. According to FIA Appendix J Art. 258A and 259 such a test looks roughly as follows. An 8 kg hemisphere of 165 mm diameter is struck on the centre of the steering wheel at a speed of 7 m/s in the axial direction of the steering shaft. The maximum acceleration shall not exceed 80 *g* for more than 3 ms. The quick release of the steering wheel shall continue to function correctly after the test.

References

1. Internetportal der FIA: http://www.fia.com/sport/Regulations/appjregs.html. Accessed on 07.02.2007
2. Heimbs, S., et al.: Crash Simulation of an F1 Racing Car Front Impact Structure. Beitrag zu 7. Europäischer LS-DYNA-Konferenz, DYNAmore GmbH (2009)
3. Incandela, S.: The Anatomy & Development of the Formula One Racing Car From 1975, 2. Aufl. Haynes, Sparkford (1984)
4. McBeath, S.: Competition Car preparation, 1. Aufl. Haynes, Sparkford (1999)
5. Murri, R., Schläppi, M.: Realitätsbezogene Abstimmung passiver Sicherheitssysteme mittels Schlittentest für Tourenwagen, Beitrag zur Tagung Race.Tech 14.-15. Okt. München (2004)
6. http://www.boltscience.com/pages/junkertestvideo.htm. Accessed on 09.01.2015
7. Grote K.-H., Feldhusen J. (Hrsg.): Dubbel, Taschenbuch für den Maschinenbau, 24. Aufl. Springer, Berlin (2014)
8. N.N.: Formula Renault 2000 Manual, Renault Sport Promotion Sportive (2001)

Cockpit

4

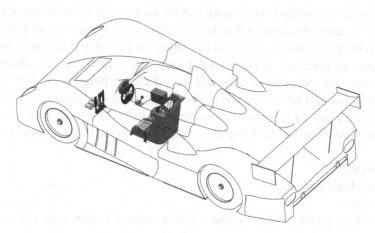

The cockpit is the driver's workplace. In this area, human influence comes into play alongside technology. Ergonomics is crucial to ensure that the driver is actually able to implement the possible driving performance of a vehicle.

4.1   Concept

The driver of a multi-track vehicle has in principle only a few control elements to influence the behaviour of his vehicle while driving:

- the steering wheel
- the foot brake
- the accelerator pedal
- the circuit
- the clutch
- the parking brake.

Whereby clutch and parking brake are not found in all vehicle types. On the other hand, in some cars there are further possibilities of influence from the cockpit, e.g.:

- adjustable brake force distribution front axle/rear axle
- adjustable stabilizer (ARB)
- adjustable limited-slip differential
- adjustable identification between accelerator pedal and motor response, i.e. torque

The driver receives feedback on the driving status mainly directly via his sensory organs and, in a few cases, via a display instrument such as a rev counter or speedometer. The driver feels the acceleration and forces of the vehicle via the seat, and the force ratios at the front wheels are perceived in the steering torque. The cockpit thus contains the most important interfaces between the driver and the vehicle, enabling the driver to move the race car at its physical limits. In addition, the cockpit creates the climatic conditions that should make it easier for the driver to concentrate on driving for longer periods of time.

Basic considerations for the design of a cockpit concern the following points:

- Regulations
- Open or closed cockpit
- Driver size(s)
- Ergonomics: visibility, operating forces, climate control, fatigue, entry and exit.

Open Cockpit
An open cockpit, Fig. 4.1, facilitates access to the driver's seat, which is advantageous for time-relevant driver changes, for example in an endurance race, where the drivers replace each other every two hours or so. The required cockpit widths vary in some regulations. One of the reasons for Audi's choice of an open sports prototype when designing the R8 was the advantage of the narrower cockpit compared to the closed version (1100 mm instead of 900 mm prescribed). With the car's width fixed, this leaves more room for the side heat exchangers and allows wider tyres to be used. This makes softer tires or longer change intervals possible.

In addition, the dazzling effect of a dirty windshield at night is eliminated. The air conditioning of the open cockpit is also much easier, which is advantageous in high outside temperatures, but also in damp, cool conditions (tarnishing of the windscreen inside). Especially when it is hot and humid at the same time, the driver's performance suffers

Fig. 4.1 Vehicle with open cockpit (Lola). The vehicle is two-seater, at least for the regulations. In fact, one person sits in the open cockpit

noticeably in a closed car. Reliable operation of the windscreen wipers also becomes a problem at the high speeds. Air resistance is not a differentiating factor in an open cockpit compared to the closed version. The thing that is the main cause of drag is approximately the same in both types of cars, namely downforce. For example, until 2008, cars with closed cockpits had only won the 24 Hours of Le Mans twice in 10 years, even though the regulations allowed these cars to have more engine power.

The safety for the driver is lower with the open cockpit. Wheels and other detached vehicle parts can easily hit the head. Serious accidents in recent times have led to the following regulations. In the open cars of Formula 1, 2 and E a ring-shaped cage (*halo*, because of its appearance) covers the cockpit area and in IndyCars a transparent screen covers the cockpit edge (see also Chap. 3 *Safety*).

Closed Cockpit

A closed cockpit, Fig. 4.2, facilitates the representation of a flexurally rigid vehicle body by including the roof in the load-bearing structure. Doors have to be provided for this purpose, which increases the design effort compared to an open cockpit, Fig. 4.3. The aerodynamics of a closed contour are in principle more favourable and the inflow conditions for the rear wing are at least theoretically better than in an open car with roll bar. In practice, the conditions are more complicated because the rear wing may not be higher than the roof in Le Mans cars, for example. In open cars, a certain height away from the ground is regulated, which means that the wings of executed cars are mounted higher and therefore have favourable flow conditions.

Fig. 4.2 Closed cockpit (Pro Sport 3000). The door extends over the driver's head and the side sill. The roof is thus reduced to a narrow connecting web between the front and rear roll bars

Fig. 4.3 Closed cockpit (Bentley EXP Speed 8). View through the left door opening. The door is hinged far to the outside and covers the side air supply to the heat exchangers, which leads around the cockpit (visible in the foreground). The result is a wide roof, which has a positive effect on vehicle rigidity. The disadvantage is the limited access to the cockpit

Ergonomics

Only an ergonomically designed cockpit enables the full potential of a vehicle to be exploited over a longer period of time by the driver. A great deal of force is required to operate actuators, strong vibrations, considerable noise levels, high temperatures and humidity cause a person to tire more quickly and concentration also deteriorates more rapidly. Elastic decoupling elements are standard in production vehicles to increase comfort. They are found, among other things, in the chassis and in the engine and powertrain mounts. Hardly any vibration-damping elements are found in racing cars because they lead to energy loss and indirectness. This leaves practically only the seat to keep strong vibration stress away from the driver.

Some regulations require it anyway, but it should still be mentioned that a targeted fresh air supply and also an air exhaust in the cockpit should be provided.

4.2   Driver's Position

The aim is to make it easier for the driver to control the vehicle while still keeping the centre of gravity as low as possible. The mass of the driver makes up a not inconsiderable part of the total mass of light vehicles. Formula Renault 2000 cars, for example, have a total mass of 485 kg [1]. A driver weighing 80 kg is therefore equivalent to one sixth.

Cockpits of racing cars of the highest categories are usually built for a specific driver. His dimensions and preferences (posture, steering wheel diameter, operating forces, switch positions ...) can be recorded directly in that case. This is no longer possible with racing vehicles that are produced in small series. Also in the case of vehicles for endurance races, the cockpit must be suitable for several drivers who share it in a "flying" change. This is where the experience gained in large-series passenger car design comes in handy. Statistical studies provide dimensions of people divided into so-called percentiles, see Fig. 4.4. If, for example, the dimensions of the column for the 95% man are used in the design of a cockpit, 95% of the male population fit into it and only 5% of the male population are too tall.

In addition to the dimensions of the human body, the masses of individual body parts are also of interest for packaging, and in particular for influencing the position of the overall centre of gravity, Table 4.1.

The driver's position can be used to influence the position of the vehicle's centre of gravity horizontally and vertically. The most favourable arrangement from a technical point of view is a reclining seating position (cradle position) of the driver. However, this is not accepted equally by all drivers. In the case of the sports prototype for Le Mans Audi R8, an upper body tilt of up to 45° was approved by the drivers [4]. A similar sitting posture is also found in formula cars with a raised nose, Fig. 4.5.

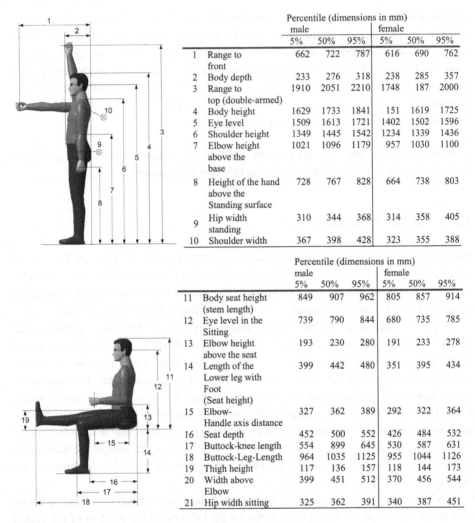

		Percentile (dimensions in mm)					
		male			female		
		5%	50%	95%	5%	50%	95%
1	Range to front	662	722	787	616	690	762
2	Body depth	233	276	318	238	285	357
3	Range to top (double-armed)	1910	2051	2210	1748	187	2000
4	Body height	1629	1733	1841	151	1619	1725
5	Eye level	1509	1613	1721	1402	1502	1596
6	Shoulder height	1349	1445	1542	1234	1339	1436
7	Elbow height above the base	1021	1096	1179	957	1030	1100
8	Height of the hand above the Standing surface	728	767	828	664	738	803
9	Hip width standing	310	344	368	314	358	405
10	Shoulder width	367	398	428	323	355	388

		Percentile (dimensions in mm)					
		male			female		
		5%	50%	95%	5%	50%	95%
11	Body seat height (stem length)	849	907	962	805	857	914
12	Eye level in the Sitting	739	790	844	680	735	785
13	Elbow height above the seat	193	230	280	191	233	278
14	Length of the Lower leg with Foot (Seat height)	399	442	480	351	395	434
15	Elbow-Handle axis distance	327	362	389	292	322	364
16	Seat depth	452	500	552	426	484	532
17	Buttock-knee length	554	899	645	530	587	631
18	Buttock-Leg-Length	964	1035	1125	955	1044	1126
19	Thigh height	117	136	157	118	144	173
20	Width above Elbow	399	451	512	370	456	544
21	Hip width sitting	325	362	391	340	387	451

Fig. 4.4 Body dimensions, statistics according to DIN 33402 [2]

In Formula 1 cars, everything is subordinated to aerodynamic efficiency. To ensure that sufficient air can flow into certain areas of the vehicle as unhindered as possible, the nose of the vehicle is raised as far as the driver's view allows. As a result, the driver lies even flatter in the cockpit, Fig. 4.6.

Figure 4.7 gives indicative values for a seating position such as might be planned in a touring car

Table 4.1 Masses of parts of the human body [3]

Body part	Mass, kg	Tolerance
Head	5.08	±0.05
Shoulder – Chest	18.82	±0.73
Abdomen – Pelvis – Upper part of the thighs	16.28	±0.68
Leg – Thigh (each)	8.35	±0.32
Leg – Shin (any)	3.13	±0.14
Foot (each)	1.27	±0.05
Arm (each)	2.18	±0.09
Forearm (any)	1.54	±0.05
Hand (any)	0.64	±0.05
Total mass	74.4	±1.4

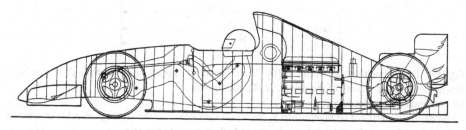

Fig. 4.5 Longitudinal section through a Formula 1 vehicle [5]. The driver is positioned as low as possible. His feet are behind the front axle and considerably above the buttocks. The steering wheel is close to the driver's upper body. Large steering wheel turns are only possible to a limited extent

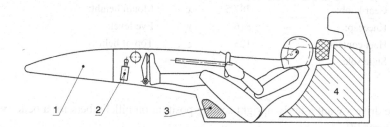

Fig. 4.6 Seat position in monocoque with raised nose. 1 impact absorbing element (nose), 2 Suspension parts (spring, damper, steering), 3 Fire extinguisher, 4 fuel tank bag. The position of the driver is exhausted to the extent that the view to the front is just given. The vehicles are used on circuits with a known course. For a rally car, for example, such a recumbent position is unusable

Recommendations for the dimensions given, mm:

a	Back Clearance:	76	h	Top of the steering wheel:	533
b	Hip point:	533	j	Top edge of shoe sole:	279.5
c	Steering wheel center:	750	k	Pedal surface center:	203

(continued)

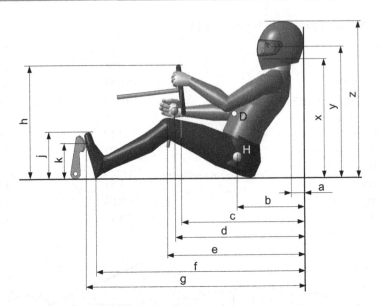

Fig. 4.7 Driver position, according to [6]. The dimensions apply to a driver with 1727 mm body height. *D* Centre of gravity of the driver, *H* Hip joint point
Of outstanding importance are:
– Dimension *h*: must not be higher than *x*, otherwise visibility over the steering wheel is obstructed
– Position of the gear lever knob in each gear and elbow room to the rear
– Dimension *k* and the radius of movement of the pedals

Recommendations for the dimensions given, mm:					
d	Gear knob:	787.5	*x*	Mouth height:	559
e	Kneecap:	876	*y*	Eye level:	635
f	Heel contact:	1257	*z*	Top of helmet:	762
g	Shoe sole:	1372			

Some drivers want a fairly upright seating position on hillclimbers for a better view in hairpin turns [7].

Ever since aerodynamics gained a dominant role in racing car design (mid-1980s), the same philosophy that applies to production cars has taken hold in the design process: Design from the outside in, i.e. the external shape is first determined according to aerodynamic aspects as well as preliminary wind tunnel tests, and all parts must then be accommodated within this as cleverly as possible. However, a certain degree of ergonomics must not be lacking in the driver's seat design. Many of the dynamic advantages of a vehicle cannot be exploited if the driver cannot sit without cramps for a full race distance [5]. What is often done in 2-seater cockpits (e.g. LMP) to achieve a streamlined contour in

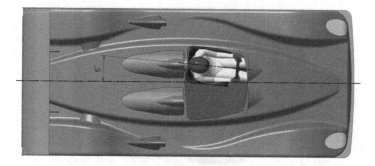

Fig. 4.8 Rotated seat position for a sports prototype

Fig. 4.9 Top view of a driver's seat (Formula 1 Ferrari)

Fig. 4.10 Upper cockpit closure of a Formula 1 car (Williams FW 18, 1997). The air intake for the engine sits above the driver's helmet. The contour underneath hugs the helmet. The cockpit rim rises towards the rear and thus follows the course of the visor cutout

plan view is to swivel the seat by up to 5° to the vehicle's centre plane, Fig. 4.8. This moves the footwell closer to the vehicle's centre and the nose area can be made slimmer.

The footwell should have a clear width of about 305 mm with three pedals as the lower limit [8].

The development begins with the construction of a dummy seat in which the driver sits to check the space conditions. It is often found that drivers are initially satisfied with their seating position and only encounter problems when driving the real vehicle [5].

When designing the cockpit of single-seaters (Figs. 4.9 and 4.10, cf. also 4.55), it must be borne in mind that in most formulae the driver must be able to leave the vehicle from the

Fig. 4.11 Cockpit in a Le Mans racing car (Audi R8S). The open cockpit is divided by a central strut so that the driver finds the necessary lateral support, as in a monoposto. The right leg facing the centre of the vehicle is also supported. The thigh support is so pronounced that a large part of the transverse but also the longitudinal forces can be absorbed by it

driving position (seat belt fastened, hands on the steering wheel) without assistance within 5 s (5-second rule). Although this is not mandatory for sports prototypes, in endurance races short standstill times in the pits and thus quick driver changes are crucial. The "24 Hours of Le Mans" are mostly lost in the pits, so to speak – or won by giving away the least amount of time there [9].

Due to the high lateral forces, a constricted legroom is even advantageous for the driver. Especially in the area of the thighs and knees, sufficient lateral support is required, Fig. 4.11.

Field of Vision
In addition to seat position, accessibility of the control levers and air conditioning, the driver's view is probably the decisive criterion in the design of the driver's seat and its surroundings. Of course, the minimum field of vision depends on the type of competition. On the circuit the view to the front, to the apex of the corner and into the rear view mirror is sufficient. The course is well known, only the position of the other competitors changes during the race. On the circuit, these can also be communicated to the driver via the on-board radio by having an observer stand on an elevated point or watch the action via a monitor. Rally cars are driven on unfamiliar roads and sometimes with large body sideslip angles (drifting) and race alone against the clock during the day and at night. By and large, however, the eye tracking of all drivers is similar and digital. On the straight, he focuses on the braking point, then moves to the apex of the corner. Out of the corner of his eye, he keeps an eye on the gauges and rearview mirror. Figure 4.12 shows these essential areas for a single-seater. The unused areas (2, 4) can be used for the placement of the rear view mirrors.

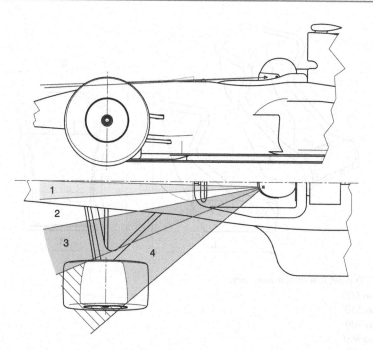

Fig. 4.12 Driver's field of vision areas, according to [10]. The left half of the vehicle is shown in plan and elevation. 1 View area straight towards braking area (red), 2 unused area, 3 view towards corner apex (green), 4 area blocked by wheel. Vision area 1 is limited downwards by the nose of the vehicle (the wind deflector at the front end of the cockpit is transparent) and intersects the road at a distance of approx. 15 m from the driver's eye. In doing so, it sweeps a total (2 eyes) of one vehicle width (2 m). The middle view to the apex of the corner (3) hits the road at approx. 13 m

4.3   Seat

A rigid attachment of the seat to the frame is important so that the driver can dose sensitively, especially during braking, and can generate the maximum actuation force [11]. Due to the extreme accelerations in a racing vehicle, the driver needs an individually adapted seat so that he can still control the vehicle accordingly.

Typical seat dimensions are shown in Fig. 4.13. The mass of such a seat is in the region of 9 kg. Seats for factory drivers are narrower and therefore lighter than those for customer sport. The athletic, wiry professionals require less wide seat shells than typical gentleman drivers [12].

The seat is usually bolted to the frame or to the body floor assembly by means of two rails, Figs. 4.14 and 4.15.

A minimum standard for the mounting of a seat is prescribed by the FIA for touring, GT and production cars, Fig. 4.16. The seat must be mounted in at least four places – two at the

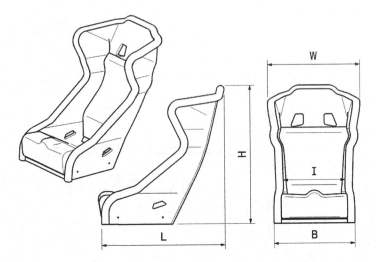

Fig. 4.13 Typical seat dimensions, mm
W = 550 to 600
B = 450 to 550
L = 610 to 700
H = 550 to 890
I = 350 to 400

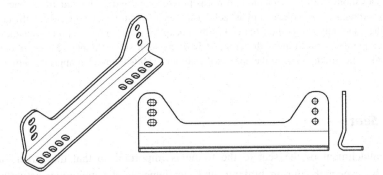

Fig. 4.14 Fixing rail for seats. The rail has elongated holes on one side towards the seat. This allows tolerances to be compensated. Likewise, the screw holes towards the frame are designed as slotted holes, so that the width tolerance of the seats is taken up

front and two at the rear – with M8 bolts or larger. The contact area between the bracket and the counter plate, which must be at least 60 mm long, must be at least 40 cm² per fixing point. Each fixing point must be able to withstand a force of 15,000 N, regardless of the direction in which it is applied. The wall thicknesses of the brackets and the counter plates are specified as 3 mm for steel and 5 mm for light metal, depending on the material.

There are also seat mounts that have elastic elements between the rail and the seat so that shocks are transmitted to the seat in an attenuated form.

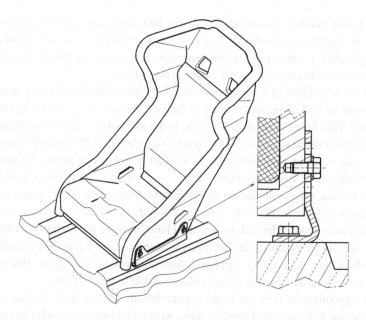

Fig. 4.15 Fastening a seat. Two rails hold the seat. The rails themselves are screwed to the frame or the floor. The rails have several holes at the screw connection points and thus enable a stepped adjustment of the seat in height and in the direction of travel

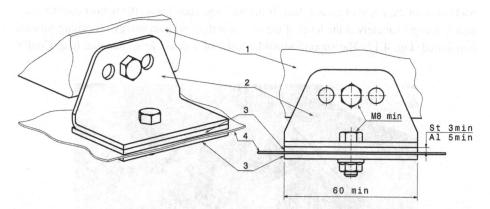

Fig. 4.16 Seat mounting according to FIA Appendix J Art. 252. *1* Seat. 2 Mounting. *3* Counterplate. *4* Body/frame

In raid vehicles, the load during landing after jumps can be so great that damage to the health of the driver and passenger is the result. The seat anchorage must therefore have predetermined breaking points which break in the event of overstressing and thus absorb part of the energy.

For races under high ambient temperatures, seats with ventilation are installed in factory vehicles for endurance races. The backrest and the seat surface are provided with numerous holes that establish a connection to an air chamber located behind them. This air chamber is formed by a shell that hugs the outside of the seat shell at a distance of about 10 mm. A hose connects the chamber to the climate box or a fresh air supply. This seat ventilation is a considerable relief for the drivers in the insulating racing suits and helps to maintain their performance and concentration over a longer period of time.

In the tub-like cockpit of a monoposto, an inherently stable seat is rarely needed. It is more common to find some sort of liner to balance the area between the driver and the cockpit wall. This lining is made as follows. In a seat bucket, which is connected to the vehicle, the driver sits down on an 80-litre plastic bag filled with PU foam. The cured seat is cut to size and tested on the racetrack, e.g. whether there are any pressure points. Teams with a low budget tape this seat and use it in the race. A much more expensive solution is to mould this PU foam seat with carbon fibre laminate. Such a seat is stiffer and lighter than the PU version, but also harder. Which means that if there is a pressure point, a completely new seat has to be made. The carbon seat shell has a mass of just under one kilogram.

In any case, it is important with a reclining seating position that the lower back (lumbar spine) of the driver is precisely supported (lumbar support). Otherwise, this leads to a hunched driver posture, which becomes uncomfortable in the long run.

Recent regulations for Formula 1 cars require that the driver can be lifted out of the car together with the seat, so that vertebral injuries to the driver can be avoided in the event of an accident recovery. For this purpose, the seat shell has five points where lifting straps can be attached.

Some seats include a head restraint to prevent the cervical vertebrae from extending backwards in the event of an accident. If the backrest ends below the helmet contact area, which is approximately at the level of the ears, a surface with an energy-absorbing support is required, Fig. 4.17. The support should consist of a dense foam material (e.g. Confor

Fig. 4.17 Head protection on a production sports car (Osella PA 20 S). A head guard is mounted on the frame side above the actual seat. In the picture you can also see the seat shell. This is foamed and covered with adhesive tape

CF42 or CF45, BSCI EIS W50), which does not release any elastic forces when compressed, but on the contrary deforms quasi plastically and thus prevents the head from rebounding. This behaviour reduces the risk of spinal injuries in the event of an accident.

4.4   Steering Wheel

4.4.1   Position of Steering Wheel

The steering wheel is one of the most important control elements (and at the same time a sensor for the road) in the vehicle and must be easy for the driver to operate. In this context, the position of the steering wheel in relation to the seat is of particular importance. Figure 4.18 shows the most important standardised dimensions for passenger cars. The normal position of the hand is such that the lower edge of the hand is in the middle of the steering wheel.

All dimensions refer to the R-point. The R-point (seating reference point) is fixed to the vehicle and corresponds to the H-point (hip joint point) of the passengers in the rear third of the adjustment range of the seat [13].

For the position of the steering wheel in racing cars, see Fig. 4.7. The steering wheel can also be designed to be axially adjustable to suit different drivers.[1] In most cases the seat cannot be adjusted.

For passenger cars and rally cars, the following applies: The steering wheel should be able to be grasped by the driver at the 12 o'clock position with one hand without leaning forward.

---

[1] See Racing Car Technology Manual vol. *4* chassis chap. 5 *steering*.

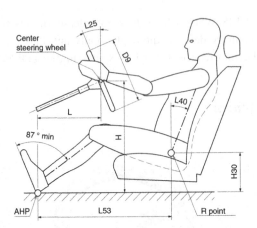

Fig. 4.18 Steering wheel position, passenger car, according to [3] and DIN 70020
R-point: seat reference point
AHP: *Accelerator Heel Point*
L53 = 130 min
H30 = 130 to 520
Steering wheel diameter D9 = 330 to 600
Steering wheel angle L25 = 10° to 70
L = 152 to 660
H = 530 to 838
L40 = 9° to 30
Dimensions in mm

4.4.2 Steering Wheel Dimensions and Types

Steering wheels have various shapes derived from the original circular shape, Fig. 4.19. For example, D- and U-shaped steering wheels are found in a wide variety of cockpits. D-shaped steering wheels either provide space for the driver's legs (flat spot at the bottom) or give visual space over the steering wheel (flat spot at the top). A closed steering wheel rim is essential for vehicles that are driven with large steering angles (rally, raid …).

Steering wheels in which the hub mounting is lower than the rim (potted design) have safety advantages. In the event of a frontal impact, the head or helmet cannot hit the hub [14]. On some steering wheels there is a marking at the 12 o'clock position. This is useful in rally cars, among other things, so that the driver can announce a value for the steering angle to the co-driver who is recording the course notes during training.

The following dimensions are encountered: Formula vehicles: diameter approximately from 250 mm to 285 mm.

Touring car, rally: diameter approximately from 310 to 350 mm. Passenger car: diameter 330 to 600 mm (see also Fig. 4.18).

Wreath dimensions: For round cross-section, the diameter is 28–35 mm; oval cross-sections fit into a rectangle of 35 × 27 mm.

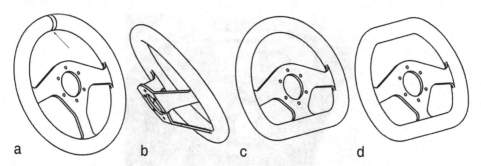

Fig. 4.19 Basic steering wheel shape. (**a**) round shape, flat spokes. (**b**) potted design. (**c**) flattened at the bottom. (**d**) flattened on both sides. For quick orientation about the steering wheel position, some steering wheels have a marking on the steering wheel rim (arrow for variant a)

Fig. 4.20 Steering wheel with quick release (Formula BMW). The principle is the same as found, for example, in quick couplings on compressed air lines. Actuation is via the collar (2) of the sliding sleeve, which relieves the locking balls (1). The steering torque is transmitted via a toothed profile (3)

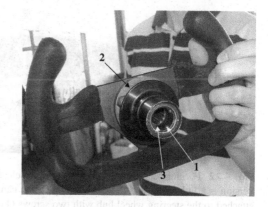

The steering wheel is screwed directly or via a defo element onto the hub with three (formula cars) or six (touring cars, rally, . . .) screws (usually M5 with a bolt circle diameter of around 70 mm).

The hub itself can be detached from the steering shaft by means of a quick-release fastener, Fig. 4.20.[2]

The maximum steering wheel torque should not exceed 10 Nm, otherwise the driver will tire too quickly.

Materials

The star and the spokes are made of aluminium or CFRP laminate.

The tubular steel rim is PU-foamed and covered with suede leather.

---

[2] For the connection of the steering wheel with the steering shaft see Racing Car Technology Manual vol. *4 chassis* chap. 5 3 *steering shaft*.

Fig. 4.21 Formula 1 steering wheel (Ferrari, 2003). From the driver's seat, not only the ignition can be switched off, but also, among other things, the differential setting (locking effect), the engine's mixture composition and the traction control can be influenced. See also Fig. 4.25

Fig. 4.22 Formula 1 steering wheel with gear selector, top view (BMW Williams, 2001). This version is a little older and therefore the operation can be seen well. The rocker switch is rotatably attached to the steering wheel hub with two screws (1). It actuates the two shift pins (2)

Fig. 4.23 Steering wheel gearshift (Formula 1, Ferrari). In the view from below, the switches for the clutch can be seen in the foreground. Behind this is the rocker switch for the gearshift (see also Fig. 4.22)

In the increasingly cramped cockpits of formula cars over the years, one solution has been to move gauges away from the dashboard and directly into the steering wheel, Fig. 4.21.

Another improvement was shifting gears without having to take the hands off the steering wheel, Figs. 4.22, 4.23, and 4.24.

Fig. 4.24 Steering wheel gear shift (Formula 1, Ferrari). In the view from above into the cockpit the release ring of the quick release can be seen. Above the steering shaft the gearshift rocker is mounted. The switches for the clutch are located under the centre of the steering wheel, but are still clearly visible (see also Fig. 4.23)

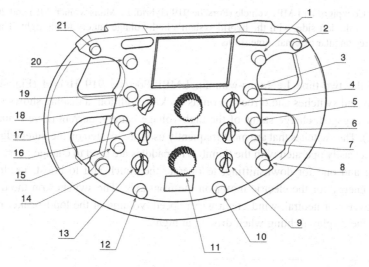

Fig. 4.25 Functions of a Formula 1 steering wheel. *1* Speed limiter for pit lane. *2* Starting aid. *3* Intercom. *4* Corner exit differential. *5* Display control up-scroll. *6* Engine brake lower speed range. *7* Clutch engagement. *8* Safety vehicle strategy. *9* Engine speed engagement. *10* Telemetry switch. *11* Multifunction switch. *12* Engine off. *13* Traction control. *14* Auxiliary oil pump. *15* Fuel filler flap. *16* Differential corner entry. *17* Downward scroll indicator control. *18* Differential lock. *19* Automatic start (launch control). *20* Idle activation. *21* Multiple switch

This development has made the steering wheel one of the most important interfaces to the driver in the vehicle, with steering still being the simplest function, Fig. 4.25. Especially in vehicles with hybrid drive (KERS, ERS), the importance of the steering wheel as an interface continues to increase. Because the size of the steering wheel cannot increase any further (cockpit dimensions, steering speed, ergonomics), less important switches, i.e. those that are rarely operated, have to be relocated to the dashboard. The steering

Fig. 4.26 Cockpit of an LMP1 vehicle (Porsche 919 Hybrid, Le Mans winner 2015 and 2016). The driver sits on the left side, with the hybrid system's high-voltage battery to his right. The steering wheel has rectangular external dimensions

wheel of the two-time Le Mans-winning LMP1 Porsche 919 Hybrid (Fig. 4.26) has 24 buttons and switches and 6 rockers on its back. In addition, a display shows important data (gear engaged, charge status of the high-voltage battery, current drive management, speed, . . .). The switches that are most frequently used are located on the outer edge, so that they can be easily operated with the thumb. The paddles are used to change gears – one for upshifting and one for downshifting, to operate the clutch and to boost, i.e. to request electrical energy for the electric motor on the front axle. The switches on the dashboard include reverse or neutral, windscreen wiper speed, volume of the loudspeaker radio and dimming the display lighting when driving at night.

4.5   Foot Lever Mechanism and Pedals

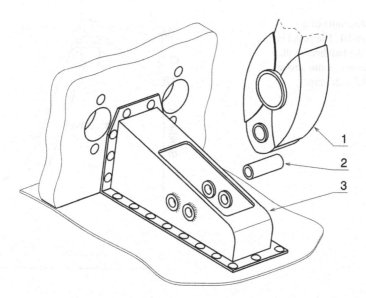

Fig. 4.27 Bracket for brake pedal, according to [15]. *1* Brake pedal. *2* Sleeve. *3* Bracket in box construction. The bracket is riveted to the floor and bulkhead. The pedal runs on a sleeve (2) which is 0.25 mm longer than the receptacle in the pedal. This maintains this lateral play when the screw connecting this sleeve to the bracket is pretensioned

Foot lever gears are mounted suspended (e.g. Fig. 4.28) or upright (e.g. Fig. 4.34). The upright version is preferred for formula vehicles. This ensures a low centre of gravity due to the low-slung master cylinders including reservoirs. The mounting of the pedals also remains as low as possible.

The pedal assembly of racing vehicles must be able to withstand forces of 4000 N without noticeable deformation. This avoids spongy brake pedal feel with associated imprecise braking [8].

The pedal suspension must be rigid, as must the seat mounting, so that the driver has good control, especially when braking. The bolting is done with M8 bolts.

The pedals may be stored in individual brackets. These should be made of at least 2 mm steel plate and fixed to the floor and to the bulkhead, Fig. 4.27. The brackets can be of box construction or consist of two lateral bearing walls with folded edges (Fig. 4.28).

Fig. 4.28 Example of a
suspended pedal. The pedal is
screwed to the bulkhead with a
bracket. The maximum stroke
for this pedal is 28 mm

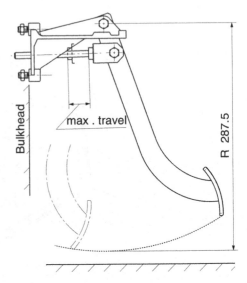

Fig. 4.29 Heel stop. A beveled
plate is screwed to the floor.
Several holes (see detail) allow
the heel position to be adjusted.
The angle of the plate is
designed so that the sole rests in
the heel area and not on the
upper edge of the plate

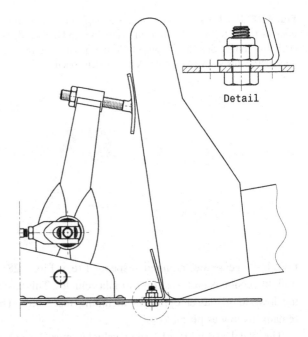

The pedals must be adaptable to the driver. This is made possible, for example, by
different bearing points or adjustable foot plates, Fig. 4.27.

Foot lever assemblies mounted in one unit allow easy adjustment to different driver
sizes by simply moving the lever assembly forward or backward, Fig. 4.37.

For standing pedals, a stop for the heels is recommended. This must also be adjustable in
the longitudinal position. Figure 4.29 shows a simple and easy solution. In certain driving

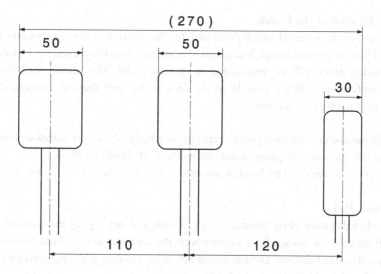

Fig. 4.30 Example of dimensions of a foot lever mechanism of a small passenger car. This example can be used as a first estimation when it comes to the footwell design

situations this stop can help to support the driver in the cockpit. When designing the heel support, it is best to start from the "working position" of the feet, i.e. the accelerator pedal depressed and the brake pedal pressed. In this position of the foot, the sole of the shoe must not rest on the upper edge of the heel support. This only unnecessarily reduces the effective pedal force and irritates the driver.

The actual footrest surface of the pedals should be rough to prevent slipping. A simple option is coating with epoxy resin and sand. On fixed sides, pedals can also have a support plate that supports the foot laterally. This is possible in any case with the accelerator pedal on the right side. If there are only two pedals (brake and accelerator pedal), both pedals can have support plates on the left and right.

Space Requirement

Particularly when designing the front of a single-seater, but also in the case of sports prototypes with a front diffuser, the first question is how narrow the footwell can be. The less space the lever system and the driver's feet take up, the more freedom there is for the monocoque or installations in the front of the vehicle. Figure 4.30 therefore shows the main dimensions of a pedal system.

A support for the left foot in the footwell is advantageous.

In a pure two-pedal arrangement – when clutching and shifting gears via paddles on the steering wheel – it can be more ergonomic and aerodynamic to follow the natural V-shaped position of the feet when the driver is lying down. The heels are closer together and the tops of the feet gape apart. The footwell and thus the nose of a single-seater can be made slimmer towards the bottom, which improves the airflow to the rear below (heat exchanger, splitter, diffuser, ...) and reduces drag.

Relative Position of the Pedals

The position of the actuated clutch pedal dictates the position of the other pedals, because this pedal has the greatest travel. The actuated brake pedal (at the pressure point) should be approximately level with the unactuated accelerator pedal. This allows for intermediate throttle application with the outside of the foot or the heel (heel-toe technique) while braking with the balls of the feet.

The lateral distance between pedals depends on whether the driver brakes with the left or right foot. In any case, the paths of the alternating foot should be small.

If the clutch is operated by hand or automatically, only two pedals are required.

Accelerator Pedal

The throttle valve must close automatically (fail-safe system), e.g. by means of an external spring, if the cable or linkage fails. Furthermore, the actuating device should not hinder the butterfly valve from closing. Double-acting actuation systems allow the driver to actively close the throttle, which increases safety.

The applied brake pedal should be at the same level as the accelerator pedal, which facilitates downshifting with "heel-toe" (heel-toe technique). The amount of spring return depends on the personal preferences of the driver. However, the principle of throttle position over pedal travel is best, as shown in Fig. 4.31 identifier (3). At 50% accelerator pedal travel, a throttle opening of about 15° should be reached first. The more powerful the engine, the more helpful such a progressive gear ratio is for the driver. To design the lever mechanism required for this or the contour of the pulley, the relationship between engine full-load torque and throttle valve angle (or slide position) must be known. The decisive factor is ultimately the objective: the effective engine torque should be linearly dependent on the accelerator pedal travel. Thus, the engine's response to the driver's desired input corresponds to his expectations. For rain races, a less aggressive layout in high-powered vehicles proves helpful to the driver. A simple solution for adjustment is offered by two different mounting points for the Bowden cable or linkage either on the accelerator pedal or on the throttle actuator. If you have access to the electronic engine management system, the desired accelerator pedal characteristics can be elegantly set via the software. The most elegant way of realising all the above wishes is with an electronic accelerator pedal (drive-by-wire). Here, the accelerator pedal is completely separated mechanically from the throttle device. Springs provide the pedal feel. The driver's wish is detected by an angle or travel sensor[3] on the accelerator pedal and transmitted by the engine control unit to the actuator of the throttle unit. The control unit knows, among other things, the gear engaged, the engine speed and the driving speed. This means that the accelerator pedal recognition can be

---

[3] In fact, two sensors are installed for safety reasons. For fault detection, both work independently and differently (redundancy).

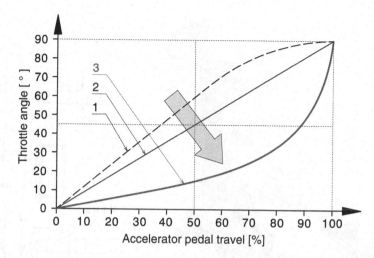

Fig. 4.31 Curve of the throttle position over the accelerator pedal travel. The diagram shows three principal identifiers. 1 degressive course, 2 linear course, 3 progressive course. The driveability increases in the direction of the grey arrow. A progressive characteristic is particularly helpful with strong engine power for sensitive operation of the throttle. A degressive characteristic makes low-powered vehicles appear powerful, at least for the driver

changed depending on the situation. For example, less engine torque will be requested in first gear at the same pedal position than in higher gears so that tire slip is not excessive. A drive-by-wire can also be used for numerous other functions (traction control, automatic intermediate throttle – blipper, ESP, . . .).

The accelerator pedal should have an end stop which is not taken over by the transmission device to the throttle valve. A pressure stop at the level of the football avoids a bending moment in the pedal.

Clutch Pedal

A clutch pedal is no longer found in all footwells. In some vehicles, the starting element is operated by a lever on the steering wheel and the clutch is automatically disengaged or the traction is interrupted by the engine control system during gear changes.

If there is a pedal, it is used for starting and downshifting. If the clutch is actuated hydraulically, there is a risk that the release arm will move too far and the clutch will be damaged. Therefore, a mechanical stop must be provided at the piston rod of the master cylinder or directly for the pedal.

Brake Pedal

The brake pedal not only has to bear the greatest forces, it usually also distributes the braking force between the front and rear axles. This is usually done by means of a balance

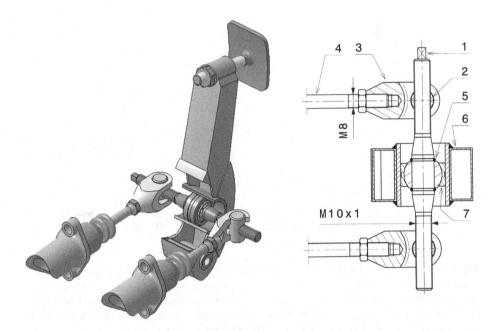

Fig. 4.32 Brake pedal with balance beam. The pedal is shown partly cut open in the left part of the picture. *1* Spindle as balance bar with square as drive for cockpit adjustment. *2* threaded pivot. *3* clevis. *4* Piston rod. *5* circlips. *6* Brake pedal. *7* spherical bearing

bar (*brake bias bar*), Fig. 4.32. A spherical plain bearing (7) slides in a tube section in the brake pedal (6). The spherical plain bearing is axially secured on a spindle (1) by two round snap rings. The spindle carries forks (3) on both sides which are hinged together with round nuts (2). The forks act directly on the two piston rods (4) of the brake master cylinders. When the spindle is turned, both round nuts move along the spindle thread in the same direction. This changes the lever ratios between the two piston rods and the brake pedal and thus the individual brake pressures with the same foot force.

Figure 4.33 shows a simple version of how the balance beam can be adjusted from the driver's seat while driving.

The piston rods to the brake cylinders are subjected to buckling and must be dimensioned accordingly. Long, slender piston rods must be avoided at all costs. With the usual dimensions (approx. Ø8 mm), lengths over 150 mm inevitably lead to problems (Fig. 4.34).

Pedal Ratio
(leverage ratio). The pedal ratios for clutch and brake pedals are approximately 5.0 to 6.25: 1 (value range approximately 3:1 to 6.5:1). The pedal travel (measured in the middle of the footrest) when operating a brake pedal is approx. 20 to 100 mm at usual ratios. To limit the maximum pedal travel, adjustable stops are provided on the clutch and accelerator pedals

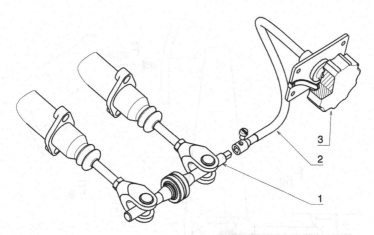

Fig. 4.33 Manual adjustment system of a balance beam. *1* Balance beam with connection . *2* flexible shaft . *3* adjusting knurl at the cockpit end of the dashboard. By turning the adjusting wheel (*3*), the balance beam (*1*) is rotated via the flexible shaft (*2*). The rotation causes the two round nuts, which actuate the brake cylinders via the balance beam, to change their position relative to the brake pedal and thus the force distribution of the balance beam. The adjusting wheel has a stepped locking device so that the setting is maintained

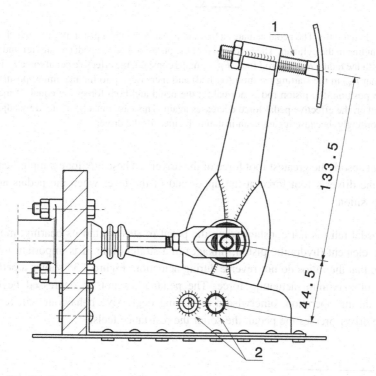

Fig. 4.34 Standing brake pedal, according to [15]. The pedal has a lever ratio of 1:3. The foot plate (*1*) is adjustable to the driver's foot via a thread. Likewise, the pedal can be installed in different bearing positions (*2*) in the console. Of course, the length of the piston rods for the brake cylinders must also be adjusted

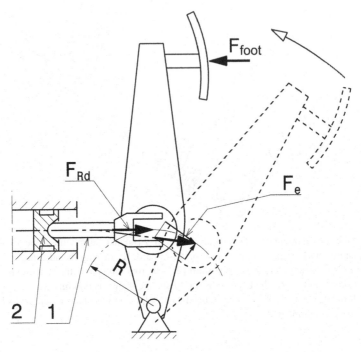

Fig. 4.35 Pedal ratio. The pedal actuates a hydraulic piston (2) via a piston rod (1), which is guided in a straight line in the cylinder. The pivot point of the piston rod on the pedal rotates around the pedal bearing with the radius R. If the rod force F_{Rd} remains constant, the effective pedal force F_e is initially smaller than F_{Rd} in the starting position (dashed) and increases up to the maximum position. In the maximum position, the piston rod is normal on the pedal and both forces are equal. If the pedal is turned further, the effective pedal force decreases again. This case must be avoided by adjusting the pedal accordingly, because it causes an irritating feeling for the driver

which can absorb the greatest foot force of the driver[4] . These are, for example, screws that support the driver's foot force at the upper end of the lever when the pedals are in the upright position.

The pedal ratio is not constant when the pedal rotates around its bearing, moving the actuating element (hydraulic piston, linkage) in a straight line. It is important for human sensation that the ratios do not reverse during actuation. Figure 4.35 shows a pedal in the position of maximum actuating force. The pedal must not be depressed beyond this position during operation, otherwise a disturbing degressive behaviour will result. The more the driver presses the pedal, the softer the resistance feels.

---

[4]The greatest foot force occurs at the brake pedal, see Racing Car Technology Manual Vol. *4 Chassis* Section 6.5 *Forces*.

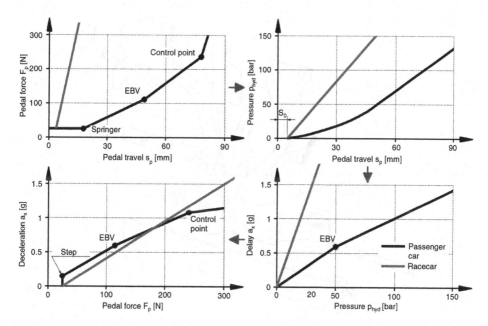

Fig. 4.36 Pedal identifiers for passenger cars and racing vehicles, partly after [16]. From the identification pedal force over – away follows the pressure over pedal travel. Further the deceleration via pressure and finally the deceleration via pedal force
s_0 Empty path
Step: Pressure increase at start of actuation
EBV: electronic brake force distribution
Control point: End of brake force support

Pedal recognition (also "pedal feel") is an important characteristic in passenger cars that supports brand-specific characteristics. When a brake booster is used, it is characterised, among other things, by response force, idle travel, jumper, gain as well as pedal travel and pedal force at the control point. Figure 4.36 shows schematically the basic conditions for a passenger car with vacuum brake booster and for a racing car. The pedal characteristic perceived by the driver is described by the force-displacement characteristic (top left). Initially, empty travel in the system – caused by clearance and closing travel in the brake master cylinder – must be traversed. The foot force does not change, but the pedal travel increases. In the case of the passenger car, this causes the pressure from the brake booster to increase disproportionately and there is already a noticeable deceleration. This engagement allows the driver to subjectively perceive the response better. As the pedal is depressed further, the pedal force builds up over the travel. From the point of intervention of the electronic brake force distribution (EBV), the gradient of the curve increases. When the steering point is reached, the maximum support of the brake booster is reached: from then on, it no longer provides any auxiliary force and the driver feels this as a mechanical stop. The steering point is generally selected in such a way that it does not have to be exceeded

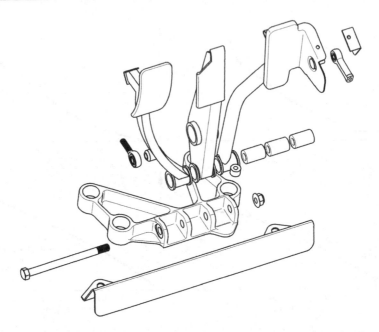

Fig. 4.37 Complete foot pedal of a Monoposto in exploded view (Dallara F306). The pedals are housed in a common bracket and run on bushings that are bolted with an M8 screw. The accelerator pedal and clutch pedal are connected to the actuators with rod ends. The brake pedal accommodates a balance beam. The accelerator pedal has a side support plate to prevent the driver from sliding off the pedal. A heel support is screwed in front of the pedal bracket

even during full deceleration. The combination of brake pedal lever and brake cylinder diameter results in a hydraulic line pressure (top right) due to the foot force. Depending on the mass of the vehicle and the dimensioning of the brake system, a certain deceleration follows for a brake pressure (bottom right). Finally, the driver perceives the actual identification as deceleration via pedal force (bottom left). The braking force distribution between the front and rear axles is changed via the ABS valves by pressure reduction at the rear brakes so that it remains close to the ideal one (function of EBV – electronic brake force distribution).

No brake booster is used on the racing vehicle. After a short system-related idle travel, the foot force rises steeply in a linear manner and leads to an equal increase in pressure. The driver perceives this identification as subjectively confidence-inspiring: the pedal hardly moves and an approximately linearly dependent deceleration is built up, which facilitates the modulation of the braking force. In contrast, a spongy, compliant pedal has a disruptive effect on a braking maneuver at the limit of the tire's circumferential force. When racing on very uneven tracks that cause correspondingly large relative movements between driver and vehicle, some drivers prefer a brake pedal with large travels because the brake is easier to dose under these circumstances.

All pedals can also be pre-assembled in one unit and bolted to the floor. Figure 4.37 shows a lever mechanism mounted in a common bracket. The actuated hydraulic cylinders

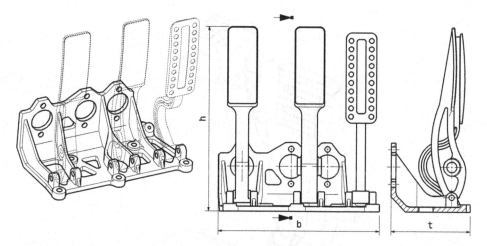

Fig. 4.38 Example of an upright lever mechanism as a pre-assembled unit. The accelerator pedal is curved to allow the flexible adjustment shaft to pass by the balance beam of the brake pedal
Dimensions: $h = 294$ mm, $w = 278.5$ (214.5 with 2 pedals), $t = 133$ mm
Pedal ratio: 4.85:1
The brake pedal is shown in brake position, the clutch pedal in middle position. The accelerator pedal position is adjusted individually

and cables are attached to the chassis. The lever unit in Fig. 4.38 accommodates all hydraulic cylinders in addition to the pedals.

Figures 4.39, 4.40, and 4.41 show examples of completed pedals including footwell.

Materials
Pedals are made of steel, aluminium and magnesium alloys. There are also pedals made of CFRP.

Metal pedals are built as sheet metal bent or stamped parts or milled from the solid.
Materials for brackets: steel, titanium alloys, cast aluminium and magnesium alloys.

4.6   *Linkage*

Accurate transmission of driver inputs in all driving conditions is important. Poorly or incorrectly engaged gears and excessively long shift travels cost time. The components of the external gearshift, i.e. the linkage or the cables, must be routed with sufficient freedom of movement so that they only transmit the desired movements and these also under the influence of large acceleration forces and deformations of the vehicle [11]. The fact that linkages also have to compensate for relative movements between the gearbox and the frame/chassis, as is the case with production cars, rarely occurs in racing cars. The engine-

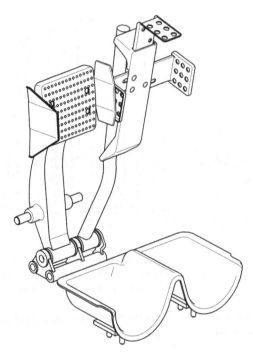

Fig. 4.39 Pedals of a Formula 1 cockpit (Ferrari F1–2002). The clutch pedal is not required because the clutch is engaged via a switch on the steering wheel. The two pedals are therefore brake pedal and accelerator pedal. The pedals are individually adjusted to the wishes of the driver. The side plates prevent the feet from slipping off the pedal. The driver's heels are also guided in tub-like recesses. There are also drivers who want to operate both pedals with one foot, for which the two inner lateral plates between the pedals are removed. In this vehicle, the braking force is distributed by means of a balance beam located on the fixed side of the brake cylinders and moved by a hydraulic cylinder. On the pedal side, rigid mounts of the two rod ends are therefore sufficient

transmission connection is usually a supporting link in the structural chain of the car, which rules out such movements.

Linkages for pure push/pull movements are connected with backlash-free ball joints. If rotational movements also have to be transmitted, universal joints take their place, e.g. Figure 4.42.[5] Figure 4.42 shows the complete outer shifting of a racing vehicle with H-shift. The linkage is made up of several individual elements which are welded or bolted together. The linkage is guided in suitable rod ends (front) or spherical bearings (rear) when passing through a (bulkhead) wall. The connection to the shift shaft of the gearbox is made via the last pipe section with a cross screw.

---

[5] See also Racing Car Technology Manual Vol. *4* Chassis, Chap. 5, Fig. 5.28.

Fig. 4.40 Legroom of a Formula BMW car. The balance beam for the brake force distribution front/rear at the brake pedal (the flexible shaft for adjustment is attached to the left) as well as the support angle for the driver's heels can be seen well. Through the opening in the front bulkhead the reservoirs of the master cylinders can be seen. Above in the middle runs the steering shaft. A cross tube connects the articulation points of the upper wishbones and thus contributes to direct power transmission

Fig. 4.41 Legroom of a Formula 1 car (Toyota 2007). The two pedals are brake and accelerator pedal. The steering gear sits relatively low. The steering shaft runs between the pedals, but this is not a problem because the feet do not change the pedals

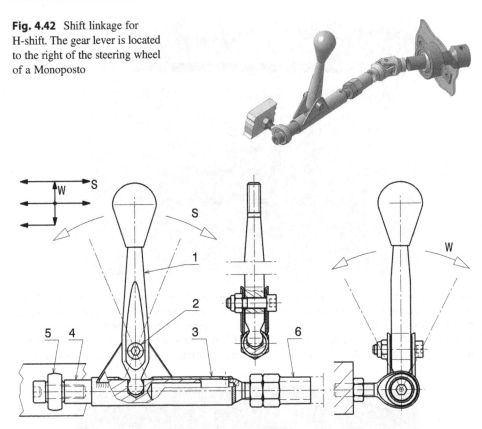

Fig. 4.42 Shift linkage for H-shift. The gear lever is located to the right of the steering wheel of a Monoposto

Fig. 4.43 Gearshift lever for H-shift. This arrangement is used for the shift system in Fig. 4.42

The function of the gear shift can be seen from the illustration of the lever bearing in Fig. 4.43. The driver's wish is initiated by the knob of the shift lever (1). A distinction must be made between a selection movement W (selection of the shifting lane) and the actual shifting movement S (connection of a specific gearwheel with the gear shaft). During the shifting movement, the lever (1) turns around the screw (2). The screw itself is inserted in lugs of the shift shaft (3). As the lever extension is held in a bore in the guide shaft (4) by a ball pin, the shift shaft is moved forwards or backwards during shifting. The guide axle itself is bolted to the frame via a rod end (5). A weld-in piece with thread (6) allows the linkage to be adjusted. During the selection movement, the lever together with the shift shaft (3) is swivelled around the guide axle.

Wire rope hoists which can transmit tensile and compressive forces are advantageously used. Figure 4.44 shows an example of an application.

A lever (1) is mounted on a pivot (2), which is screwed to the cockpit wall. The end of the lever is connected to a rod end (3). The push-pull bowden cable is screwed directly into the nut thread of this rod end. On the lever side a clamp (4) and on the gearbox side an angle

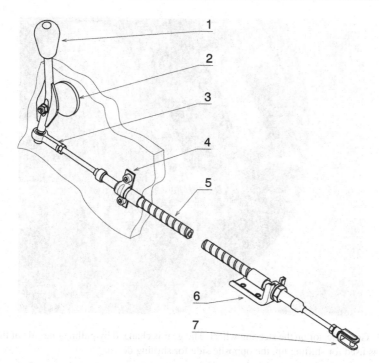

Fig. 4.44 Gear shift by push-pull cable. 1 shift knob. 2 Lever pivot bearing bolted to frame/chassis. 3 Rod end with female thread. 4 Clamp. 5 Push-pull cable. 6 bracket. 7 clevis

piece (6) form the abutment. Both systems allow easy adjustment. The design of the connection to the gearbox lever depends on its shape. Angle heads, rod ends, clevises and similar can be screwed onto the end of the Bowden cable and secured by means of a lock nut.

The gearshift lever should be as close as possible to the steering wheel. This means that the path for the hand to change gears is short and the time required remains short. Steering and shifting must of course be possible without restriction. The hand force required for shifting gears in passenger cars should not exceed 120 N and is generally in the range of 80–120 N. The transmission efficiency of the gearshift system must be taken into account. The transmission efficiency of the entire gearshift must be taken into account, which is often less than 70% [17]. The usual transmission ratios between the gearshift lever and the gearshift sleeve are in the range 7:1–12:1.

If the gearshift is performed sequentially via a lever, it is advantageous that upshifts are performed by pulling and downshifts by pushing. The acceleration forces acting on the driver thus support the shifting movement [18].

Reverse and neutral can be activated on actuated transmissions by a separate button on the instrument panel. With manually shifted sequential transmissions, reverse gear and neutral are blocked by a device in the transmission. This blocking must be deliberately

Fig. 4.45 Gear selector on the steering wheel. The gear is changed by pulling one side of the rocker. One side is used for shifting up, the opposite side for shifting down

released by the driver via a lever in the cockpit, and only then can he engage the desired gear via the gearshift lever. This ensures that the driver does not get into the wrong gear in the heat of the moment.

The operation of sequential shifts can also be displayed directly on the steering wheel via buttons, separately for upshifting and downshifting, or via shift paddles, Fig. 4.45.

4.7 Dashboard

The dashboard contains the most important displays and switches. Especially in single-seaters, a neat arrangement of the instruments can be achieved with a horseshoe-shaped dashboard like in an aircraft cockpit. Elastic mounting of the dashboard or individual instruments has proved successful. Rigid mounting to the frame can unacceptably shorten the service life of instruments due to vibrations.

The following gauges can be found on executed vehicles, among others: tachometer, shift timing, oil pressure, coolant temperature, oil temperature, fuel pressure, ignition-on indicator light, oil pressure warning light.

The displays are supplemented by these switches: main switch, fuel pump, starter actuation, tail light, engine rev limiter, fire extinguisher actuation.

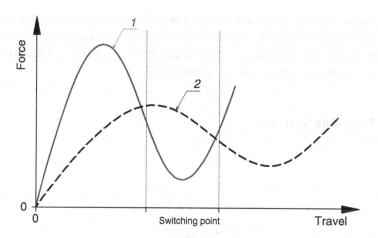

Fig. 4.46 Force-displacement diagram of probes, [19]. *1* short-stroke pushbutton, *2* silicone switching mat. Operation (*1*) characterizes a concise circuit typical of a short-stroke pushbutton. Characteristic (*2*) is a harmonic circuit as realized with silicone switching mats

Switch

Switches must be designed in such a way that they can be operated without restriction when wearing racing gloves. The actuating forces should, on the one hand, provide clear tactile feedback about the switching process that has taken place (Fig. 4.46) and, on the other hand, prevent unintentional switching processes – for example through the effect of acceleration forces or vibrations on the arm. Switches should be arranged in logical groups, e.g. according to function (headlamp, rain lamp, ...) or according to operating sequence (fuel pump, ignition, starter, ...). Mechanical finger guides by means of bars or brackets between switches, as known from the aviation industry, have also proved their worth. In this way, switches can be operated in a targeted manner even in the dark or in the event of vibrations. Toggle switches indicate the switching status (on/off) by their end position. Membrane switches can be combined with an LED and thus also indicate the switching status to the driver at a glance. Membrane switches are available in industrial quality and can be freely programmed. In addition to other advantages, such as low construction effort, this makes their use in racing vehicles attractive.

Switches must be labelled clearly and legibly. This is done using symbols or (mostly English) text.

Display

Displays can also be installed in the cockpit to provide the driver with specific information, see also Chap. 3 *Safety*, marshalling display. However, reading the contents of a screen can take about 1 second. This averting of the driver's gaze poses a risk at high speeds and in unclear racing situations. Head-up displays can theoretically provide a remedy. They project the desired data onto the lower area of the windshield in such a way that the driver does not have to avert his gaze when reading. The driver has the impression that the display is floating in front of the vehicle. The use of current systems in racing cars fails due to the installation space required for the projection device, the image quality and the susceptibility of the image generation components to vibration [20].

4.8   Restraint Systems

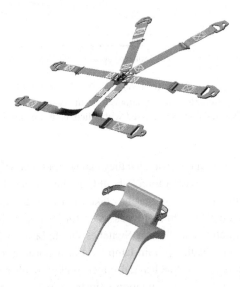

Restraint systems have the task of firmly connecting the driver to the seat. This is necessary during extreme driving maneuvers so that the pilot can operate the controls and does not get caught on them. However, the restraint function also becomes important in the event of an accident, because without a restraint system the combination of flexible crash element and rigid safety cell becomes ineffective for the person inside. Seat belts and airbags are recognized restraint systems. While the airbag is standard in passenger cars, its use in motor sports is extremely rare. There are several reasons for this. Firstly, it only works once. For example, it is useless in the event of multiple overturns. Secondly, motor sportsmen wear helmets. This can lead to jaw fractures in the event of an unfavourable collision with the airbag. Additional systems have also been developed for racing applications, e.g. the HANS system. HANS stands for Head and Neck Support. HANS prevents the head (additionally "weighted down" with the helmet mass) from being pulled forward too much in the event of an impact. The rest of the body is held back by the straps. HANS therefore relieves the neck in the event of an impact. For the use of HANS, the helmet requires its own mounting. The two retaining straps are clipped into both sides of the helmet, Fig. 4.47.

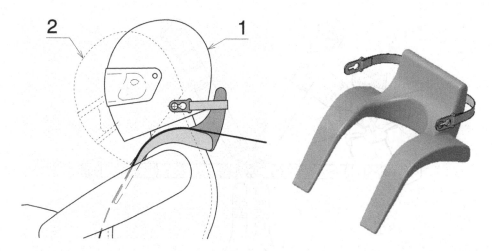

Fig. 4.47 HANS system
Two positions of the head at impact are shown:
1 with HANS. 2 without HANS
The HANS system stabilises the head/helmet bandage so that the neck is relieved in the event of an impact. The system consists of a yoke that the driver slips over the shoulders. Between the shoulders and the yoke, an air cushion conforms to the driver. The shoulder straps are placed over the yoke. In the neck area there is an elevation over which a retaining strap runs. This strap is anchored on both sides of the helmet with a quick-release fastener

Table 4.2 Safe combinations of restraint systems, according to [14]

Restraint system combination			
Belt	Airbag	HANS	Helmet
3-point	Yes	No	No
6-point	No	Yes	Full visor

Table 4.2 shows safe combinations of restraint systems for motor sports.

Seat belts, together with a rigid rollover structure in racing cars, form the survival cell for the driver. Belts are anchored to the frame, the roll bars and the body. Basically all available belt systems are the same, they only differ in the number of attachments. The number of attachment points determines the designation. The greatest protection is provided by six- and seven-point belts, which are mainly used in monoposto cars, Fig. 4.48. In other racing classes, three- and four-point belts are also used. The decisive factor here is the driver's posture. A prone position calls for a 6- or 7-point harness. A reclining position is defined as soon as the backrest deviates more than 30° from the vertical.

When arranging the belt parts in the vehicle, it is important that they can be adjusted to different driver dimensions. Not only the shoulder belt but also the lap belt must be easily adjustable. The belt forces are absorbed by the pelvic bone and the central buckle therefore

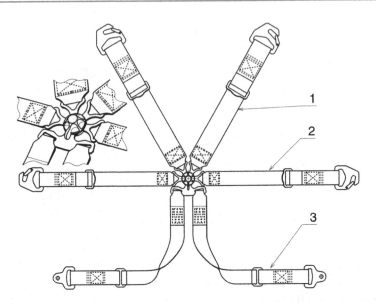

Fig. 4.48 6-point harness six-point harness. 1 shoulder strap. 2 Lap strap. 3 Leg or crotch strap *(sub-belt)*. According to the FIA, the shoulder belt (1) must be 3 inches (75 mm) wide and the lap belt (2) 2 inches (50 mm) wide. The third strap is the leg strap. The buckle is a central twist buckle

rests against the lower abdomen. In touring cars, the same tensioning system as for the shoulder belt is sufficient for lap belts. In formula cars, where the helpers can only tension the belt parts from above, systems that are tensioned in tension are more manageable.

For open vehicles, arm restraints are also used to prevent the driver's arms from hitting the edge of the cockpit in the event of a rollover. These loops are threaded onto the lap belt so that the driver does not have to open an additional latch to exit the vehicle (5 second rule!).

For racing vehicles, the FIA prescribes minimum specifications for safety belts in Annex J, Article 253.

Belts must not be attached directly to the seat, but must support the forces directly on the frame or body. The areas in which attachment points are recommended result from the recommended positions of individual belt parts in relation to the seat, Fig. 4.49. In the case of the shoulder belt, it should be mentioned that the horizontal position, which appears ideal at first glance, is only ideal for a frontal impact. In a rollover, the belt running downwards keeps the driver in the seat much more efficiently. The recommended range is therefore a compromise of both cases.

Each attachment point must be able to withstand a force of 1470 daN or 720 daN for crotch straps. Two belt ends may also be attached to one attachment point (e.g. crotch straps). This must then be able to withstand the sum of both forces. If counter plates are required for body panels, these must have a minimum thickness of 3 mm and ensure a contact surface of at least 40 cm^2.

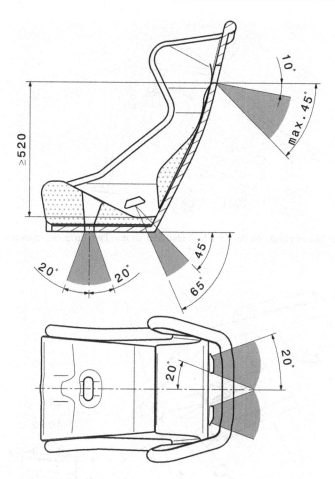

Fig. 4.49 Areas for seat belt attachment points according to FIA. The seat-related recommended areas (shaded grey) for the individual belt components indicate the locations where the belt ends must be secured in the vehicle

The mounting points in the vehicle are summarised in Fig. 4.50.

For the ends of the two shoulder straps there are separate regulations for fastening, Figs. 4.51 and 4.49.

If there are no frame tubes – such as in a monocoque – an alternative is available for shoulder harnesses, Fig. 4.52.

Eyebolts or cap screws can be used at all body attachment points, Figs. 4.53 or 4.54. The screws can be screwed into the original belt mounts on production cars. As with any bolted connection, the bolt should produce a frictional force directed transversely to its axis via its pre-tensioning force. This frictional force is the actual holding force of the belt lug. The screw should therefore not be pulled by the belt force.

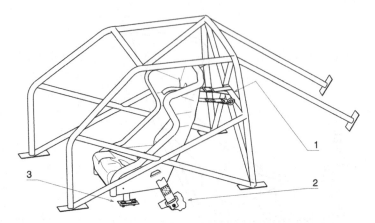

Fig. 4.50 Attachment points on the vehicle. 1 shoulder straps. Two different types are shown. 2 Lap belt. 3 Crotch strap

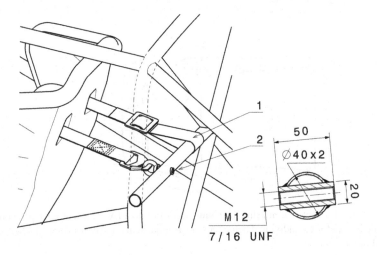

Fig. 4.51 Types of shoulder strap attachment. Shoulder straps may be mounted on the roll cage or an appropriately sized reinforcing brace
1 Loop fixing. 2 Eyebolt
The reinforcing strut shall be 38 × 2.5 or 40 × 2 mm tube. The seamless cold drawn tube shall be made of carbon steel with $R_m \geq 350$ N/mm^2. If a screw is used (variant 2), it must be screwed into a welded sleeve with the specified dimensions. These screws must also be at least M12-8.8, as with other fastening points. An alternative fastening can be seen in Fig. 4.52

For the belt ends on crotch straps there is a simple method of anchorage to the bodywork without sheet metal lugs, Fig. 4.55.

Regulation No. 14 of the German Road Traffic Licensing Regulations (StVZO) "Anchorage of safety belts in passenger cars" stipulates, among other things [3]:

Fig. 4.52 Alternative attachment of a shoulder strap. This mounting is required if there are no frame tubes or no roll cage available. The mounting bracket itself must be bolted to the body with a counter plate

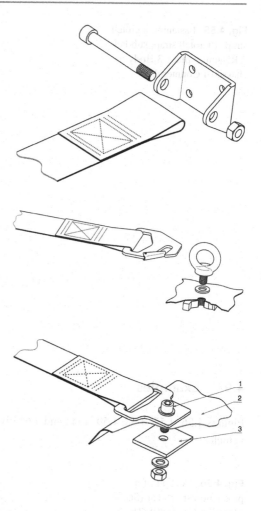

Fig. 4.53 General attachment point for a belt end with eyebolt. The eyebolt can also be screwed into the original mount of the belt on production cars

Fig. 4.54 General design of a fastening point with cap screw. **1** Bolt min. M12-8.8 or 7/16" UNF. **2** Body or frame. **3** Counter plate. The counter plate must be at least 3 mm thick and have a contact area of 40 cm^2

- the minimum number of anchorages to be provided (two lower and one upper for the outer seats and two lower for all other seats)
- the position of the belt anchorages
- the resistance of the anchorages (this is tested by means of a traction device. The upper and the opposite lower anchorages and, at the same time, the two lower anchorages shall withstand a tractive force of 13,500 N each and, if lap belts are used, the two lower anchorages shall withstand a tractive force of 22,250 N over 0.2 s).
- the dimensions of the threaded holes of the anchorages.

Fig. 4.55 Fastening a crotch strap. 1 Crotch strap, sub-belt. 2 Retaining plate. 3 Body or chassis. 4 Counterplate

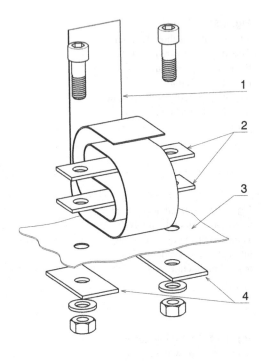

4.9   Examples

Figures 4.56, 4.57, 4.58, 4.59, 4.60, and 4.61 give an impression of the cockpits of different vehicles.

Fig. 4.56 Cockpit of a production sports car (Sauber Mercedes C9, 1989). The gearshift is operated by a gear lever next to the steering wheel. The driver's thighs are held in position laterally by belts

Fig. 4.57 Cockpit of a Formula BMW car. The shift lever (1) for sequential shifting is located to the right of the steering wheel. The rotary switch (2) is used to adjust the brake force distribution front/rear

Fig. 4.58 Driver's position of a Formula Renault racing car. The steering wheel is removed with the quick release, the end of the steering shaft is clearly visible. A 6-point harness holds the driver in the individually foamed seat. This consists of two parts, i.e. the driver's buttocks touch the cockpit floor. The battery is located under the driver's knees

Fig. 4.59 Cockpit of a production sports car (Faust P94). The cockpit is two-seated for the regulations, when in fact only one person can sit in it. Note the lateral support of the pilot and the tub-like seating position (distance between driver's knee and steering wheel)

Fig. 4.60 Cockpit of a touring car. The vehicle is based on a production car with all unnecessary internals removed and a roll cage added. The steering wheel is removed and dangles from a hook next to the right front door

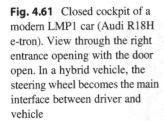

Fig. 4.61 Closed cockpit of a modern LMP1 car (Audi R18H e-tron). View through the right entrance opening with the door open. In a hybrid vehicle, the steering wheel becomes the main interface between driver and vehicle

References

1. N.N.: Formula Renault 2000 Manual. Renault Sport Promotion Sportive (2001)
2. Niemann, G., Winter, H., Höhn, B.-R.: Maschinenelemente, Band 1 Konstruktion und Berechnung von Verbindungen, Lagern, Wellen, 3. Aufl. Springer, Berlin (2001)
3. Henker, E.: Fahrwerktechnik. Vieweg, Wiesbaden (1993)
4. Appel, W.: Development of the chassis fort the R8. AutoTechnol. **3**, 56–59 (2003)
5. Tremayne, D.: Formel 1, Technik unter der Lupe. Motorbuch, Stuttgart (2001)
6. Staniforth, A.: Race and Rallycar Source Book, 4. Aufl. Haynes Publishing, Sparkford (2001)
7. McBeath, S.: Firehawk ascending. Racecar Eng. 10, 54 ff (2002)
8. Incandela, S.: The Anatomy & Development of the Formula One Racing Car from 1975, 2. Aufl. Haynes, Sparkford (1984)
9. Paefgen, F.-J., Gush, B.: Der Bentley Speed 8 für das 24-Stunden Rennen in Le Mans 2003. ATZ. **4**, 281–289 (2004)
10. Newey, A.: How to Build a Car. HarperCollins Publishers, London (2017)
11. McBeath, S.: Competition Car Preparation, 1. Aufl. Haynes, Sparkford (1999)
12. Treusch, M.: Auslegung eines Fahrerplatzes für ein GT-Rennfahrzeug unter dem Aspekt motorsportspezifischer Anforderungen. Diplomarbeit: FH Joanneum, Graz (2014)
13. Braess, H.-H., Seiffert, U.: Vieweg Handbuch Kraftfahrzeugtechnik, 4. Aufl. Vieweg, Wiesbaden (2005)

14. Murri, R., Schläppi, M.: Realitätsbezogene Abstimmung passiver Sicherheitssysteme mittels Schlittentest für Tourenwagen, Beitrag zur Tagung Race.Tech 14.–15. Okt. München (2004)
15. Smith, C.: Tune to Win. Aero Publishers, Fallbrook (1978)
16. Breuer, B., Bill, K.-H. (Hrsg.): Bremsenhandbuch, 1. Aufl. GWV Fachverlage/Vieweg, Wiesbaden (2003)
17. Lechner, G., Naunheimer, H.: Fahrzeuggetriebe, 1. Aufl. Springer, Berlin et al. (1994)
18. Ludvigsen, K.: Mercedes Benz Renn- und Sportwagen, 1. Aufl. Motorbuch, Stuttgart (1999)
19. Pfeffer, P., Harrer, M. (Hrsg.): Lenkungshandbuch, 2. Aufl. Springer Vieweg, Wiesbaden (2013)
20. Bermes, K.: Untersuchungen zum Einsatz von Head-Up-Displays im Motorsport. Master Thesis: Fachhochschule Trier (2012)

Aerodynamics

5

Whereas at the beginning of racing car development the influence of the air on a vehicle was of neglected or secondary importance, today aerodynamics is the determining development tool in most racing series, to which everything else is subordinated.

5.1   Introduction

Aerodynamics deals with the flow of air around an object, in this case around the vehicle. The influences are manifold and are not limited to air resistance. This is indeed a component of the development work, it is mainly responsible for the achievable top speed, but at the same time it is tried to keep the downforce as large as possible, so that the grip of the tires is supported. However, it is not enough to keep the air flowing around the vehicle as efficiently as possible, it must also be directed through the vehicle to dissipate heat from the brakes and engine. Furthermore, cool air is equally important for the driver

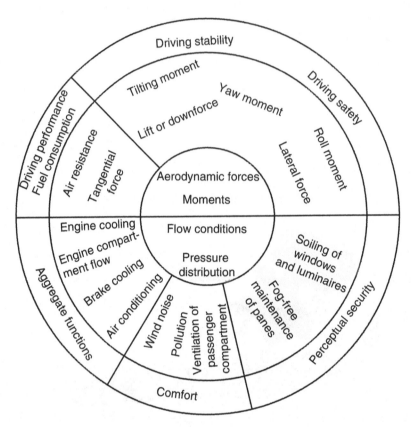

Fig. 5.1 Aerodynamic influences on vehicle functions [1]. The effect of air on a moving vehicle is manifold and concerns not only air resistance and downforce

and the engine to function. The air in the form of wind and as an existing medium also influences the driving stability of a vehicle, especially at high driving speeds, Fig. 5.1.

The immense influence of downforce on the driving performance of racing vehicles can be seen in Fig. 5.2. Above all, the drivable lateral acceleration and the braking capacity increase strongly with increasing driving speed. But also the traction and thus the acceleration capacity increase initially with speed and naturally reach an end when the engine tractive force and the driving resistances are equal.

High downforce provides much greater possible lateral acceleration at the expense of top speed.

The most difficult thing in development is probably the fact that no component acts alone in the airflow, but on the contrary, all parts influence each other. A wing, which is flowing on its own in the wind tunnel, generates a significantly different downforce than the same wing at the end of a vehicle, which is in the airflow influenced by the vehicle. Of course, this does not only apply to parts on one car. If two cars drive behind each other, for

Fig. 5.2 Influence of
aerodynamic measures on the
driving performance of racing
vehicles, after [2], cf. also
Fig. 2.12. Low downforce
results in higher top speed, but
lower drivable lateral
accelerations

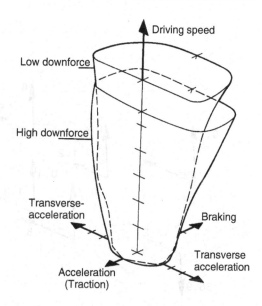

example, both influence each other's aerodynamic behaviour. In addition, the aerodynamic
behaviour of a vehicle changes in the case of an oblique flow, i.e. an inflow deviating from
the longitudinal axis of the vehicle. This is the case with crosswinds or when cornering.
When cornering, a slant flow results from the car's sideslip angle (in the order of magnitude
of up to 10°, in Formula 1 cars max. 5° [3]), which in turn is a consequence of the tire slip
angle. Due to a slant flow, the aerodynamic drag of the car increases and the downforce
decreases.

Figure 5.3 and Table 5.1 show examples of individual contributions of vehicle parts to
air resistance and downforce or lift for a Formula 1 car.

This also reveals those areas that deserve special attention in aerodynamics development. These include the undertray, which causes only 10% of the aerodynamic drag and
makes a significant contribution of 41% to downforce. Its aerodynamic efficiency, in terms
of downforce that costs little power, is expressed in the very high ratio (10.9) of downforce
to drag. Likewise, the front wing works very effectively. The rear wing exhibits similar
efficiency to the overall vehicle, and therefore lends itself to being an adjustment element at
the track for a desired downforce-to-drag ratio. The wheels, on the other hand, not only
generate unwanted lift, but also contribute significantly to drag.

5.2 Air Resistance

Air resistance (drag) directly influences the achievable top speed and fuel consumption.
Nevertheless, its importance must not be overestimated, especially with sufficient engine
power. Even racing cars do not always drive at top speed and downforce, braking

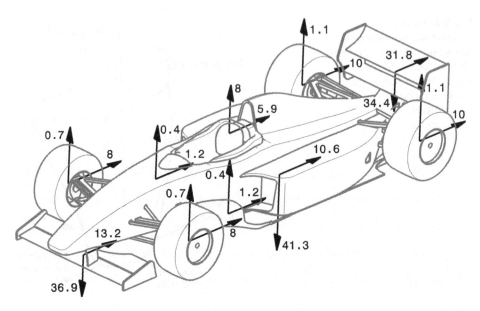

Fig. 5.3 Contributions to drag and lift of vehicle parts (Ferrari F1-2000), after [2]. Figures in %. The values refer to a medium downforce setting and 16/46 mm front/rear ground clearance. The numbers on the driver's helmet represent all other parts of the vehicle

Table 5.1 Contributions of vehicle components to aerodynamic drag and downforce [2]

Component	Air resistance		Downforce		c_A/c_W
	c_W	Share [%]	c_A	Share [%]	[−]
Front wing	0.123	13.2	0.9699	36.9	7.859
Rear wing	0.297	31.8	0.899	34.4	3.029
Undertray	0.099	10.6	1.080	41.3	10.911
Front wheels	0.150	16.0	−0.038	−1.4	−0.251
Rear wheels	0.187	20.1	−0.061	−2.3	−0.326
Barge boards	0.023	2.4	−0.020	−0.8	−0.889
Rest	0.055	5.9	−0.210	−8.0	−3.793
Total	0.934	100	2.617	100	2.802

Note: Negative signs mean lift

performance, heat dissipation, etc. also determine lap times. Figure 2.10 illustrates this clearly.

The air resistance depends mainly on the dynamic pressure caused by the vehicle in the air and on the cross sectional area:[1]

---

[1] For more details see Racing Car Technology Manual vol. 3 *Powertrain*, Sect. 4.1.

$$F_{L,X} = \frac{1}{2}\rho_L \cdot c_W \cdot A_V \cdot v_L^2 \tag{5.1}$$

$F_{L,X}$ Air resistance, N
ρ_L Density of air, kg/m³. $\rho_L = 1.199$ kg/m³ at a temperature of 20 °C, an air pressure of
        1.013 bar and a relative humidity of 60%
c_W Drag coefficient
A_V Cross sectional (frontal area) area of the vehicle, m²
v_L Incoming flow velocity, m/s. When there is no wind, $v_L = v_V$. With v_V Driving speed

The incident flow velocity v_L transferred to the moving vehicle is the geometric sum of
the wind velocity and the driving velocity v_V acting against the direction of travel.

The air density for other temperatures and pressures can be determined approximately
with the ideal gas equation:

$$\rho_L = \frac{p_0}{R_L \cdot T_L} \tag{5.2a}$$

p_0 Air pressure, bar
R_L Gas constant of air, J/(kgK). $R_L \approx 287$ J/(kgK)
T_L (absolute) temperature of the air, K. $T_L = \vartheta_L + 273.15$. ϑ_L ...temperature in °C

More precise values for the air density are provided by the relationship via the air
humidity:

$$\rho_L = \frac{349 \cdot p_0 - 131 \cdot p_e}{T_L} \tag{5.2b}$$

p_e Partial pressure of the water vapour contained in the air, bar.
        $p_e = p_{e,\,max} \cdot \varphi/100$
$p_{e,max}$ maximum steam pressure at T, bar
φ relative humidity, %

To reduce the air resistance, the vehicle can be influenced in the process:

- Drag coefficient c_w as a measure of aerodynamic shape quality
- projected area (frontal area)

Changes to the c_w value can be achieved by individual measures such as undertray
panelling, spoilers, seals, etc. The changes that can be achieved are not absolute but relative
to the original condition of the vehicle. The achievable changes cannot be stated in absolute
terms but relative to the initial condition of the vehicle, Table 5.2.

The projected area is influenced by the concept or the regulations:

Table 5.2 Changes in the c_w value achievable through individual measures for passenger cars [1]

Impact of	Δc_w [%]	Impact of	Δc_w [%]
Level lowering 30 mm	Approx. −5	Flow through heat exchanger and engine compartment	+4 ... + 14
Smooth hubcaps	−1 ...−3		
Wide tires	+2 ... + 4	Brake cooling	+2 ... + 5
Outside panes	Approx. − 1	Interior ventilation	Approx. +1
Sealing of gaps	−2 ...−5	Open windows	Approx. +5
Floor coverings	−1 ...−7	Open sunroof	Approx. +2
Folding headlights	+3 ... + 10	Surfboard roof transport	Approx. +40
Exterior mirrors	+2 ... + 5		

Fig. 5.4 Air brake on a historical racing car (Mercedes). The flap lies against the bodywork during normal driving. When braking, it is moved into the position shown and thus increases both the drag coefficient and the chip area enormously

open – enclosed wheels, open – closed cockpit, track widths front – rear, arrangement of heat exchangers, tyre dimensions, roll bars, position and size of wings.

Air resistance is so great, especially at high driving speeds, that it contributes greatly to the braking deceleration of a car. In a Formula 1 car, the deceleration due to driving resistance at high speed is about 1 g (!).

In the past, this led to the idea of using air resistance specifically to relieve the wheel brakes, Fig. 5.4.

NACA *Duct*

This opening is the recommended air inlet design of the NACA (*National Advisory Committee for Aeronautics*) at the body surface, which causes hardly any impairment of the air resistance, Fig. 5.5. Such an inlet thus represents a viable alternative to scoops or similar. Laminar air flowing on the body surface is deflected into the inlet by a suction effect caused by edge vortices.

Fig. 5.5 NACA inlet. Brakes, oil coolers, engines, etc. can be supplied with air in a streamlined manner via NACA inlets

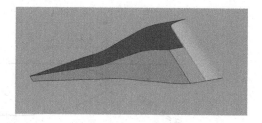

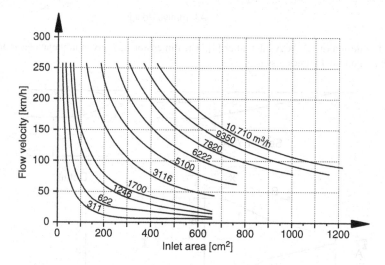

Fig. 5.6 Relationship between air velocity and inlet area for NACA inlets, after [4]. From a desired air flow rate in m³/h follows from the average flow velocity the desired inlet area of the NACA opening

Design of NACA Inlets

1. From the air requirement at a given speed for the engine, heat exchanger, etc., the inlet cross-sectional area A_1, follows theoretically from Fig. 5.6.
2. The actual inlet cross-sectional area $A_1 = b \cdot h$ is chosen to be about 2× larger because the flow coefficient of a NACA inlet is about 0.5.
3. from the area A_1 follow from the favourable height/width ratios height h and width b of the inlet: $h/b = 1{:}3.5$ to 5.5.
4. The thickness t of the lip should be about $0.5 \cdot h$. Lip design see Fig. 5.7.
5. the total length l follows from the recommended ramp angle α of 5 to $11° \rightarrow l = (h + t)/\tan \alpha$
6. the inlet width at the beginning results from the course according to Table 5.3 to: $b_{\text{Beginning}} = 0.083 \cdot b$.
7. the course of the boundary curve is shown according to the values below.

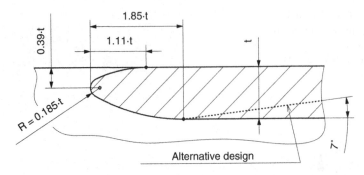

Fig. 5.7 Lip design of a NACA inlet, after [4]. The dimensions are given in parameter representation, i.e. based on the lip thickness *t*

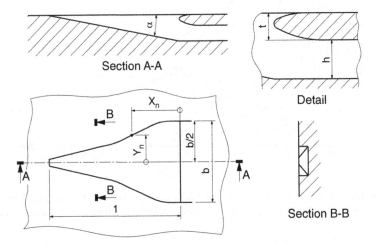

Fig. 5.8 NACA inlet design data, after [4]. Ramp angle α: 5 to 11°, total length *l*, width *b*, height *h*, lip thickness *t*, X,Y point of the edge curve. Section B-B shows that the corners are sharp-edged

The actual coordinates (x/y) for a point at location *n* follow from the chosen values for the length *l* and the width *b* from this:

$$x_n = l \cdot X_n$$

$$y_n = \frac{b}{2} \cdot Y_n$$

x_n, y_n Coordinates of the point at location *n*, mm
l Length of NACA inlet, mm
b Width of NACA inlet, mm

Table 5.3 Trace/schedule of the boundary curve of a NACA inlet in parameter representation [4]. For the position of the parameter pairs (X, Y) see Fig. 5.8. The curves of two sources are shown

X_n	Start Inlet cross section											
		1.0	0.9	0.8	0.7	0.6	0.5	0.4	0.3	0.2	0.1	0
Y_n	RAS[a]	0.083	0.158	0.236	0.312	0.388	0.466	0.614	0.766	0.916	0.996	1.0
	C. Smith[b]	0.084	0.140	0.204	0.276	0.356	0.454	0.590	0.754	0.920	0.992	1.0

[a] According to the Royal Aeronautical Society
[b] According to Caroll Smith: [5]

Fig. 5.9 Air intake and safety
cockpit on a Formula
1 (McLaren Mercedes 2003)

Finally, some thoughts on the design of NACA inlets [4]:

- Too little air is more of a problem for the assemblies to be supplied than too much.
- The supplied air must also be discharged. The outlet area should be about twice as large as the inlet area.
- Baffles or the like will interfere with a NACA inlet to the point of inoperability.

Safety Cockpit for Monoposti (single seaters)
The increase in safety is a measure that has had a detrimental effect on the car's aerodynamic properties since it became mandatory for the 1996 Formula 1 season, due to the increased drag of the wide, bulky centre section, Fig. 5.9. It affects rear downforce because the rear of the cockpit no longer has such a clean flow around it. But it is a very small amount.

Airbox
An airbox allows targeted routing of combustion air to the engine. Favourable intake points are above the driver's head and via flues arranged on both sides behind the cockpit. For two-seat production sports cars, the rollover structure behind the passenger is also a good choice. Theoretically, the bottom line is that the advantages of such a solution are not great at all. For one thing, a top-mounted air intake creates drag by increasing the span area and turbulence. On the other hand, the charging effect caused by the dynamic pressure is extremely small. Even at 200 km/h, the dynamic pressure of the air at 20 °C is only about 0.02 bar. This results in an additional output of roughly 0.2 kW per liter of displacement at 10,000 min^{-1}.

In practice, however, an airbox does pay off. In the 1996 season, for example, Ferrari and Benetton's Formula 1 cars had problems getting enough air into the airbox above the driver's head. As a result, at high speeds, they lost the extra power produced by the air stagnation. This was mainly due to the slightly higher seating position compared to the

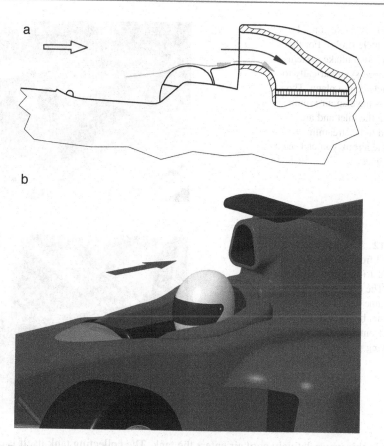

Fig. 5.10 Design of the engine air intake. (**a**) Unfavourable design: boundary layer also flows in (light blue), (**b**) Favourable design: inlet area all around clear

Williams or Jordan cars. The Ferrari drivers therefore had to tilt their heads to the side on the fast straights to avoid covering the intake [6].

While a dynamic pressure supports the incoming airflow, a complete clearance of the inlet opening additionally prevents the low-energy boundary layer flowing over the helmet and cockpit rim from entering as well, thus improving the supply of air to the engine, Fig. 5.10. The same considerations can also serve for the design of inlet openings of heat exchanger ducts, cf. Fig. 5.78b.

The inlet should not be made unnecessarily large because the frontal area of the car is increased. For a 3-liter engine, an intake area of about 50–100 cm^2 is sufficient even at the highest speeds (approx. 17,000 min^{-1}).

Especially if the regulations require an air restrictor for the intake air of the engine, the design of the air intake area is of great importance. The easiest way to do this is to design the collector to have good flow and to locate it away from the engine in the direction of

Fig. 5.11 Airboxes on a Le Mans vehicle (Lola). For the V-engine, two intake systems are arranged symmetrically to the vehicle centre plane. The collectors have an air volume limiter at the inlet and are designed to be streamlined. Under the airbox you can see a NACA inlet

Fig. 5.12 Aerodynamically designed front suspension on a Formula 1 (McLaren Mercedes 2003). The track rod is not only at the same height as the upper wishbone, but is even integrated into its front arm. The tierod also has a wing profile

travel. In this way relatively cool air enters the tank. The collecting tank itself is cooled by the air flowing around it, Fig. 5.11.[2]

Wheel Suspension

Wing-sectioned wishbones and pushrods help reduce drag and allow airflow to flow to heat exchangers and the rear wing with as little interference as possible, Fig. 5.12.

Wheels

In cars with (regulation) open wheels, they are responsible for more than 30% of the aerodynamic drag. When narrower tyres were introduced for 1993 in Formula 1 and the front cross-section was reduced, aerodynamic drag was reduced from 40 to about 33% of the total drag. But this did not significantly change the influence of drag overall on driving performance [6].

---

[2] See also Racing Car Technology Manual Vol. 3 Powertrain Chap. 1, Fig. 1.103.

Fig. 5.13 Airflow in the front wheel area of a production sports car (Norma N20). You can see the left front side of the vehicle. Part of the air entering the wheel housing is used for brake cooling and the other part flows around the fuselage. Subsequently, a part of the air flows through the heat exchanger, which is located behind the cover on the right side of the picture

A wide nose front can at least reduce the influence of the front wheels. On the Tyrell P34 (6-wheel Formula 1 car), this idea was implemented. On today's Formula cars, however, airflow to the sidepods is more important than bulkheading the front wheels. A fairing of the wheels is forbidden by the regulations. At the rear axle one helps oneself with appropriate design of the rear area (bottle neck effect, *Coke-bottle effect*), Fig. 5.15.

If fairings are permitted – as on production sports cars – the air is directed around the wheels, Figs. 5.13, 5.14, 5.15, and 5.16.

The design of the outside of the wheel also has an influence on the flow around the wheels. For example, an axial offset between the outer tyre contour and the wheel disc increases the air resistance, Fig. 5.17. The smaller the step between the surfaces of the tyre sidewall and the wheel disc, the closer the flow and thus the smaller the c_W value.

In general, however, care is taken to ensure that the tyre overhangs the rim flange. If a wheel comes into "contact with the enemy", the contact with the elastic tyre is much more harmless than a blow to the rim flange. This can deform and the air can escape abruptly, which means that this tyre can no longer transmit forces and the stability of the vehicle is endangered.

Likewise, the shape of the wheel's outer surface has an influence on turbulence as the air passes by. In terms of low drag, smooth, continuous surfaces are more favorable than pronounced spokes. In fact, a compromise will have to be found so that heat can be dissipated from brake and wheel bearings.

If the regulations allow it, continuous, smooth wheel discs are suitable, which are cut off at the top, so that a D-shaped opening is available for the cooling air to escape. The disc

Fig. 5.14 Wheel house ventilation on the front axle (Lola). With enclosed wheels, air builds up on the outgoing side of the tyre. A vent allows pressure equalization. The fins direct the air from the wheel arch to the rear via the bodywork. For LMP vehicles (Le Mans Prototypes), venting of the wheel housing has been mandatory since 2012 for safety reasons (possible climbing of the vehicles)

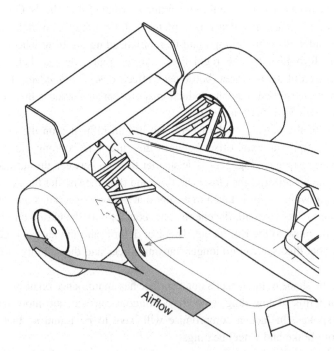

Fig. 5.15 Reduction of air resistance for monoposto with open wheels. If the rear of the vehicle is tapered like a bottle neck, the air that has built up in front of the wide rear wheel can flow past the wheel on both sides. The location (1) lends itself to an air intake. This is the area where the exhaust turbochargers of the Formula 1 cars of the 1980s drew in fresh air. The arrangement of the brake in the wheel (as opposed to the inboard brake) facilitates the desired design, as does a slim gearbox

Fig. 5.16 Wheel arch lower
part of a production sports car
(Osella PA 20 S). Pictured is the
right rear wheel with the engine
cover removed, which houses
the upper part of the wheel
housing

Fig. 5.17 Dependence of air
resistance on the offset between
tyre and wheel surface [7]. The
results were determined on 16 to
18″ wheels. The greater the
offset between the outer surface
of the rim and the sidewall of the
tyre, the greater the c_W value

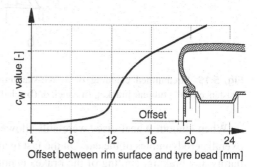

Fig. 5.18 Partially closed wheel. The rim flange carries a ring-shaped disc that creates a smooth, stepless transition to the tire bead. The remaining central opening allows the cooling air of the wheel brake to escape almost unhindered between the spokes. The wheel is pre-forged from one piece and is finished by machining. It is then powder-coated, with the exception of the functional surfaces (nut and flange support). Vehicle: Porsche 919 Hybrid. (Le Mans overall winner 2017)

itself is mounted rotatably on the wheel axle. The opening thus always remains at the top, even when the wheel is rotating. A compromise solution between low air resistance and good brake cooling is shown in Fig. 5.18.

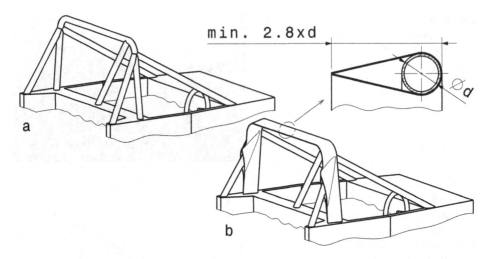

Fig. 5.19 Streamlined cladding of a free-standing roll bar. A simple cladding reduces the air resistance of the tubular bracket. (**a**) unclad, (**b**) clad

When smooth wheel discs were still allowed in Indycars, they proved helpful for the overall aerodynamics of the car. On the fast oval tracks with banked corners, where there is hardly any braking, cooling of the brakes is not critical and the wheel disc can be closed. Because the air could not escape through the wheel to the outside, it flowed under the car. At the rear axle, the closed wheel disc prevented air from flowing under the rear wing and the negative pressure there had a positive effect on the functioning of the diffuser [8].

Other Parts
Side shafts rotating directly in the airflow, as is the case with most single-seaters, disturb the flow in many ways. Friction on the surface causes a lift effect and the flow to the rear wing is deflected.[3]

Free-standing roll bars, such as those used in open sports prototypes, have a circular cross-section as the tube. By cladding, this can be approximated to a much more favourable teardrop cross-section, whereby the c_W value drops to about one tenth of the tube value, Fig. 5.19.
A vehicle undertray that is raised to the rear also affects drag. Good combinations of angle of rise, ground clearance and length of rise can reduce the drag of the vehicle, see "Diffuser" in Sect. 5.3 for more information.

---

[3]Remedial measures are shown in the Racing Car Technology Manual vol. 3 Powertrain, Section 5.5.2.

5.3 Downforce

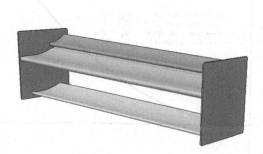

Downforce is the ideal solution when increased circumferential and lateral force is to be generated at the wheels without simultaneously lifting the mass of the vehicle. The disadvantage is that the effect is speed-dependent and that – depending on the operating principle – the position of the vehicle in relation to the road is a decisive factor, which can make the effect unpredictable for the driver.

A simple view of a mass point in a corner shows the potential of downforce.

The maximum related lateral acceleration a_y is equal to the available coefficient of friction $\mu_{W,Y}$:

$$\frac{a_{y,\,max}}{g} = \mu_{W,Y} \tag{5.3}$$

$\mu_{W,Y}$ Friction value in transverse direction
$a_{y,max}$ Max. lateral acceleration, m/s^2
g Acceleration due to gravity. $g \approx 9.81$ m/s^2

With a coefficient of friction of $\mu_{W,Y} = 1$, the maximum achievable lateral acceleration thus follows to 1 g.

If a downforce $F_{L,Z}$ is added, the drivable lateral acceleration increases:

$$\frac{a_{y,\,max}}{g} = \mu_{W,Y}\left(1 + \frac{F_{L,Z}}{m_{V,t} \cdot g}\right) \tag{5.4}$$

$F_{L,Z}$ Downforce, N
$m_{V,t}$ Total mass of the vehicle, kg

The downforce of a complete vehicle is calculated analogously to the air resistance, (5.1), to:

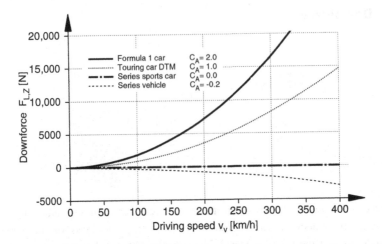

Fig. 5.20 Calculated progression of the downforce over the travel speed, according to [9]. Frontal area for all vehicles 2 m². Production vehicles do not have any downforce-generating elements and therefore "suffer" from lift. In production sports cars, care is taken to ensure that at least no lift occurs at high speeds

$$F_{L,Z} = \frac{1}{2}\rho_L \cdot c_A \cdot A_V \cdot v_L^2 \tag{5.5}$$

ρ_L Density of air, kg/m³. See (5.2)
c_A Downforce coefficient, see Table 5.1
A_V frontal area of the vehicle, m²
v_L Incoming flow velocity, m/s. When there is no wind, $v_L = v_V$, with v_V Driving speed

Figure 5.20 shows the calculated trace of the downforces of different vehicles over the driving speed. For comparison purposes, the frontal faces are assumed to be uniformly 2 m².

At 240 km/h, a Formula 1 car actually generates around 16 kN of downforce and weighs just over 600 kg, even with fuel and driver. So the downforce is about 2.7 times the weight. This means it could literally drive on the ceiling if the road suddenly twisted from bottom to top!

In the numerical example above with $\mu_{W,Y} = 1$, this increases the drivable lateral acceleration to $1 + 2.7 = 3.7$ g.

The maximum speed at which a vehicle can negotiate a flat corner follows from (5.4) and (5.5) with the relationship for lateral acceleration a_y:

$$a_y = \frac{v_V^2}{R}$$

v_V Driving speed, m/s
R Corner radius, m

$$v_{co,\,max} = \sqrt{\frac{g}{\frac{1}{\mu_{W,Y} \cdot R} - \frac{\rho_L \cdot c_A \cdot A_V}{2 m_{V,t}}}} \tag{5.6}$$

$vco_{,max}$ Maximum corner speed, m/s
$m_{V,t}$ Total vehicle mass, kg

It can be seen in (5.6) that for each vehicle there is a critical corner radius above which the maximum driving speed is no longer limited by the lateral acceleration that occurs. The speed goes towards infinity when the denominator becomes zero.

$$R_{krit} = \frac{2 m_{V,t}}{\mu_{W,Y} \cdot \rho_L \cdot c_A \cdot A_V} \tag{5.7}$$

R_{krit} Critical corner radius, m

For a Formula 1 vehicle with a mass of 600 kg, a coefficient of friction $\mu_{W,Y}$ of 2 and a coefficient of downforce c_A of 2.6, this results in a critical radius of approx. 160 m. Accordingly, this vehicle can negotiate flat (unbanked) corners with a larger radius at its maximum speed, or viewed differently: The driver only has to brake before corners with a smaller radius.

However, a downforce also improves the traction capacity in the longitudinal direction (driving and braking).[4] Thus, the adhesion limit of the rear axle $_{FW,X,A,max}$ increases with increasing speed because the decisive downforce $F_{L,Z,r}$ increases:[5]

$$F_{W,X,A,\,max} = \mu_{W,X} \cdot \left(m_{V,t} \cdot g \cdot \frac{l_f}{l} + F_{L,Z,r} + m_{V,t} \cdot a_X \cdot \frac{h_V}{l} \right) \tag{5.8}$$

FW,X,A,max Maximum tractive force rear axle, N
$\mu_{W,X}$ Coefficient of friction in the longitudinal direction
$F_{L,Z,r}$ Proportion of the downforce on the rear axle, N. Is determined by the position of the center of pressure
h_V, l_f, l_r Dimensions of the centre of gravity position, m. See Fig. 2.8
l Wheelbase, m

---

[4] For the influence on braking see Racing Car Technology Manual Vol. 4 Chassis, Chap. 6.
[5] See also Racing Car Technology Manual Vol. 3 *Powertrain*, Chap. 4 *Tractive Force Diagram*.

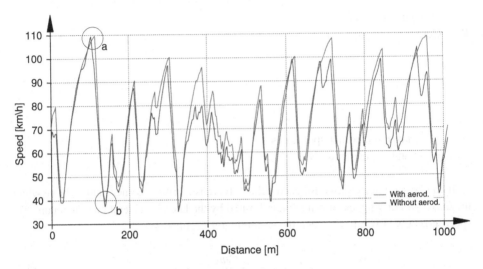

Fig. 5.21 Lap time comparison: Influence of downforce (simulation), according to [10]. A flying lap of a Formula Student car at the Endurance race (Formula Student Austria competition 2015. Software: OptimumLap) is shown. Once without and once with aerodynamic downforce aids consisting of front and rear wing including wing side boxes as well as undertray with rear diffuser. All other vehicle parameters are the same in both cases. The lap times are 52.6 s and 56.9 s respectively, which is a 4.3 s advantage for the car with downforce aids
For orientation purposes, two recurring areas are highlighted as examples: a Start of braking, b Corner apex

However, a high downforce can only be usefully implemented with a high engine output, due to the corresponding increase in air resistance.[6]

A simulation clearly demonstrates how and where the advantage of downforce affects the speed and thus the lap time, Fig. 5.21. With aerodynamic downforce aids (red), it is possible to brake later and harder (e.g. area a) or it is not necessary to decelerate at all (e.g. at 700 and 950 m). The advantage is also noticeable in the corners (e.g. area b). The lowest speed is always above that of the comparison car (blue).

Downforce is created by a pressure difference above and below a part of the vehicle. The greater the effective area of this part and the greater the pressure difference, the greater the resulting force on this part. Some principles of how pressure differences in air flows can be generated result from the consideration of physical relationships, Fig. 5.22. The energy equilibrium along a stream path of an incompressible flow describes Bernoulli's equation (here without weight members, i.e. without the share of potential energy):

---

[6] See also Racing Car Technology Manual Vol. 3 *Powertrain*, Chap. 4, Fig. 4.14.

Fig. 5.22 Basic principles of downforce generation. Digits: *0* surrounding, *1* above the part, *2* below the part

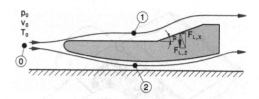

$$p_t = p_0 + \frac{\rho_{L,0}}{2}v_0^2 = p_1 + \frac{\rho_{L,1}}{2}v_1^2 = const. \tag{5.9}$$

p_t Total pressure, Pa
p_0, p_1 Static pressure at point *0* (ambient) or at point *1*, Pa
$\rho_{L,0}, \rho_{L,1}$ Air density at point *0* or *1*, kg/m^3
v_0, v_1 Flow velocity, m/s

For a downward force, the pressure above the part (location 1) must be greater and/or the pressure below the part (location 2) must be smaller. This is achieved by a low flow velocity v_1 above the part (thus p_1 large) and/or by a high flow velocity v_2 below the part (thus p_2 low). High flow velocities are achieved by reducing the flow cross-section (nozzle effect). The air density offers a further control lever. This is inversely proportional to the temperature, see (5.2). Hot air under a part can thus increase the downforce. The inertial forces of the air can also be used to generate force through the appropriate design of air control elements: A portion of the dynamic pressure force (point with flow velocity = 0) of the air against an inclined surface acts downwards:

$$F_{L,Z} = \frac{\rho_{L,1}}{2}v_1^2 \cdot A \cdot \cos\alpha \tag{5.10}$$

$F_{L,Z}$ Downforce, N
$F_{L,X}$ Resistance, N
A Area of the inclined part, m^2
α Deflection angle, °

One of the first known ground effect vehicles was the Lotus 78 Formula 1 car, which already had the key features, Fig. 5.23:

- Pointed, wedge-shaped nose (nose)
- As slim as possible fuselage, so that the side boxes as wings are as wide as possible
- As narrow a tail as possible and fully shrouded engine so that the airflow can exit almost unimpeded under the vehicle.

The wing cars were constructed as follows, Fig. 5.24. The side boxes (1) on either side of the fuselage are constructed like the wings of an airplane, but with an upside-down airfoil so that the wings create an air force that points downward. These wings are relatively short

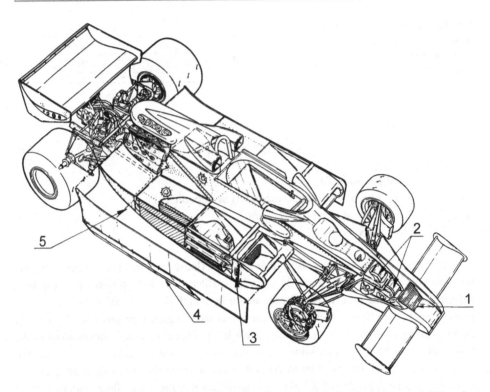

Fig. 5.23 Formula 1 car from 1977: Lotus 78 [11]. The Lotus 78 is the first wing car. 1 Oil cooler, 2 Fire extinguishers, 3 Water radiator, 4 Plastic skirt, 5 side pod

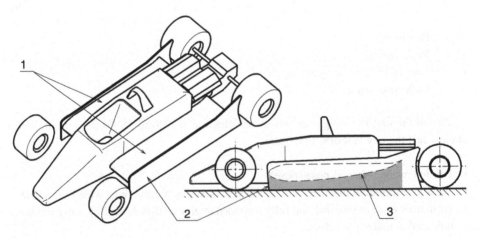

Fig. 5.24 Principle of a ground effect car [11]. 1 side pods, 2 skirt, 3 underwing profile

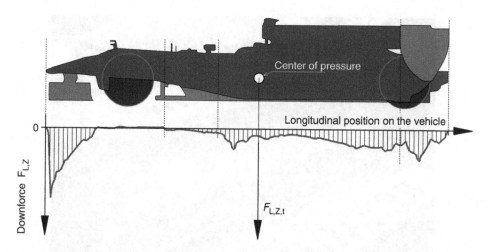

Fig. 5.25 Course of the downforce over the vehicle length, Formula 1 car Red Bull RB6 (2010) according to [13]. In addition, the total downforce $F_{L,Z,t}$ is entered, which is the resultant force from the integral of the curve and which determines the position of the center of pressure in the longitudinal direction

and therefore side plates are attached to their end. This prevents some of the airflow from the surrounding area from entering the negative pressure area between the side box and the roadway. To reinforce this barrier effect, sealing strips are used on the lower edge of the side plates, which slide on the roadway (skirts). The proximity of the wings to the carriageway further enhances their downforce effect (ground effect).

The downforce of a racing vehicle is generated by the interaction of numerous individual elements that cannot be considered independently of each other and whose contribution to the overall effect varies. Nevertheless, it can be stated that the surfaces facing the road (low pressure, high flow velocity) react more sensitively to changes (ground clearance, inclination, roughness, ...) and have a greater influence on downforce than those that can be seen from above. Accordingly, more attention is paid to the non-visible surfaces in the development [12]. In Fig. 5.25, the variation of downforce is plotted along the vehicle length of a Formula 1 car. The front wing and the diffuser in conjunction with the rear wing make the largest contributions. In the area of the front wheels there is even a slight lift. The center of pressure (imaginary point of application of the resulting air force) is in the area of the vehicle's centre of gravity and thus enables neutral handling at high speeds.

At the following, some elements for generating resp. supporting downforce are described.

Skirts

Skirts sealed the side area next to the wing-like sidepods of Formula 1 flying cars from the road. So they had to be movable to compensate for vehicle movement and ground unevenness, yet stiff enough to maintain the pressure differential between the environment

Fig. 5.26 Apron on a Formula 1 car (Renault RE20) [11]. The skirt shown is attached to the left side box and slides up and down in the guides

and the underside of the car. They were not without problems, because wear due to abrasion led to unpredictable changes in driving behavior. Accordingly, a great deal of development effort was put into improving these sealing systems. In some teams, mechanics were assigned only to the task of maintaining the function of the skirts through intensive maintenance [11]. The development of the skirts naturally influenced the design of the suspension. Thus, suspension systems were soon used that allowed the ground clearance to be varied while the vehicle was in motion. Figure 5.26 shows a solution in which stiff skirts can slide vertically up and down in a guide and are pressed onto the road surface by springs.

After numerous accidents, the design of aprons was regulated, Fig. 5.27, and eventually they were banned altogether.

Wing
A wing is an aerodynamic body that generates lift on aircraft. In motorsport, a wing is installed upside down and thus generates a downward force, which is called downforce. Of course this downforce is bought with a certain air resistance. Important characteristics of wings are shown in Fig. 5.28.

Characteristic dimensions are the chord length c and the span s. Furthermore, the aspect ratio Λ formed from them is significant.

$$\Lambda = s^2 / A_{wing}$$

Fig. 5.27 Section through a skirt on the right side of the vehicle, according to F.I.S.A. 1982, after [11] 1 flexible material attached to fuselage side, 2 ground rubbing strip

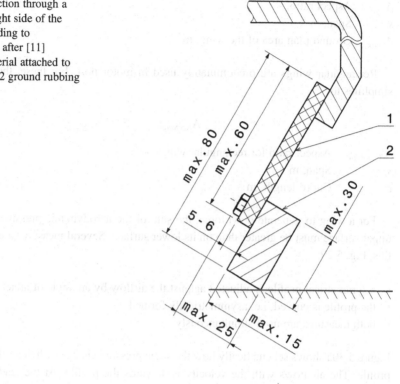

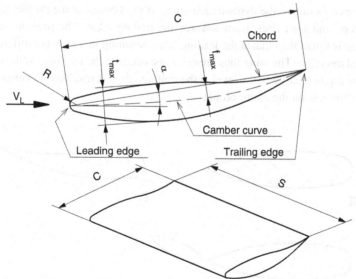

Fig. 5.28 Terms and sizes of a wing (*aerofoil, AE airfoil*): c chord length, s span, α angle of attack, f camber, t profile thickness, v_L air-flow velocity, leading edge, trailing edge, length/width ratio (*aspect ratio*)

Λ Aspect ratio, -

s Span, m

A_{wing} Ground plan area of the wing, m^2

Rectangular wings are predominantly used in motorsport. For these, the aspect ratio simplifies to:

$$\Lambda_{Rectangle} = s/c$$

$\Lambda_{Rectangle}$ Aspect ratio for rectangular wings, -

s Span, m

c Chord length, m

For a wing to generate downforce, the sum of the aerodynamic pressure forces on its upper surface must be greater than on its lower surface. Several measures are available for this, Fig. 5.29:

- a symmetrical profile is adjusted against the airflow by an angle of attack
- the profile is curved, i.e. asymmetrically formed
- both measures are used simultaneously

Figure 5.30a shows schematically how the static pressure changes when air flows around a profile. The air flows with the velocity v_L towards the profile. At the leading edge, the pressure increases due to the dynamic pressure. At the bottom of the profile, the pressure initially drops and then rises again towards the trailing edge. The pressure level at the trailing edge is lower than that at the leading edge, resulting in a pressure difference in the longitudinal direction. The wing thus creates a resistance to the air flow. At the upper side, the pressure drops steadily to the level at the trailing edge. A resulting downward pressure difference thus acts on the entire profile.

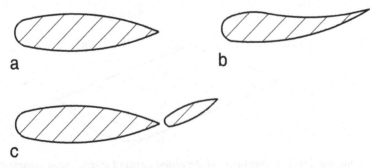

Fig. 5.29 Types of wing sections aerofoil cross sections. (**a**) symmetric aerofoil section, (**b**) cambered profile, **c** symmetric profile with camber flap

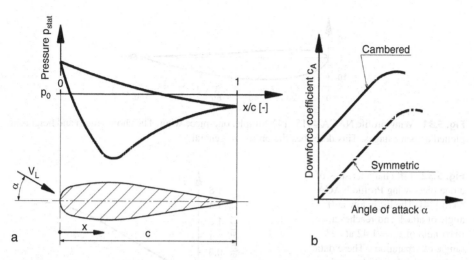

Fig. 5.30 Pressure curve at the wing surface and downforce. p_0 Ambient pressure, (a) Negative pressure prevails over the largest area of the lower side of the wing. In contrast, overpressure prevails on the upper side. The resulting force points downwards. (b) A curved airfoil generates more downforce at the same angle of attack

The angle of attack strongly influences the lift behaviour of a wing, Fig. 5.30b. It is the angle between the velocity of the airflow and the chord of the wing. It is important to note that the direction of the incoming air on a vehicle is not the same as the direction of the roadway. Especially with rear wings this will hardly be the case. The lift increases approximately linearly with the angle of attack until the airflow detaches at the bottom of the airfoil.

The downforce is calculated from the downforce coefficient:

$$F_{L,Z} = c_A \cdot \frac{\rho_L v_L^2}{2} \cdot s \cdot c \qquad (5.11)$$

$F_{L,Z}$ Downforce, N
c_A Downforce coefficient, -
ρ_L Air density, kg/m^3
v_L Air velocity, m/s
s Span, m
c Chord length, m

The main focus of the wing design is the consideration of the ratio of downforce to drag. In addition, the separation behaviour, usable range, speed range as well as the degree of turbulence of the incoming flow are also examined. Unlike in aviation, criteria such as resulting moment are of secondary importance, which is why wing profiles in motor sports

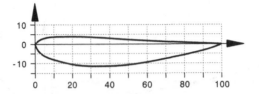

Fig. 5.31 Wing profile NACA 4415 [14]. Simple, one-piece wing. The chord length and heights are plotted as percentages. This describes the profile in general

Fig. 5.32 Lift characteristics of a one-piece wing Profile NACA 4415 [14]. The lift is low at 0° angle of attack and reaches a maximum of $c_A = 1.42$ at 12° (angle of separation). These data apply to an infinitely long wing

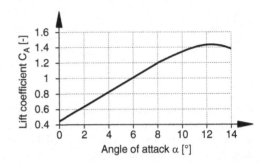

(Figs. 5.31 and 5.32) are designed differently than in aircraft construction [9]. The wings used in racing become noticeably effective from about 80 km/h [11].

The following parameters can be used as a guide for a one-piece wing for racing vehicles [14]:

- Aspect ratio $\Lambda_{rectangle} = 5$ to 8 [9].
- small angle of attack for low drag and downforce, max. Angle of attack of approx. 14 to 16°.
- low profile thickness for low resistance and downforce. Thickness up to 0.12 c for stronger downforce. Thickness is not so important in low speed ranges
- slight camber at the 0.3 c point for low downforce and greater camber further back from 0.05 c to 0.15 c at the 0.5 c or 0.6 c point for higher downforce
- Radius of leading edge about 0.01 c to 0.03 c.

Above all these guideline values, however, are the regulations, so that no generally applicable design guidelines are meaningful. With increasing wing width the downforce increases, but in general the span is limited by the regulations to the vehicle width. The further away a wing is from other parts of the vehicle, the less interference interferes with its desired flow around the wing and thus its effectiveness. This is precisely why the space in which a wing must be fitted is also prescribed by the regulations in relation to the vehicle. The same applies to the ground clearance of a front wing.

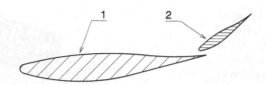

Fig. 5.33 Schematic of a two-part wing [14]. 1 wing, 2 flap. A multi-piece wing acts similar to a one-piece wing with a larger camber, thus generating more downforce. The mobility of the flap (s) results in easier adjustment possibilities

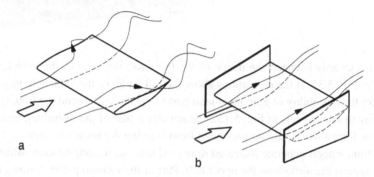

Fig. 5.34 Effect of end faces on wings. (**a**) Wing without end faces, (**b**) Wing with end faces. In case a, the air flows from the overpressure side above to the underpressure side below. In case b, this is prevented by the end faces. The end faces increase the downforce by up to 30%

A large selection of different airfoil profiles is offered by the UIUC Airfoil Coordinates Database [15]. Here, one will select airfoils with smaller Reynolds number *Re* for road vehicles.

The camber of a wing and thus its downforce can be increased by a multi-part design, Fig. 5.33. A flap is added to one wing (split wing). By adjusting the flap the camber can be varied. However, this is done before the race because aerodynamic elements are generally not allowed to move during the race.

End Plates
. The laterally attached vertical surfaces help to channel the airflow over the wing, Fig. 5.34. They prevent air from flowing laterally over the wingtips, which otherwise inevitably happens due to the pressure difference between the upper and lower surfaces. End plates thus increase the effective wing width. On the front wings they even improve the flow at the rear of the vehicle by helping to direct the flow around the front wheels towards the undertray and on to the diffuser.

In the case of the front wing of single-seaters, it must also be taken into account that the front wheels change their position in relation to the end plates during steering. If the wing width is limited by the regulations to the inside of the tyres, a curved end plate extending or

Fig. 5.35 Curved end panel of a front wing (top view). The end plate is shaped in such a way in the top view that the air flow flows in the desired direction to the rear of the vehicle even when the front wheel is turned in

pointing backwards between the inside of the tyres and the nose of the vehicle is more favourable, Fig. 5.35. If the regulations allow the full width of the vehicle to be used, end plates offer the possibility of guiding at least part of the airflow around the front tyres (in a similar way to rear wheels in Fig. 5.15). Particularly a part of the disturbing end vortices can thus be forced outwards and prevented from entering the undertray area.

Open front wheels generate increased drag (and lift) due to their rotation, which moves the tread against the airflow on the upper half. Part of the incoming flow follows the tread and is pushed away to both sides in the inlet area of the slat. The vortices damaging the downforce can be prevented by a deflector strip at the end surface of the front wing from negatively influencing the ground flow, Fig. 5.36.

The end plates act mainly where the greatest pressure differences exist, i.e. near the trailing edge of the wing and on the underside. Therefore, an efficient end plate will project further downwards than upwards (min. 3 t, t . . .airfoil thickness) and this projection will increase towards the rear.

Production

Wings are manufactured in a variety of ways. Even though they generate large downforce, the specific stress is relatively low (about 0.55 N/cm^2 [5]), so there are many lightweight design methods. One way is to form them from thin sheets (e.g. 0.5 mm aluminium) glued and riveted to frames. Another way uses the possibilities offered by fibre-reinforced plastics (GRP, CFRP). The most important thing is a smooth surface, especially on the first third of the wing that is exposed to the airflow. Openings, rivet heads, sheet metal joints, etc. cause the flow to detach and render the wing areas following in the direction of flow ineffective, cf. also Fig. 5.39.

A basic construction consists of transverse frames (2) connected to spars (1), Fig. 5.37. The frames provide the desired wing profile and the spars provide the required bending stiffness of the wing. The top layer (4) is placed over the formers and glued and/or riveted to them. If this is done with sheet metal and an end rabbet is needed, it should interfere with

Fig. 5.36 Air flow in lathe run front wheel. The air flow (light blue arrows) is deflected downwards by the front tyres (the right one in the picture) and squeezed out on both sides in the inlet area of the tyre contact patch. A deflector strip (black arrow) facing the road at the end plates prevents the squeezing vortices from penetrating into the undertray area

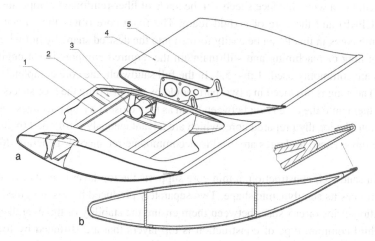

Fig. 5.37 Structure of a wing. The wing is attached to the front of the vehicle on both sides of the nose. (**a**) Axonometric view (partially sectioned), (**b**) Cross-section through the wing. 1 Spar (*cross beam*), 2 Rib, 3 Mounting bracket, 4 cover layer, 5 Adjustment plate

the flow on the upper surface as shown, i.e. it should act like a Gurney strip. The spigot (3) for attaching the wing is located at the center of pressure of the airfoil. A tube is inserted through the spigot, which makes the connection to the vehicle and transmits the wing forces. To change the angle of attack, the wing can be rotated around the tube axis. The

Table 5.4 Overview of sandwich construction methods, according to [16]

Construction	Sketch	Stiffness	Weight	Layup effort	Joining effort
Full Sandwich		++	+	++	++
Skin sandwich		+	++	+	0
Profile reinforcement		+	+	0	+

Legend: ▓ Core, ——Top layer (e.g. CFRP), ++ very good, + good, 0 satisfactory

wing is fixed with a lug (5) which has a curved slotted hole at the end or individual holes for the screw connection.

The frames are made of wood or shell-formed aluminium sheet with a thickness of 0.7–0.9 mm. The spars can be bent from sheet steel 0.9–1.2 mm thick. The top layer is formed from sheets 0.7–1.2 mm thick. If the outer skin is wound onto the wing substructure in one piece, the result is a surface without joints in the direction of flow. If a rebate is required, it should be at the trailing edge of the wing and at the top, Fig. 5.37.

Another manufacturing method is sandwich construction. Here, two tensile stiff cover layers are kept at a spatial distance by a shear stiff core, so that a high section modulus is maintained. For a wing, the face sheets can be made of fiber-reinforced composite material (GFRP, CFRP) and the core of a rigid foam. The foam core offers the advantage over honeycomb cores in that it can be easily formed into the desired shape by hot wire cutting, thermoforming or machining and still maintain the required compressive strength. Three concepts are commonly used, Table 5.4. In the full sandwich, the core completely fills the interior. The wing is produced in a two-part mould by first inserting the face sheets and then pressing them onto the core lying between them as they close. Lay-up packages can also be inserted into the locally prepared core, which are advantageously positioned by the core at the same time. These packages are used to accommodate retainers or generally to transfer loads.

In skin sandwich construction, a thin core supports the top layer over its entire surface and determines its aerodynamic shape. Two separately produced halves are glued together and additional ribs or crossbars between them ensure the stability of the overall shape.

The third common type of construction is top layers that are stiffened by foam-filled profiles (stringers).

A particularly pressure-resistant and dimensionally stable core is the PMI[7] rigid foam Rohacell. Achievable tolerances of less than 0.2 mm, creep resistance at a process temperature of 180 °C and possible consolidation in an autoclave at 10 bar [16] make Rohacell an ideal core material for wings.

---

[7] Polymethacrylimide.

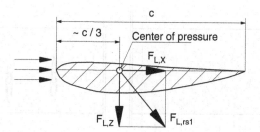

Fig. 5.38 Aerodynamic forces on the wing. All forces can be combined acting in the center of pressure. The center of pressure is located approximately in the first third of the wing chord. $F_{L,Z}$ downforce, $F_{L,X}$ air resistance, $F_{L,rsl}$ resulting total force

Rigid foam cores offer further advantages in driving operation. In the event of an impact load (stone chips, hail, vehicle parts, pylons, ...), the foam core elastically supports the cover layers and, in the event of their failure, it also absorbs plastic deformation [17]. In case of minor damage, there is a possibility of repair. The cover layers and the foam core are removed locally. After gluing in a suitable piece of foam, individual cover layers can be glued to the existing structure and rebuilt [17].

Fastening
Wings must be attached to the vehicle. The ideal would be a direct connection to the wheel carrier, so that the downforce only acts on the wheels and not also on the sprung mass of the vehicle. However, this mounting is forbidden in all regulations because this represents a moving downforce aid. Firstly, it is helpful to consider the aerodynamic forces acting on the wing which the wing attachment must absorb, Fig. 5.38. In addition, however, it is important not to forget the forces acting on the wing when the vehicle has a spin at high speeds.

The force quantities of the wing are related to its base area. The air resistance of the wing is calculated with a known drag coefficient as follows:

$$F_{L,X} = c_W \cdot \frac{\rho_L v_L^2}{2} \cdot s \cdot c \tag{5.12}$$

$F_{L,X}$	Air resistance, N
c_W	Resistance coefficient, -
ρ_L	Air density, kg/m^3
v_L	Air velocity, m/s
s	Span, m
c	Chord length, m

All aerodynamic forces can be thought to act at the *centre of pressure* without any moment acting on the wing. The attachment of the wing should therefore act in its vicinity.

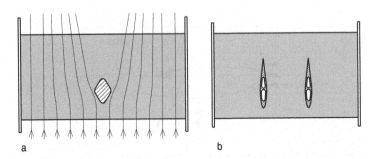

Fig. 5.39 Flow influence by elements for wing attachment. (**a**) flow separation at a column of a rear wing, (**b**) flow-favourable cross-section of two supports of a front wing

The connection to the vehicle should itself be as streamlined as possible and influence the effect of the wing as little as possible. Basically, there are two possibilities. The attachment is done with a pillar in the middle of the wing or externally by means of the two end plates.

Figure 5.39a shows schematically the effect of a disturbance of the flow by a mounting column. The flow detaches at the column and a much wider area is formed behind the column in which the wing surface is not covered by the flow, i.e. it is not effective. Similarly, openings in the wing surface also allow the flow to detach. Figure 5.39b shows a flow-favourable design of two supporting elements of a front wing. To reduce weight, the two columns are hollow.

The angle of attack should be adjustable for vehicle tuning. The pivot point of rear wings is preferably at the trailing edge of the wing. This way, it can never exceed the maximum height allowed by the regulations when being adjusted.

In addition to the aerodynamic forces, the attachment must of course also be able to withstand the inertial forces caused by the wing mass. As a rule of thumb, the wing and its attachment are sufficiently dimensioned if the vehicle can be pushed by hand at the outer wing edge.

Front Wing

Fig. 5.40. The front wing can be regarded as the most important single aerodynamic element of a racing vehicle. On the one hand it generates about a quarter to a third of the downforce, on the other hand its downstream flow influences the aerodynamic behaviour of all subsequent parts, i.e. the complete vehicle. For this reason, a compromise is made in its design and attention is not paid exclusively to its function as a downforce element.

At the end of the long straight at Estoril, for example, the downforce of a Formula 1 front wing is 5500 N [6].

The ground effect on front wings (see also Fig. 5.69) increases the pitch sensitivity of a vehicle.

Fig. 5.40 Multi-part front wing with end surfaces on a Formula 1 (Mercedes AMG Petronas F1 W05 Hybrid). The wing not only generates downforce, but also channels the airflow into the area between and around the front wheels. Among other things, it thereby acts as a vortex generator

Rear Wing

Since the diffuser has been greatly shortened by some regulations (usually it may only start from the leading edge of the rear wheels), the rear wing produces about 30% of the downforce. For a Formula 1 rear wing, this is still about 9.8 kN of downforce on the long straight at Estoril [6]. Figures 5.41, 5.42, and 5.43 show examples of designed wings.

With multi-part rear wings, the lowest wing (*beam wing*) is of particular importance. It carries the entire wing package and not only generates downforce – as do the upper wing elements – but above all enhances the effect of the diffuser by increasing the negative pressure in the diffuser outlet area. This results in increased airflow through the entire diffuser and thus amplification of the desired downforce.

Gurney Strip or Gurney Flap, AE: Wicker

The Gurney strip, or "Gurney" for short, is a narrow strip fitted to the trailing edge of a wing at right angles to its upper surface (i.e. at right angles to the direction of flow), Fig. 5.44. The height at right angles to the direction of flow is a maximum of 10–15 mm. Common designs are found in the range 3–10 mm. It increases the downforce of the wing, but also increases its drag. Since it is easy to install or remove, it is an important part of fine tuning.

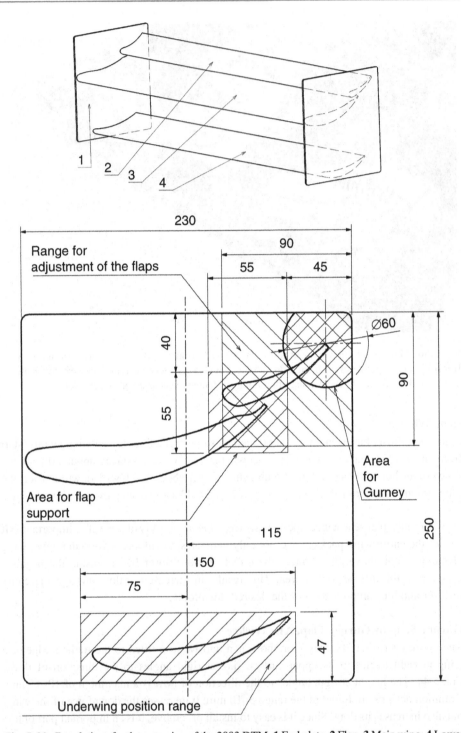

Fig. 5.41 Regulations for the rear wing of the 2003 DTM. **1** End plate, **2** Flap, **3** Main wing, **4** Lower wing. The dimensions of the end plates and the installation spaces of individual wing elements are prescribed

Fig. 5.42 Multi-part rear wing of a Formula 1 car with flat end plates (Ferrari). The wing is attached with the lowest element to the rear impact element, which in turn is attached to the end of the gearbox housing. The angle of attack of the wings and flaps is adjustable via several holes in the end plates

Fig. 5.43 Multi-part rear wing on a production sports car (Osella PA 20 S). The entire rear wing is cantilevered over the lower wing element with two plates at the gearbox end. In this way, the downforce acts on the rear axle via a lever. The two plates also provide the towing lugs

Figure 5.45 shows the measured influence of a Gurney strip compared to the original wing profile.

As expected, gurneys not only increase downforce, they also increase drag at the same time. An interesting design variant to reduce the required compromise can be seen in Fig. 5.46.

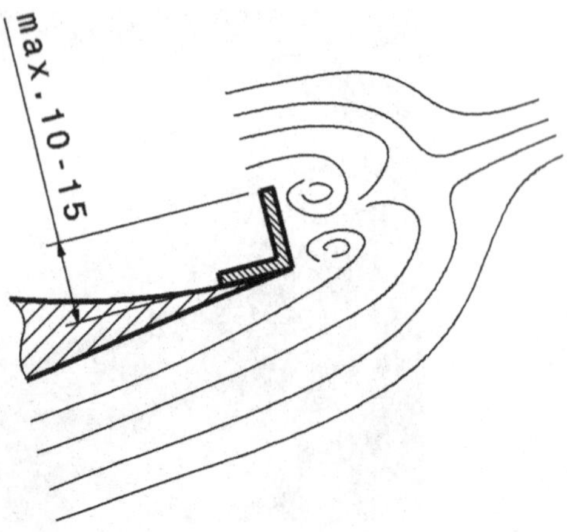

Fig. 5.44 Principle of the
Gurney bar, according to
[14]. The flow is deflected
upwards at the trailing edge.
Two counter-rotating vortices
form behind the lip. This gives
the flow an additional vertical
component. This deflects the
flow upwards and increases the
downforce

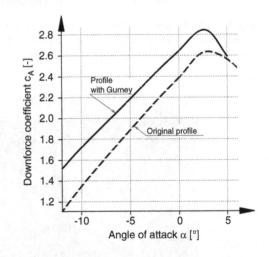

Fig. 5.45 Influence of a Gurney
on the downforce, according to
[9] Compared to the original
airfoil without Gurney, the
Gurney flap clearly raises the
downforce of this wing

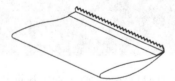

Fig. 5.46 Serrated Gurney strip. This design represents a compromise between drag and downforce.
The spikes act like a high, continuous bar in downforce, but generate only half the drag of a high,
continuous design

Despite the increase in drag of the wing, which is accompanied by a deterioration in the
efficiency of this part, Gurneys are attractive for the tuning of the overall vehicle. Especially if certain wing profiles are prescribed by the regulations or the installation space for
wings is restricted, Gurneys can show a positive influence, Fig. 5.47.

Fig. 5.47 Gurney on a front wing (Ferrari Formula 1). The picture shows the left part of the front wing in front of the front wheel. The Gurney strip is screwed onto the outer part of the offset shaped front wing

Fig. 5.48 Influence of the nose shape on drag and lift at the front axle, according to [9] (**a**) high nose, (**b**) middle position of the nose tip, **c** lowered nose with wedge-shaped nose

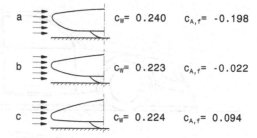

Nose

That's the whole front end of the car tapering forward. Raised noses were a consistent development on Formula One cars in the late 1990s because they help keep the front wing flowing around in relative isolation, effectively directing airflow after the wing to the undertray and diffuser at the rear.

As always, the effect of the nose does not depend on it alone, but on the interaction with the other parts. In general, however, the relative position of the nose to the road and to the rest of the car body is important, Fig. 5.48. A high nose (a) actually generates lift, due to the resulting high stagnation point. A tapered nose (b) lowers drag. If the tip of the nose is lowered further (c), the drag increases and even downforce at the front axle results.

Tail End

Similar to the nose, the rear also influences the aerodynamic behaviour, but in an even stronger form, Fig. 5.49. Long, slender rear shapes (teardrop shapes) are ideal for low aerodynamic drag, but in practice these are limited by the regulations (rear impact, rear overhang) and other criteria (stability, crosswind sensitivity).

Front Spoiler

A spoiler continues the surface of a vehicle without any interruption between the body and this device. If the surface between the bodywork and the spoiler is interrupted, one can possibly already speak of a wing [14].

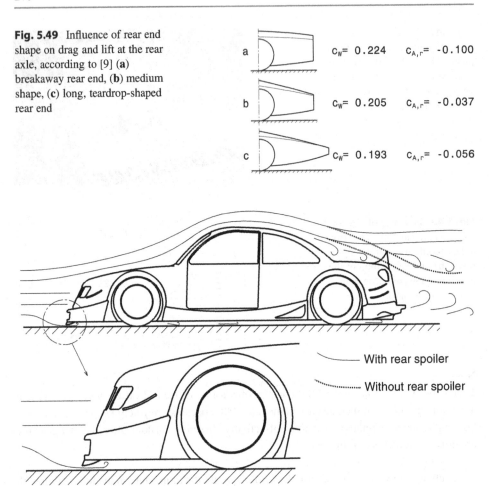

Fig. 5.49 Influence of rear end shape on drag and lift at the rear axle, according to [9] (**a**) breakaway rear end, (**b**) medium shape, (**c**) long, teardrop-shaped rear end

a $c_W = 0.224$ $c_{A,r} = -0.100$

b $c_W = 0.205$ $c_{A,r} = -0.037$

c $c_W = 0.193$ $c_{A,r} = -0.056$

—————— With rear spoiler

················· Without rear spoiler

Fig. 5.50 Airdam front spoiler and rear spoiler, after [14]. The front spoiler reduces the air mass flowing under the car. This reduces the air resistance with fissured underbodies (despite the increase of the frontal area) and the pressure under the car decreases. The rear spoiler slows down the flow and the pressure increases

Front spoilers (skirts) reduce the underflow (airdam spoiler) of the car floor and thus provide a pressure difference between the car top and bottom, which increases the downforce, Fig. 5.50.

Rear Spoiler

The rear spoiler disturbs (= spoils – name!) the flow around the rear of the car and allows the flow to separate earlier. It thus reduces the lift at the rear axle, Fig. 5.50. The spoiler must be fitted in the laminar area of the flow, otherwise it has no effect.

Fig. 5.51 Dive plate on the nose of a touring car

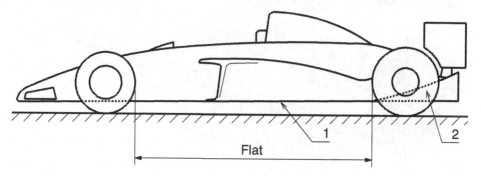

Fig. 5.52 Schematic of a vehicle with flat undertray. **1** flat section of the undertray, **2** diffuser area

A rear spoiler also increases the effect of a rear wing. If a rear spoiler is mounted in front of a wing, the downforce and thus also the drag of this wing increases.

Dive Plate

Such adjustable plates (in the eponymous manner of a depth rudder on submarines) on the bow of the car, when set at an angle, produce a certain amount of downforce which can be used for fine adjustment, Fig. 5.51.

Undertray

The removable undertray makes an extremely important contribution to downforce, especially as it allows this to be achieved with relatively low losses. In many regulations, a continuous undertray is now prescribed due to high-speed accidents, which may only deviate from it in certain areas – usually defined in relation to the wheels, Fig. 5.52. For downforce, the area near the rear wheels thus becomes interesting, starting from where the floor may be bent. In the case of sports prototypes and touring cars, the nose of the vehicle also moves to the centre of the concept considerations, because there are also possibilities for a diffuser.

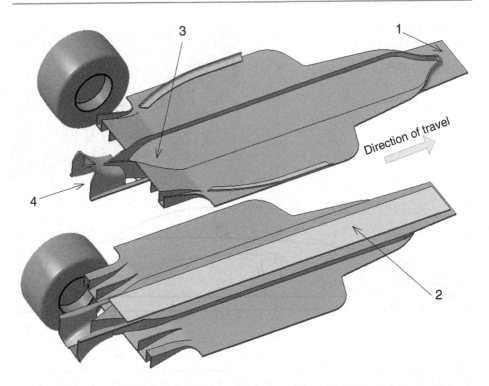

Fig. 5.53 Undertray on a formula car (top view and mirror image below). The floor is stepped due to regulations and runs flat up to the rear wheels. The left rear wheel is also shown for orientation. The lowest area in the middle of the floor forms a splitter (1) at the front and contains the floor board (2), which must not exceed a certain abrasion value after the race. In the area of the gearbox (3) the bottom runs together like a ship's hull. Subsequently, a separate diffuser (4) is attached

In the case of Formula 1 cars, the undertray must also be stepped by an additional 50 mm, i.e. it must not represent a continuous flat surface even in the basically flat area between the wheels, Fig. 5.53.

A diffuser in the rear of the vehicle and an undertray that rises slightly towards the rear accelerate the air flow in the front and in the floor area and, if designed correctly, provide downforce and (!) lower air resistance.

Diffuser

In a diffuser, in contrast to a nozzle, the flow cross-section increases over the length. As a result, the flow velocity decreases when flowing through the diffuser and, due to the conservation of energy, the static pressure increases.

In vehicles, diffusers are used at air intakes and in the undertray area. In the undertray area, the effective surfaces of the diffuser are formed by the vehicle and the road surface, Fig. 5.54. Such a diffuser affects the pressure distribution under the entire (!) car floor. The

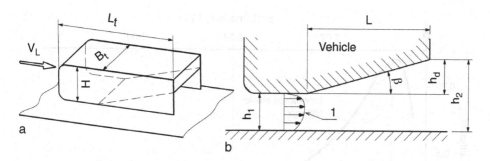

Fig. 5.54 Principle of a rear diffuser. (**a**) oblique section simple vehicle model, v_L air velocity, L_t, W_t, H vehicle dimensions. (**b**) section with dimensions, 1 asymmetric airfoil, h_d diffuser height

static pressure in the area of the floor decreases and thus downforce is created thanks to the floor area and the pressure difference to the top of the car.

An undertray diffuser is asymmetrical. The widening part of the channel is formed by the vehicle floor and the counter surface represents the road surface, which additionally moves relative to the vehicle. A ratio that is decisive for the design is the area ratio k_A. It determines the velocity ratio and thus the pressure ratio between the inlet and outlet cross-sections of a diffuser. The following applies to a diffuser as shown in Fig. 5.54 with parallel side walls [9]:

$$k_A = \frac{h_2}{h_1} = 1 + \frac{L}{h_1} \cdot \tan\beta$$

k_A Area ratio, -
h_1 Ground clearance, mm
h_2 Outlet height, mm
L Diffuser length, mm
β Diffuser angle, °

In order for this theoretical ratio to be achieved as well as possible by practically designed diffusers, the air flow must not detach from any boundary wall. This alone sets limits to the diffuser angle and the pressure increase cannot be increased at will for a given diffuser length.

An asymmetrical flow profile in the inlet caused by frictional influences also worsens the diffuser effect. Furthermore, a non-uniformity of the flow in the diffuser itself is increased.

In general, the following parameters should be kept in mind when designing diffusers [9]:

• the aspect ratio k_A. It determines the theoretical upper limit of the pressure recovery.
• the referred diffuser length L/h_1. It is a measure of the flow resistance.

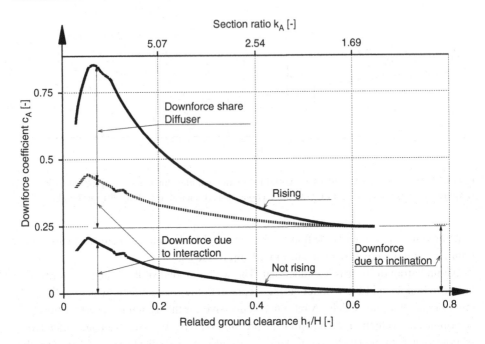

Fig. 5.55 Composition of the downforce by the undertray, after [9]. The measurements are based on a vehicle model according to Fig. 5.54a with the following numerical values. $L_r/H = 2.4$, $W_r/H = 1.29$, $W_r/h_1 \approx 20$

- the blockage of the inlet flow. It is partly responsible for a non-uniform flow profile.

Investigations on a simple vehicle model (Fig. 5.54a) break down the complex downforce generation in the undertray area. Even though the model is very simple, the results can at least be applied to all vehicles with a smooth undertray. For single-seaters, the flow conditions are somewhat different and therefore the following statements only apply to a limited extent to this vehicle category.

Two designs of the undertray were compared during the investigation. One vehicle has the undertray raised at the rear and the other has no slope to the floor, Fig. 5.55.

It can be seen that the undertray without a rise in the rear does not generate downforce at high ground clearances ($h_1/H > 0.7$), but the raised version does. If the distance to the ground is reduced, the interaction with the road surface leads to the generation of downforce for both variants. If the distance to the road is too small, the downforce decreases again due to toughness effects. If you now subtract the share of downforce due to the interaction from the downforce curve with raised undertray, the share that is due to the diffuser alone remains.

So downforce by undertray is composed of three independent effects:

1. Downforce due to interaction with the road surface with low ground clearances
2. Downforce due to the lifting of the rear undertray
3. Downforce through the diffuser.

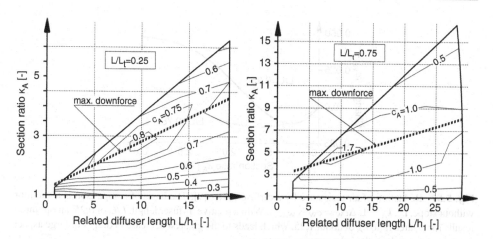

Fig. 5.56 Total coefficient of drag c_A for two relative diffuser lengths, after [9]. These measurement results refer to the vehicle model from Fig. 5.54a. Next to the lines of equal downforce, the curve of maximum downforce values is plotted. It is a straight line

Based on the findings of the mentioned investigation, a first dimensioning of the diffuser can be done by means of Fig. 5.56. The diagrams show for two relative diffuser lengths L/L_t (= diffuser length in relation to the vehicle length) the course of the downforce coefficient, which arises in sum from the three effects. $L/L_t = 0.25$ corresponds approximately to the ratio, which is given by most regulations, namely that the diffuser may begin only in the area of the rear wheels. For comparison, the map of an extremely long diffuser with $L/L_t = 0.75$ is compared.

Given ground clearance h_1, which will generally be the case, the diffuser length L follows from the diagram by the desired downforce coefficient c_A on the straight line of maximum downforce and by feasible aspect ratios k_A. The diffuser length may also be dictated by geometric and regulatory constraints, giving the ratio L/h_1. The design possibilities result in the case from the diagram by the aspect ratio, from which the exit height h_2 follows.

Because the constant downforce curve in the diagrams represents the contour lines of a conical hill, there are several possible diffuser designs for a given downforce. When choosing the most suitable diffuser, it is helpful to consider not only geometric considerations but also the fact that a raised undertray also influences drag. Figure 5.57 illustrates this again for two relative diffuser lengths.

Compared to a vehicle with a flat undertray without a rise, the drag decreases when the undertray is raised at the rear, i.e. the exit height h_2 is increased. With a further increase of the aspect ratio k_A the drag decrease reaches an extreme and from there on the effect decreases until it finally leads to an increase in drag.

As an example of the dimensions of a diffuser in the rear section, Fig. 5.58 shows the undertray for two-seater racing sports cars as prescribed by the FIA.

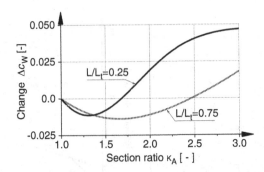

Fig. 5.57 Reduction of air resistance by a diffuser, after [9]. The basis of comparison is a subfloor without slope, i.e. $k_A = 1$. $\Delta c_W = c_W - c_{W,base}$. With a relative diffuser length L/L_t of 0.25, an optimum results for an area ratio k_A of about 1.25, which leads to the greatest decrease in drag. For large aspect ratios (approx. >1.6), the air resistance increases

In past racing seasons, when the diffuser was allowed to start well ahead of the rear axle in Formula 1, it was responsible for up to 70% of the total downforce. Today, this share is about "only" 40%. Figure 5.59 shows in the view from the rear the extension of the undertray and thus the diffuser of a Formula 1 car. In the area of the rise of the floor there are vertical guiding elements which guide the air flow and reduce the disturbing effect of the edge vortices coming from the rear wheels.

Diffusors do not necessarily have to be used only at the rear of a vehicle. If the regulations permit, raised underbodies can also be used in the nose area to generate downforce. The exhaust air from the diffuser area is used for flowing through heat exchangers or it leaves the vehicle upwards or to the side in areas of displaced external flow, Fig. 5.60.

Figure 5.61 shows schematically the air flow on the underside of a sports prototype. The lateral inflow of air into the diffuser (2) in front of the rear axle is striking. The air flowing over the sharp edges of the tunnel forms vortices which support the flow on the upper side of the diffuser. These two vortices (4) are also present after the air leaves the diffuser. The suction effect of the diffuser also pulls air sideways into the tunnel area behind the rear axle. This flow forms a smaller vortex that increases both downforce and drag [19].

Winglets

These are small additional wings that can be fitted wherever the regulations have left a gap. They are usually mounted high on the side panels and immediately in front of the rear wheels and are intended to improve downforce, Fig. 5.62.

Downforce-generating elements work as desired if the flow is in the intended manner. Deviations occur, for example, when driving in the slipstream of a vehicle in front or in the event of a strong crosswind. The most dangerous situation is when the vehicle changes its

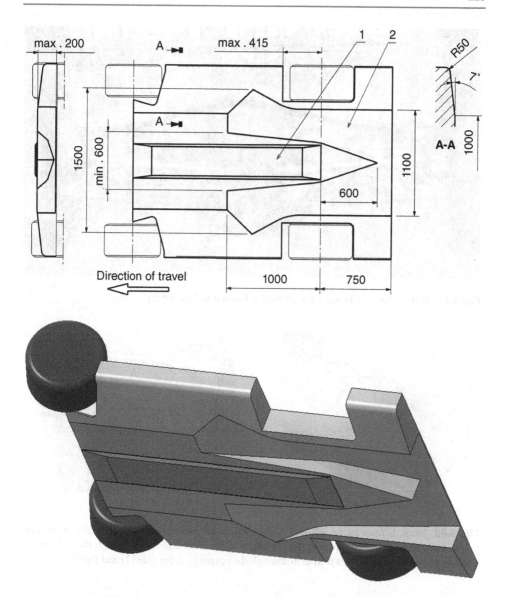

Fig. 5.58 Undertray according to FIA [18]. **1** skid block, **2** rear diffuser. The diffuser may be 1750 mm long and max. 200 high. The rest of the subfloor is flat, except for the skid plate and a 7° bevel on the edge

intended position in relation to the road. At high speeds, a loss of pressure in a rear tire may be enough to cause the rear undertray to sag and the resulting inflow of air to literally lift the vehicle. Similarly dangerous situations can occur when the car is spinning at high speed: the vehicle is then hit from behind or from the side. There are devices that reduce the lift of

Fig. 5.59 Diffuser on a Formula 1 car (Benetton Renault B 195, 1995)

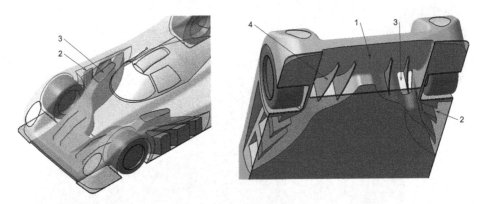

Fig. 5.60 Front diffuser (*nose diffuser*). Left: View from above with air duct. Right: View from below, base plate shown transparent. The air enters at the front and flows through the diffuser (1). Baffles (4) direct the airflow, which exits through the openings at the side (2) and top (3)

a vehicle in such cases. In the North American Winston Cup, this resulted in such strong lift that the vehicles took off. Safety flaps and strips now prevent this phenomenon, for more details see Sect. 3.3 *Protective Devices*.

Drag Reduction System

(DRS). In order to make the races more interesting again by overtaking manoeuvres, in 2011 Formula 1 deviated from the principle that no downforce-generating element may be adjustable while driving. In certain situations, the driver is now allowed to reduce the drag

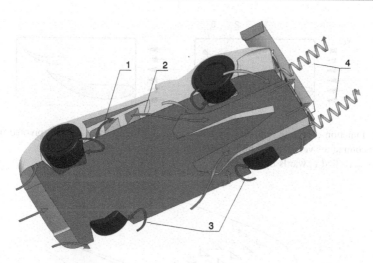

Fig. 5.61 Airflow on the underside of a sports prototype, schematic, after [19]. 1 Cooling air outlet, 2 Lateral diffuser inlet, 3 Flow separation, 4 Diffuser edge vortex

Fig. 5.62 Winglets on a Formula 1 car (Toyota). At the air intake of the engine behind the cockpit there are additional wings which generate lift. The central T-shaped element is a housing prescribed by the FIA in which on-board cameras can be housed. It is deliberately designed in such a way that it does not bring any aerodynamic advantages

of the rear wing on a certain straight line so that he can attack the car in front. For this purpose, the rear wing has a pivoting flap, Fig. 5.63, which can be pivoted by an actuator – controlled from the cockpit. The axis of rotation is located near the trailing edge of the flap so that in the event of a failure of the adjustment actuator, the flap is forced into the downforce position by the incoming air. The reduced drag allows a vehicle to build up a speed advantage of about 4–5 km/h on a typical straight line.

One disadvantage of DRS is that only the rear wing changes its downforce behavior. The aerodynamic balance of the car is shifted towards the front axle when DRS is activated because the front wing generates downforce undisturbed. This altered handling behaviour

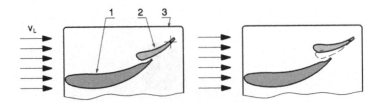

Fig. 5.63 Function of a drag reduction system on a rear wing. 1 main wing, 2 pivoted flap, 3 end plate, v_L Incoming air velocity. (**a**) high downforce position, (**b**) low drag position
The flap is swivelled upwards by 4 to 10°

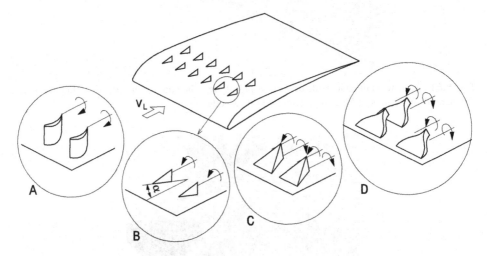

Fig. 5.64 Vortex generator on a wing, after [19]

poses a challenge for the driver when braking and especially entering corners. Some Formula 1 teams have therefore developed systems that use ducts to direct air from the rear to the front wing when DRS is activated and also reduce its downforce. This maintains the balance and makes the car easier for the driver to control.

Vortex Generator
These are small devices that generate vortices. If such are attached to the wing surface in the area where the flow could detach, the flow energy of the boundary layer is increased with the help of the air flow and detachment takes place only later. Therefore, vortex generators are designed only slightly higher than the boundary layer thickness. The use of vortex generators increases the downforce of the wing and reduces the drag at high downforce values. Vortex generators have the shape of small wings or more complex shapes. Some possible designs are summarized in Fig. 5.64. Another simple option is to roughen the surface in the desired area.

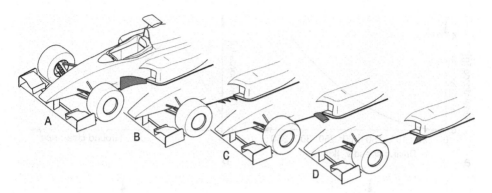

Fig. 5.65 Vortex generator for the extension of vortices under the vehicle, after [19]. **A** lateral deflector, **B** row of delta wings, **C** mini wing, **D** delta wing

Figure 5.65 shows other known vortex generators on formula cars, which help to extend the vortex distance under the car even with low ground clearance. Variant A is a deflector (barge board) which generates vortices at the lower end and thus contributes to the total downforce. Variation B, sawtooth-shaped flat wings, represents one of the first Indy Car applications. Variation C uses the end vortex of a mini-wing to affect the flow on the underside of the car. Variation D creates an edge vortex on the outside of the side box.

Summary
Downforce is a multi-layered phenomenon that has become a decisive competitive advantage, especially in high-performance vehicles. However, downforce is dependent on many influencing variables and this in a non-linear manner, which makes it difficult for the driver to control in several ways. Figure 5.66 provides an overview of the basic course of the downforce caused by the most important influencing variables.

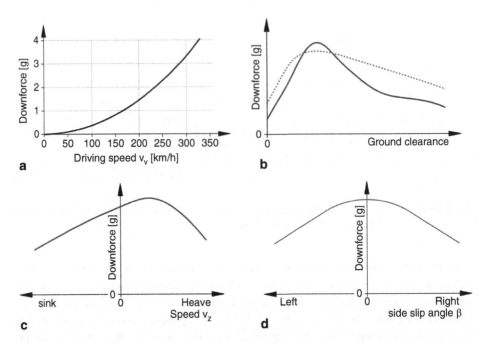

Fig. 5.66 Basic course of the downforce as a function of important vehicle sizes. The downforce is represented as a force related to the vehicle mass, i.e. an acceleration. (**a**) Driving speed: The downforce increases with the square of the driving speed because it is caused by dynamic air forces. (**b**) Ground clearance: The downforce reaches a maximum at a certain static ground clearance (depending on the undertray design and angle of attack). If the ground clearance is too low or too high, the effect decreases strongly. (**c**) Vertical speed: A change in ground clearance increases the negative pressure when the ground is raised, or compresses the air under the car when it is lowered. (**d**) Sideslip angle: If the car shows a sideslip angle, the air flows towards it at a corresponding angle and the downforce-generating elements only act through an angular portion of the flow

5.4   Wing Calculation

In the following, a selection of a wing configuration, i.e. front and rear wing, will be made [14]. From wind tunnel studies comes the knowledge that in single seaters the front wing hardly raises the drag of the vehicle. Thus, a simple design option arises because only the rear wing lifts the power requirement and only this is included in the considerations. The top speed of the vehicle without wing depends primarily only on the air resistance, which must be overcome by the engine:

Table 5.5 Efficiency of the drive train [4]

Vehicle type	η_{Drive}
Single-seater with rear engine and cold or narrow tyres, e.g. Formula Ford, hill climb racer	0.91
Single-seater for circuit racing with warm and wide tyres, e.g. Formula 1, Formula 3000	0.875
Sedan for racing, sports car with engine above the drive axle, e.g. Le Mans car, Imp, Mini	0.85
Racing cars with front engine and rear-wheel drive, e.g. Clubmans	0.82

$$v_{\max} = \sqrt[3]{\frac{P_{M,\max} \cdot \eta_{drive}}{c_W \frac{\rho_L}{2} A_V}} \tag{5.13}$$

$v_{\max}$ Theoretical maximum speed without wings, m/s
$P_{M,\max}$ Max. motor power, W
η_{Drive} Efficiency of the drive train, -. See Table 5.5
c_W Drag coefficient of the vehicle, -
ρ_L Air density, kg/m³. $\rho_{L,\,medium} \approx 1.22$ kg/m³
A_V Cross sectional area (frontal area) of the car, m²

The efficiency of the drive train depends not only on the detailed design of its components but also on the basic arrangement.[8] Table 5.5 can be used for an initial consideration of the choice of wings.

If a rear wing is now attached to the car, its additional drag causes the achievable top speed to drop while engine power remains the same. So you have to decide how much of the originally possible top speed you can or want to give up in favour of the wing (and thus for downforce). The required power for the selected top speed with wing follows to:

$$P_{W,requ} = c_W \frac{\rho_L}{2} v_{with-wing}^3 \cdot A_V \tag{5.14}$$

$P_{W,requ}$ required power for reduced maximum speed with wing, W
$v_{with-wing}$ Maximum speed with wing, m/s. $v_{with-wing} < v_{\max}$

Here, the power "absorbed" by the rear wing:

$$\Delta P = P_{M,\max} \cdot \eta_{Drive} - P_{W,requ} \tag{5.15}$$

ΔP Share of power caused by the rear wing (primarily), W

With the main wing dimensions, the upper limit for its drag coefficient is thus fixed:

---

[8] See Racing Car Technology Manual Vol. 3 *Powertrain*, Sect. 5.1 *Power Transmission*.

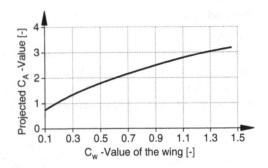

Fig. 5.67 Example of the relationship between lift and drag of a wing, after [14]. The diagram provides an orientation about the relationship between the two coefficients

$$c_{W,wing,r,\max} = \frac{\Delta P}{\frac{\rho_L}{2} v_{with-wing}^3 \cdot A_{wing,r}} \tag{5.16}$$

$cW_{,wing,r,max}$ Maximum permissible c_W value of the rear wing, -
$A_{wing,r}$ Wing area of rear wing, m^2. $A_{wing,r} = s_r \cdot c_r$
s_r Span of the rear wing, m
c_r Chord length of the rear wing, m

The downforce coefficient is related to the c_W value via the wing profile:

$$c_{A,r} = f(c_{W,r})$$

$c_{A,r}$ Downforce coefficient for rear wing, -
$c_{W,r}$ Drag coefficient for rear wing, -

For rough orientation, Fig. 5.67 shows this relationship between the aerodynamically decisive coefficients graphically.

This $c_{A,r}$ value is used to select a wing from wing catalogues. First the basic configuration (one, two or more parts) is determined and then a specific airfoil with the required angle of attack is selected.

$$F_{L,Z,r} = c_{A,r} \cdot \frac{\rho_L}{2} \cdot v_V^2 \cdot A_{wing,r} \tag{5.17}$$

$F_{L,Z,r}$ Rear wing downforce, N
v_V $= v_{with-wing}$, m/s

The basic handling of the car should remain the same even with wings, so the same pitch characteristics are aimed for. This in turn means equal spring travel changes front and rear so that the pitch angle of the car remains unchanged. So the conditions for the suspension travel are:

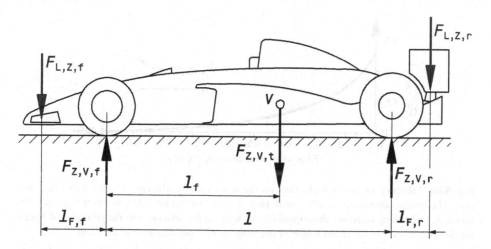

Fig. 5.68 Forces and dimensions on the vehicle. $F_{L,Z,f}$, $F_{L,Z,r}$, $F_{Z,V,f}$, $F_{Z,V,r}$ Forces, see text, l Wheelbase, $l_{F,f}$, l_f, $l_{F,r}$ Dimensions

$$\Delta s_f = \Delta s_r \rightarrow \frac{F_{W,Z,f}}{c_f} = \frac{F_{W,Z,r}}{c_r} \rightarrow \frac{F_{W,Z,f}}{F_{W,Z,r}} = \frac{c_f}{c_r} = const = \Phi_{Sp} \qquad (5.18)$$

with: $\quad \Delta s_f, \Delta s_r$

Suspension travel changes at the front or rear, m

$F_{W,Z,f}, F_{W,Z,r}$ Front and rear wheel contact forces, N

c_f, c_r Wheel-related spring rates of the front or rear suspension, N/m

Φ_{Sp} Ratio of wheel related spring rates front/rear, -

From the equilibrium of forces and moments on the complete vehicle (Fig. 5.68), a correlation between the downforce forces of the two wings at the front and rear is obtained. It can be seen that a rear wing which does not act directly on the rear axle but is offset to the rear generates a greater downforce on the rear axle. However, this is at the expense of the front axle load, which becomes smaller as a result. The front wing thus becomes a necessary balancing element in such a configuration, so that the front axle has sufficient axle load even at high speeds.

$$F_{L,Z,f} = \frac{F_{L,Z,r}(B \cdot D - 1) + F_{Z,V,t}(B \cdot l_f - 1)}{1 + B \cdot l_{F,f}} \qquad (5.19)$$

$F_{L,Z,f}$ Required downforce of the front wing to balance $F_{L,Z,r}$, N

B, D Auxiliary sizes:

$$B = \frac{\Phi_{Sp}+1}{l} \text{ and } D = l + l_{F,r}$$

$l, l_{F,f}, l_f, l_{F,r}$ Lengths, m. See Fig. 5.68

$F_{Z,V,t}$ Total weight of the vehicle, N

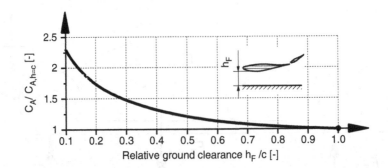

Fig. 5.69 Influence of the ground effect on the downforce coefficient of a two-part wing, after [14]. The relative downforce coefficient is plotted above the relative distance to the ground. The downforce of a wing increases disproportionately with smaller distances to the ground. The base is the distance $h = c$, i.e. the chord length of the wing. h_F Ground clearance of the wing

From the downforce of the front wing $F_{L,Z,f}$ determined with the equation above follows the downforce coefficient of this wing:

$$c_{A,f}^* = \frac{F_{Z,L,f}}{\frac{\rho_L}{2} v_{with-wing}^2 \cdot A_{wing,f}} \tag{5.20}$$

$c_{A,f}*$ Downforce coefficient of the front wing without ground influence, -
$A_{wing,f}$ Wing area of the front wing, m²
     $A_{wing,f} = s_f \cdot c_f$

Due to the ground effect, the actual downforce behaviour of the front wing may be significantly different depending on the distance to the ground. The distance of the wing to the ground must therefore be taken into account:

$$c_{A,f,\mathrm{requ}} = f(c_{A,f}^*, h_{F,f}) \tag{5.21}$$

$c_{A,f,requ}$ Actual required downforce coefficient of the front wing which, with ground influence, produces the desired downforce $c_{A,f}$, -
$h_{F,f}$ Ground clearance of the front wing, m

Example values for the effect of the ground effect with a front wing can be taken from Fig. 5.69. The values also show in passing why pitching of vehicles with front wings can become such a serious problem. At small ground distances the wing shown achieves more than twice the downforce of the value at distance $h = $ chord length and thus "supports" the pitching motion even more.

From the boundary conditions specified by the regulations and the installation space, a ground clearance h_F can now be selected for the front wing. This ground clearance in turn determines for a given wing the ratio $c_A/c_{A,h}=0$ with which the theoretical downforce coefficient is amplified. With this ratio the required downforce value for the wing selection follows to:

$$c_{A,f,requ} = \frac{c_{A,f}^*}{c_A/c_{A,h=0}}$$ (5.22)

$c_A/c_{A,h}=0$ Ratio of downforce change due to ground influence, -
           See, for example, Fig. 5.69

A suitable front wing with $c_{A,f,requ}$ can now be selected from a wing catalogue. The wing is mounted with the appropriate distance h_F to the roadway.

5.5 Air Deflectors

Baffles, Deflector Shields (Barge Boards)

They have a similar function to the end-plates on the wings. They first appeared in Formula 1 in 1994 after the "end-plates" had to be radically reduced in size according to the regulations. Barge boards can be horizontal (Fig. 5.71), but most often vertical. Their function is to influence and smooth the airflow behind the front wing before the air continues to flow towards the rear, Fig. 5.70. The outer surfaces of vertical baffle boards, push the wake of the front tyre deflected by the end vortex of the front wing towards the outside of the car. This prevents the wake from getting under the car and severely disrupting the diffuser effect. Just how important they are is demonstrated by the deflection elements on the 1997 Formula 1 Ferrari F310B with their strongly bent upper edge. They were an effective means of combating the strong understeer tendency that impaired handling in the first tests [6].

Fig. 5.70 Baffles on the front suspension of a Formula 1 car (Ferrari). You can see the right side of the vehicle between front wheel and radiator inlet

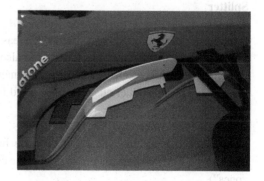

Fig. 5.71 Baffles on the nose of
a touring car (Abt-Audi, DTM).
The left front of the vehicle is
shown. The u-shaped element is
bolted to both sides of the nose.
It directs the air around the front
part of the wheelhouse. In
addition, the splitter is well to be
recognized on this photo

Fig. 5.72 Deflector in front of
the rear wheel of a Formula 1 car
(Renault R25). The deflectors
help to direct the air effectively
around the rear wheels

Scallops

These are deflectors or small curved mouldings in front of the rear wheels. They are fitted
where the "bottleneck effect" begins at the constriction of the bodywork, Fig. 5.72. This is
another means of smoothing the air vortices, i.e. influencing the flow around the rear tyre
and increasing the diffuser effect.

Splitter

A splitter splits the airflow (name from *split* = *to* divide, separate) and thus helps to
increase the desired effect of subsequent areas or to reduce disturbing effects, see also
Fig. 5.78b. A splitter at the vehicle nose can also increase the downforce at the front of the
vehicle because it is located in the zone of dynamic pressure. The pressure difference with
the area below the splitter creates a downward force, Figs. 5.73, 5.74, and 5.75.

Side Panels or *Sidepods*

Most of them have fluid lines, but they are by no means merely decorative means to
disguise the water coolers; they also protect the car sides as deformable elements ("crumple
zones").

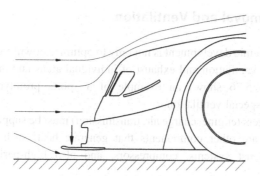

Fig. 5.73 Effect of a splitter in the nose area. The splitter is attached to the undertray in the dynamic pressure area of the bow. The pressure difference between above and below the splitter results in a downward force. The size of this force can be influenced within certain limits by the length of the splitter. The length is limited, among other things, by the ground clearance required for pitching movements of the carriage

Fig. 5.74 Splitter on a Formula 1 car (BMW Williams). The splitter is located under the raised nose of the vehicle and splits the airflow between the undertray and the inlet to the heat exchangers. It also uses the dynamic pressure in this area to generate downforce. The edge at the front is rounded in a teardrop shape

Fig. 5.75 Front spoiler with splitter on a Le Mans prototype car (Bentley EXP Speed 8)

5.6   Heat Removal and Ventilation

The task of aerodynamic development is not only to optimize downforce and air resistance, but also to provide ventilation and exhaust of individual areas and thus for targeted heat dissipation. Figure 5.76 shows an example of a sports prototype, showing which assemblies require special ventilation.

All heat exchangers (engine, charge air, transmission) must be supplied with cooling air. In addition, there are other components that generate heat, such as brakes, exhaust turbochargers, exhaust systems, compressors, and also high-performance electronic components.

Generally, it is attempted to spatially combine assemblies with a similar maximum permissible temperature. Then an air flow can be used "several times". If the temperature of an exhaust air stream of the preceding subsystem is below the usable temperature of the following subsystem, the air stream can be forwarded directly. Such a useful temperature gradient results, for example, in the arrangement of charge air cooler after the air inlet followed by heat exchanger for transmission oil or exhaust air via duct for brake cooling.

Usually, stowage areas are preferred for inlet openings because this allows the largest pressure drops. Typical stowage areas are found at the front of the vehicle, in flared side boxes and in attached chimneys. The required outlets after the flow through heat exchangers, etc. are located on the upper side or flanks of the bodywork or on the undertray

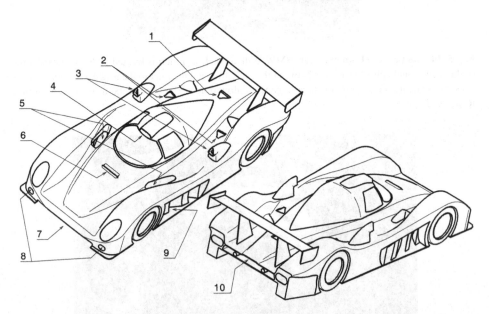

Fig. 5.76 Ventilation and exhaust openings on a racing vehicle. **1** Inlet transmission cooler, **2** Inlet rear axle brakes, **3** Inlet turbocharger (engine), **4** Ventilation engine compartment, **5** Outlet front diffuser, **6** Ventilation cockpit, **7** Inlet front diffuser/splitter, **8** Inlet front axle brakes, **9** Outlet nose area, **10** Ventilation engine compartment

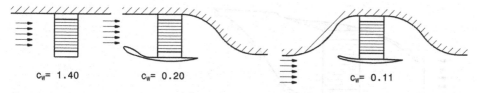

Fig. 5.77 Air resistance of selected heat exchanger arrangements, after [5]

or in the rear area. Aim of design of inlets and outlets is their high efficiency and naturally most small negative influence on downforce. Inlet openings for heat exchangers should not be located near the road surface, because air temperatures near the asphalt on hot days are much higher than a few decimetres above it.

Heat Exchanger

A shaft provides the necessary flow through the heat exchanger. This is much cheaper than simply hanging the heat exchanger "into the wind", Fig. 5.77.

The free-standing heat exchanger is the simplest, but at the same time the most unfavourable solution.

In a shaft, not only the air inlet but also the air outlet is important for an effective pressure gradient. The outlet area of the air should be about twice as large as the inlet area [4]. In extreme cases, only three times the area has led to satisfactory results. Furthermore, it is important for the design of the inlet that there is also an oblique flow, i.e. the air does not flow towards the opening at a right angle "as planned" in every driving condition. In order to avoid shadowing effects due to flow separation in the case of oblique inflow, the inlet area is rounded and designed with a drop profile in cross-section, Fig. 5.78. The rounding radius of the inlet lip should be as large as possible and at least in the order of 6–12 mm [5]. The inlet cross-section is designed with at least 25% of the heat exchanger area in new designs. However, if the design of the shaft is unfavourable to the flow, up to 60% may be required.

The heat exchanger represents a flow obstacle, therefore the air should flow through it at low velocity. This is achieved by the shown design of the shaft as a diffuser. This diffuser also directs the flow at a right angle onto the heat exchanger mesh. In order for the air to generate as few losses as possible when leaving the shaft, its velocity must be accelerated back to the previous level after the heat exchanger. This task is performed by the nozzle section of the shaft. Another measure to increase the efficiency of the inlet opening is to separate the low-energy boundary layer (see appendix). To prevent the boundary layer from flowing into the opening, the opening is moved away from the surface of the rest of the vehicle (Fig. 5.78b, left variant) or a splitter separates the air flow in an advantageous manner (Fig. 5.78b, right variant).

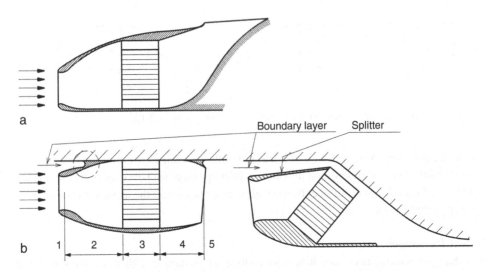

a

Boundary layer   Splitter

b   1   2   3   4   5

Fig. 5.78 Shaft design for heat exchangers, according to [5]. (**a**) bow arrangement (vertical sectional view), cf. Fig. 5.79, (**b**) lateral arrangement (two variants, horizontal sectional view). A streamlined shaft is divided into 5 sections: **1** inlet, **2** diffusor, **3 heat** exchanger (flow resistance), **4** nozzle, **5** exit

Fig. 5.79 Heat exchanger in the nose of a monoposto (Lotus 49 R6, 1970). The inlet at the front of the vehicle is rounded. The area for the heat exchanger behind it is higher and wider than the inlet. The exhaust air escapes through two openings at the top of the fuselage, cf. Fig. 5.78a. Between them is a cowl for the ventilation of the cockpit. Moreover, note the sheet metal angle on the top of the bow near the inlet opening. It acts similar to a Gurney strip on a wing

The relatively sensitive heat exchanger networks are even protected from stone chips in circuit vehicles with a protective grille in the inlet area of the air shaft. Figures 5.80 and 5.81 show examples of heat exchangers in operation.

Fig. 5.80 Favourable arrangement of a heat exchanger (Formula Renault, 2000). The vehicle shown has two heat exchangers symmetrically on both sides of the cockpit. The heat exchanger under consideration is located to the left of the cockpit. It is positioned at an angle to the direction of travel and thus has a relatively small projected span area. The design of the surrounding area ensures that air nevertheless flows through the heat exchanger over its entire surface. Both connections (flow and return) are located at the bottom of the heat exchanger. The lower water box thus contains a separating plate so that water flows through the cooler twice at its height. The engine control unit is located above the air intake. It is guided by two angles and fixed by an easily removable elastic band

Fig. 5.81 Air duct to heat exchanger (Dallara Formula 3). The air enters from the side next to the vehicle fuselage and is slowed down in a diffuser-like shaft and directed onto the slanted heat exchanger. The exiting air flows out the back through a nozzle-like shaft (not shown). The heat exchanger is housed in the side box and hardly widens the fuselage because of its slanted position

Design of Heat Exchangers

Components whose heat must be dissipated have a certain mass and thus a certain heat storage capacity. For this reason, the heat exchanger does not have to be designed for the

Fig. 5.82 Principle of a cross-flow heat exchanger. The air (index L) flows through the tubes conducting the coolant (index C) and absorbs heat in the process. Entry: Index 1, Exit: Index 2

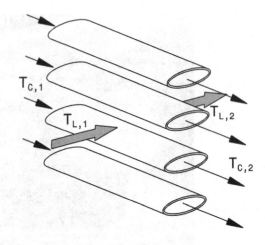

possible maximum heat generated. In the case of an engine, the maximum power is known from measurements. A Formula 1 gasoline engine has an overall efficiency of approximately 26% [2], i.e. 74% of the energy supplied by the fuel must be dissipated via the cooling system and the exhaust tract. Roughly, a third of the energy content in the fuel can be assumed for all gasoline engines. The same amount of engine (useful) power must therefore be dissipated through the cooling system and the same amount is dissipated to the environment via the exhaust gas.

The cooling system is designed for the average amount of heat generated within a lap or run. The heat-absorbing air flow results from the average vehicle speed, more precisely from the air speed through the heat exchangers. This is approximately 15% of the speed at which the vehicle is flowing [2]. In rallies, the average speed on most routes is about 80 km/h [20].

Of course, compromises must also be made in this area. On the one hand, the cooling system should be as small as possible for reasons of low weight and air resistance; on the other hand, a voluminous cooling system increases the (thermal) stability of the engine.

Heat exchangers that transfer heat to the air are usually designed as cross-flow heat exchangers, Fig. 5.82.

The coolant flows through the tubes primarily at right angles to the direction of travel and transfers heat to the air flowing through the heat exchanger network. The temperature of the coolant thus drops and the air is heated at the same time:

$$\Delta T_C = T_{C,1} - T_{C,2} \geq 0 \tag{5.23}$$

$$\Delta T_L = T_{L,2} - T_{L,1} \geq 0 \tag{5.24}$$

ΔT_C Temperature difference coolant, K
ΔT_L Temperature difference air, K

Fig. 5.83 Block volume and
mass flow density of a
cross-flow heat exchanger
From the mass flows
$\dot{m}_L$ bzw.$\dot{m}_C$ of air and coolant
respectively, their mass flow
densities follow:

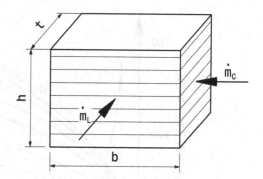

$$\dot{M}_L = \frac{\dot{m}_L}{b \cdot h}$$

$$\dot{M}_C = \frac{\dot{m}_C}{h \cdot t}$$

The heat transfer capacity of a heat exchanger is:

$$\dot{Q} = k \cdot A \cdot \overline{\Delta T} \qquad\qquad (5.25)$$

$\dot{Q}$ Heat flow through the cooler, W
k Heat transfer coefficient, W/(m^2 K)
$\overline{\Delta T}$ mean temperature difference, K

The heat flow related to an average temperature difference of 1 K (i.e. $k \cdot A$) is almost
independent of the cooler size in relation to the block volume (Fig. 5.83) of a cross-flow
heat exchanger.

$$\frac{k \cdot A}{V} = f(\dot{M}_L, \dot{M}_C) \qquad\qquad (5.26)$$

kA/V Specific heat exchanger capacity, W/(K dm^3)
V Block volume of the cooler, dm^3. $V = b \cdot h \cdot t$.
       b, h, t dimensions, dm; see Fig. 5.83.
$\dot{M}_L, \dot{M}_C$ Mass flow density of air or coolant, kg/(m^2 s); see Fig. 5.83
$\dot{m}_L$, $\dot{m}_C$ Mass flow of air or coolant, kg/s

The usual block thickness t is about 50 mm upwards. The face dimensions (b, h) depend
on the installation conditions.

The specific heat exchanger capacity kA/V is determined by measurement and shown in
diagram form over the two mass flow densities of air and coolant, Fig. 5.84. This allows the
specific heat exchanger capacity of a cooler to be determined.

For the determination of the heat exchanger capacity, the operating characteristic
diagram is now used. The heat capacity currents are used for the input variables of this
diagram:

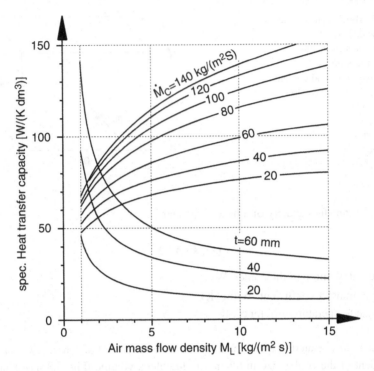

Fig. 5.84 Heat exchanger diagram [9]. The diagram is based on the following conditions: Coolant: water-glycol mixture 50/50%, average fluid temperature 355 K, average air temperature 310 K. From the mass flow densities $\dot{M}_L$, $\dot{M}_C$ of air and coolant follows the specific heat exchanger capacity kA/V of the cooler

$$\dot{W}_L = \dot{m}_L \cdot c_{p,L} \qquad (5.27)$$

$$\dot{W}_C = \dot{m}_C \cdot c_{p,C} \qquad (5.28)$$

$\dot{W}_L$, $\dot{W}_C$ Heat capacity flow of the air or coolant, W/K

$c_{p,L}$, $c_{p,C}$ specific heat capacity at constant pressure of the air or coolant, J/(kgK)

From this follows $\dot{W}_{min}$ and thus in turn the input variables for the operating characteristics diagram:

$$\dot{W}_{min} = \mathrm{Min}\left(\dot{W}_L, \dot{W}_C\right) \qquad (5.29)$$

From the operating characteristic diagram, Fig. 5.85, the operating characteristic Φ determined by the ratio of the heat capacity currents $\dot{W}_L$ and $\dot{W}_C$ and the heat flux kA can be read.

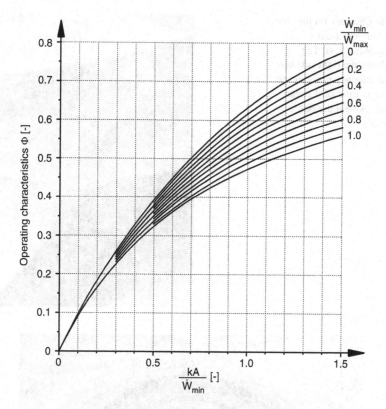

Fig. 5.85 Operating characteristic diagram of a cross-flow heat exchanger [9]
Via the input variables $\dot{W}_{min}/\dot{W}_{max}$ and $kA/\dot{W}_{min}$ Φ can be read off

The heat exchanger capacity $\dot{Q}$ can thus be completely determined:

$$\dot{Q} = \Phi \cdot \dot{W}_{min} \cdot \Delta T_1 \qquad (5.30)$$

Chimneys are found on the sidepods of some formula cars to vent the engine compartment and thus also dissipate some of the heat to the environment, Fig. 5.86.

Ventilation is also often provided at the wheel arch (*AE: fender*) of vehicles with enclosed wheels. Louvre-like outlet slots (*louvres*) on the upper side of the wheel arch allow the air introduced for brake cooling to escape to the rear without disturbing the external flow too much, or the accelerated external flow reduces the pressure in the wheel arch and can thus also cause downforce, Fig. 5.87.

With all the technical considerations regarding heat dissipation and ventilation, the cockpit must not be forgotten. Particularly when the vehicle is closed, care must be taken to ensure that the interior is adequately air-conditioned. As the temperature rises, the driver's performance and ability to concentrate decreases, which is a decisive criterion, especially in

Fig. 5.86 Chimney on the side box of a Formula 1 car (McLaren-Mercedes). The picture shows the right side box. Under the chimney you can see the splitter in front of the rear wheel and above it the bottle neck intake

Fig. 5.87 Venting a wheel arch on a touring car (Mercedes). At the top of the wheel arch, lamella-like openings are directed towards the rear. The air flowing around the outside thus creates a negative pressure in the wheel arch, which can even generate downforce

the case of long-distance vehicles. In addition, some regulations (e.g. for the 24 Hours of Le Mans) now contain provisions regarding the maximum air temperatures in the cockpit. Some vehicles – especially those with front-mounted engines – thus actually feature an air-conditioning system that is otherwise only known from production cars. When choosing the ventilation opening, rainwater has to be considered in addition to possible heat sources. There have been long-distance vehicles whose cockpits have been flooded when driving in the rain. Cockpit air conditioning is important not only from a human point of view, but also from a technical one, because it can prevent the windshield from fogging up.

5.7   Dimensioning and Setup

The aerodynamic design of a racing vehicle is a compromise between air resistance, which should be as small as possible, and downforce, which should be as large as possible for braking, acceleration and above all cornering (cf. also Fig. 5.90).[9] If the downforce is increased (e.g. by changing the angle of attack of a wing) the drag increases and vice versa, cf. Fig. 5.88.

The selected setting thus depends on the track and is optimized based on the lap times. On narrow courses, a higher air resistance is accepted because better lap times are achieved due to the higher traction. Remarkable at this point is that professional drivers feel a wing adjustment of already 1° (!).

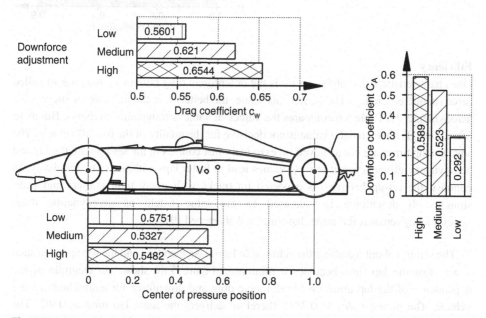

Fig. 5.88 Mutual influence of important aerodynamic variables on a monoposto (Formula Renault), after [21]. For three different downforce settings, center of pressure position, downforce and drag coefficient are mentioned. V Vehicle centre of gravity

---

[9] See also Racing Car Technology Manual Vol. 5 *Data Analysis, Tuning and Development*, Chaps. 5 and 6.

Table 5.6 Aerodynamic efficiency of selected downforce generating elements

Designation	Aerodynamic efficiency c_A/c_W [–]	Comment
Diffuser	6 to 7:1	Especially at high speeds even greater
Multipart wings	5 to 6:1	
Wing	2.5 to 3:1	
Splitter	1.5 to 2:1	

Fig. 5.89 Lap times as a function of downforce and drag coefficient of a generic racing vehicle at the Hockenheimring [9]. Vehicle model data: Vehicle mass $m_{V,t} = 1100$ kg, frontal area $A_V = 2.0$ m^2, weight distribution front to rear axle 47/53. Standard racing tyres, track width, wheelbase, etc. like standard touring cars

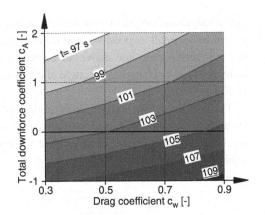

Efficiency

One characteristic value of the set-up is the downforce: drag ratio or c_A/c_W, the so-called aerodynamic efficiency. The higher this value, the better; it is the measure of success for aerodynamicists. Table 5.6 compares the values of some aerodynamic elements. But as so often, a single characteristic value is not decisive for the quality of the overall vehicle. The smallest lap times are not achieved on extremely fast courses in the set-up with the highest aerodynamic efficiency, but low c_W values lead to faster laps with the same efficiency. In contrast, the set-up behaves as expected on tracks with many tight corners and short straights: If downforce is increased to the same extent as aerodynamic drag, i.e. efficiency remains the same, lap times are shortened [9].

The setup is about matching the achievable lateral acceleration and the top speed in such a way that the lap time becomes a minimum. Figure 5.89 shows an example of the dependence of the lap times on aerodynamic drag and downforce for a simulated racing vehicle. The tuning $c_W/c_A = 0.3/0.5$ therefore delivers the same lap time as 0.9/2. The course of the lines of equal lap time shows that high downforce and low aerodynamic drag deliver short lap times in any case. However, the slopes of the lines are not constant: With increasing drag, the amount of downforce change must increase to improve the lap time by the same amount.

The observation from the point of view of the lap time is made in Fig. 5.90. The calculated lap times are plotted for different downforce settings. In addition, the

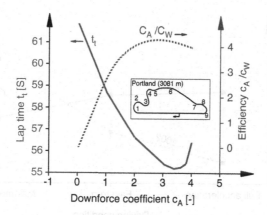

Fig. 5.90 Influence of downforce on lap time (simulation) [19]. Lap times in Portland (USA) are plotted for a generic race car for different downforce settings. The drag coefficient was determined according to the following relationship: $c_W = 0, 04(c_A + 0, 2)^2 + 0, 3$. The pole position was achieved on this circuit in 1991 with a time of 56.4 s, i.e. $c_A = 2.15$. The aerodynamic efficiency is near its maximum in the range

aerodynamic efficiency is plotted. It can be seen that the minimum lap time is achieved with high downforce and that too much downforce leads to decreasing efficiency, which causes the lap time to increase again.

For Formula 1 cars, the values of c_A/c_W range from 2.54 to 3.13 [22]. Le Mans prototypes achieve values of 4:1 [23]. This shows the specific design in the direction of downforce. For vehicles with enormously high engine power, other development goals arise than just reducing aerodynamic drag. The engine power allows theoretical top speeds, which cannot be reached on the relatively short straights of race tracks anyway. Typical straights on circuits are about 1000–2000 m long. This makes it easy for the manufacturers to convert this excess power into downforce.[10]

Statistical studies have found the following relationship between drag and downforce on race cars [19]:

$$c_W = k \cdot (c_A + c_{A,0})^2 + c_{W,0}$$

k Correction constant, -
$c_{W,0}$ Drag coefficient of the vehicle without downforce aids, -
$c_{A,0}$ Downforce coefficient of the vehicle in the condition with $c_{W,0}$, -

---

[10] See also Racing Car Technology Manual Vol. 3 *Powertrain*, Chap. 4, Fig. 4.14.

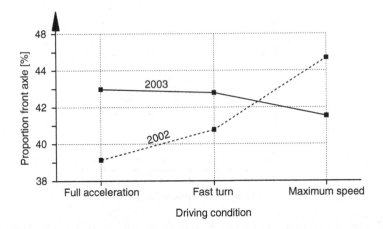

Fig. 5.91 Downforcesplit of two generations of a production sports car (Bentley EXP Speed 8), after [23]. The respective share of the total downforce on the front axle is plotted for different driving conditions. The newer car (2003) has a much more balanced distribution than its predecessor (2002)

This represents a second degree polynomial. The curve is therefore also called a *polar curve*. If you plot a polar curve on a diagram like Fig. 5.89, you can find an optimal configuration for this circuit.[11]

Formula 1 teams solve the conflict between downforce and aerodynamic drag in such a way that the downforce increases just up to the speed of the fastest corner through a targeted wing design and then the flow breaks off. This reduces drag on the following straight. A further increase in downforce would not be helpful at this point anyway.

Balance

In general, the aim should be to achieve a design that ensures balanced driving behaviour over the entire speed range and, above all, in different driving conditions. This makes the set-up easier for the driver and the vehicle engineer. The decisive factor here is the aerodynamic balance (*aero split*), i.e. the percentage distribution of downforce between the front and rear axles, e.g. $c_{A,f}/c_A$. Figure 5.91 illustrates the development of the downforce split of a production sports car on two generations. The older model (2002) produced the most downforce on the front axle at top speed. Ground clearance in that condition was only 15 mm and the front diffuser worked correspondingly effectively. Conversely, under full acceleration in second gear, the axle load shifted to the rear and the front end was lifted. Additionally, in such a situation, the downforce distribution also shifts to the rear. Both increase the understeer tendency and prevent the driver from exploiting the potential of the tyres when exiting a corner. The successor (2003) – the Le Mans double winner 2003 by the way – shows a much more balanced interaction of vehicle centre of

---

[11]In the Racing Car Technology Manual vol. 5 *Data Analysis, Tuning and Development*, Chap. 5 *Tuning there is an* example of this.

gravity and aerodynamic balance. In addition it is to be noted that the adjustment possibilities are closely limited particularly for sport prototypes at the finished vehicle. At the rear, the rear wing offers the usual adjustment options for racing cars, but in the nose area, until the LMP regulation change in 2014, at best minor corrections could be made with additional parts (deflectors, etc.). If more serious changes were required, this meant a new vehicle in the worst case! Now wings with adjustable flaps are also allowed in the front area.

The *centre of pressure* is the point on the vehicle through which the lines of action of all aerodynamic forces theoretically pass. Its position therefore determines on the one hand the downforce distribution between the two axles and on the other hand the yaw effect of a crosswind.

Ideally, the downforce forces on the front and rear axles should correspond to the static axle loads. As a result, downforce forces do not change the basic handling of the vehicle [24]. The ideal downforce aids would have to act directly on the wheels and would thus not additionally load suspension parts (springs, dampers, bearings, . . .). The center of pressure must therefore be close to the vehicle's centre of gravity for balanced driving behaviour. For the vehicle in Fig. 5.88, for example, it should always be between 50 and 60% of the wheelbase. In practical designs, the tuning is complicated by the fact that the downforce at the ends of the vehicle does not increase to the same extent with increasing speed, but that the front of the vehicle finds more favourable conditions than the rear (apart from driving in traffic). With increasing speed the downforce of the front wing increases and the decreasing ground clearance reinforces this tendency. This leads to (extremely unpopular) oversteer at high speed. To remedy this phenomenon, the wings are adjusted so that the downforce of the rear wing increases more than that of the front wing. Generally, the aim is for the center of pressure to move backwards as the speed of the car increases. Drivers are in fact faster at low speeds with oversteering vehicles and at high speeds with understeering ones.

In general, it is found that extremely increased downforce is accompanied by a narrow-band characteristic. Aerodynamic driveability suffers as a result and it becomes more difficult for drivers to stay within the (peak) range of the maximum. The development goal is therefore to generate downforce over a wide range of ground clearances, pitch and roll angles.

However, the center of pressure is not only of interest for the vertical dynamics of the car, but it also significantly influences the lateral dynamics. For stable handling in crosswinds and at large yaw angles, the center of pressure must be behind the centre of gravity, Fig. 5.92.

If the center of pressure is behind the centre of gravity, the air force generates a stabilising, reverse yaw moment. This favourable shift of the center of pressure towards the rear is influenced by large end plates on the rear wing, tail fins, but also high engine covers and high-mounted air scoops, as seen on many monoposti. In the meantime, tail fins have already become mandatory in the regulations of various racing series. At Le Mans, for

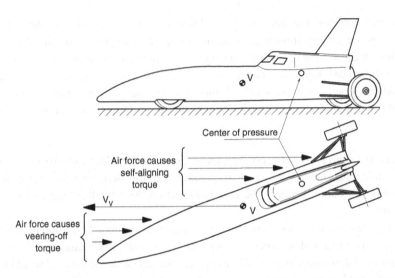

Fig. 5.92 Stabilizing effect of a tail fin. The surface of the large tail fin shifts the center of pressure (air attack point) to the rear in favour of directional stability. *V* Centre of gravity of the vehicle

example, vehicles in the LM P1 and P2 categories must be fitted with vertical air deflectors on the engine cover, see Sect. 3.3 *Safety*.

In general, downforce is easier to achieve in the nose area of the car, where the airflow is still little affected, than in the rear area, where the driver, engine covers and various superstructures have deflected and disturbed the flow.

The means for this are airdam spoilers, splitters and wings.

The first development step at the rear of high-speed vehicles is the reduction of lift, for example by means of spoilers or wings.

In addition to testing on the race track, investigations are carried out in the wind tunnel, especially during the development phase. Leading teams use 1:1 wind tunnels with conveyor belts that allow the relative movement between the vehicle and the track to be taken into account.[12]

The most targeted way of gaining knowledge is by means of simulation, Fig. 5.93. Modern computer programs, together with three-dimensional CAD vehicle models, make it possible to calculate the flow around the vehicle. The test and the measurement complement each other here. The wind tunnel provides the magnitude of values and the simulation calculation explains the qualitative influences of changes to the vehicle. CFD simulations (see appendix) are thus used in advance development, the (model) wind tunnel represents the design tool and the final decisions are made on the basis of real track tests. The analogy observation is supported by pressure taps in interesting areas of the bodywork of the real

---

[12]See Racing Car Technology Manual Vol. 5 *Data Analysis, Tuning and Development*, Chap. 6 *Development*.

Fig. 5.93 CFD (Computational Fluid Dynamics) model of a racing vehicle (OpenFOAM software). Current threads are displayed for a selected area. The scale shows the flow velocity in m/s. The inflow occurs at 70 km/h (green), below the front wing the flow accelerates (red), above it decelerates (blue)

vehicle. By comparing the measured values obtained in this way with the calculated variables, the CFD calculation becomes more accurate. A realistic behavior of the mathematical tire models used (deformation and contact-patch shape) is important for useful results of CFD calculations. CFD calculations also provide information on the wing loading of components. This can be used to dimension wings and their mounts. For production vehicles, CFD analysis tends to be used in the early stages of vehicle development, e.g. to optimize the vehicle shape. In later phases, CFD is used to investigate detailed flows.

Pitch Sensitivity

The sensitivity of a racing car to movements around its lateral axis (pitching (see appendix) or pitching) largely determines its handling. If the aerodynamic balance remains within the desired range despite the pitching forces acting on it, this is referred to as low sensitivity. Achieving low sensitivity is not so easy, as the downforce changes considerably with changing ground clearance for wings and diffusors. The more central the downforce is to a vehicle, the less sensitive it is to pitch. Wings mounted at the end of the vehicle are the least favorable in this regard, while diffusers that extend far below the undertray are much better.

The less sensitive the car reacts to changes in ground clearance, the softer the suspension set-up can be, and vice versa. A soft set-up allows the driver to safely drive the shorter and thus faster line over the track boundary (*curbs*).

Sensitivity to pitching has a greater influence on a car's handling than that to yaw (about the vertical axis) and to roll or roll (about the longitudinal axis). To some extent, pitching can also be counteracted by the chassis design. By arranging the pitch poles of the front and rear axles accordingly, a brake and acceleration pitch compensation can be installed.[13] Ideally, the car floor then remains at the desired angle (rake) to the road. The design of the chassis to support the aerodynamic properties of a vehicle is thus a central development objective.

The current Formula 1 cars with the stepped floor are unlikely to pitch.

Slipstream
In the development of a racing vehicle that is on the track together with several vehicles (circuit), the consideration of slipstreaming also becomes significant. In the slipstream of a vehicle, the flow conditions of a car change and reduce the downforce forces considerably depending on the longitudinal and lateral distance from the car in front. This applies especially to the downforce on the front axle, which can approach 0 (!) when driving under 3 m longitudinal distance without lateral offset [25]. However, the air resistance of the pursuing car is also reduced, which is why it can drive faster. But also the car in front is not unaffected. For it, the air resistance can be reduced up to 30% (!) [9]. As a result, cars can go faster in a dense bulk than on their own. Rear downforce tends to decrease the closer the chaser gets. Similar effects also occur in production cars, but the distances between the cars are (usually) much greater, so they are hardly noticeable.

Naturally, these influences have different effects in different racing series, not least because the regulations specify different vehicle categories, Fig. 5.94. At the right-hand end of the graph are racing vehicles such as karts. They do not generate downforce and have a relatively high aerodynamic drag due to their streamlined shape. The classic behaviour of drivers in a competition is to drive close up in the slipstream and overtake in the braking zone before corners. Formula 1 cars represent the other extreme. They are designed for enormous downforce, which is bought with a correspondingly large air resistance (which is compensated in this case, however, by the engine power). The lap time advantage of such cars results from the high cornering speeds and short braking distances. The drivers of these vehicles are faster alone than in the bulk. Therefore they try to keep a distance – lengthwise or crosswise – to vehicles in front in the competition. Between the two extremes there is a design where the advantage of low drag on the straight is offset by the disadvantage of low downforce in the corner (indicated by curved brackets). For comparison, typical passenger cars are also shown. These even show lift at high speeds and have a not inconsiderable air resistance due to their size.

The different aerodynamic set-ups also have an effect on the typical driving behaviour in the race.

---

[13]For more details see Racing Car Technology Manual Vol. 4 Chassis, Chap. 2 *Wheel Suspension* 2.2.

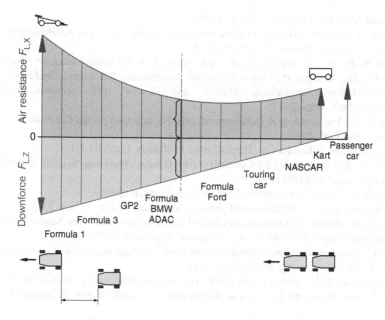

Fig. 5.94 Downforce and air resistance of different vehicles (schematic)

Testing

In addition to the wind tunnel tests mentioned above, tests are also carried out with the vehicle on the track or at a suitable test site. Similar to the wind tunnel, the function of components is examined. For example, measurement runs are carried out to validate aerodynamic measures. These include coast-down tests to determine the aerodynamic drag, measurements of the ground clearance to determine the downforce (air force pushes the car downwards) and the aerobalance, and runs with glued-on wool threads (*tuft testing*) or with dyed oil to visualize the flow.[14]

References

1. Bosch: Kraftfahrtechnisches Taschenbuch, 22. Aufl. VDI, Düsseldorf (2001)
2. Wright, P.: Ferrari Formula 1. Under the Skin of the Championship-winning F1-2000, 1. Aufl. David Bull Publishing, Phoenix (2003)
3. Tremayne, D.: The Science of Formula 1 Design, 1. Aufl. Haynes Publishing, Sparkford (2004)
4. Staniforth, A.: Race and Rallycar Source Book, 4. Aufl. Haynes Publishing, Sparkford (2001)
5. Smith, C.: Tune to Win. Aero Publishers, Fallbrook (1978)
6. Tremayne, D.: Formel 1 Technik unter der Lupe, 1. Aufl. Motorbuch, Stuttgart (2001)

---

[14]For more details, see Racing Car Technology Manual Vol. 5 *Data Analysis, Tuning and Development,* Chap. 6.

7. Der neue Audi A6, ATZ MTZ extra März (2004)
8. Bennett, N.: Inspired to Design. F1 Cars, Indycars & Racing Tyres: The Autobiography of Nigel Bennett. Veloce Publishing, Poundbury (2013)
9. Hucho, W.-H. (Hrsg.): Aerodynamik des Automobils, 5. Aufl. Vieweg Verlag, Wiesbaden (2005)
10. Amhofer, Th., Inzinger, P.: Correlation of the Aerodynamic Behaviour. Abschlussbericht der Lehrveranstaltung "Engineering Methods and Design", FH Joanneum, Institut für Fahrzeugtechnik, Graz (2017)
11. Incandela, S.: The Anatomy & Development of the Formula One Racing Car From 1975, 2. Aufl. Haynes, Sparkford (1984)
12. Newey, A.: How to Build a Car. HarperCollins Publishers, London (2017)
13. Rendle, S.: Red Bull Racing F1 Car 2010 (RB6), Owners' Workshop Manual, 1. Aufl. Haynes Publishing, Sparkford (2011)
14. McBeath, S.: Formel 1 Aerodynamik. Motorbuch, Stuttgart (2001)
15. http://m-selig.ae.illinois.edu/ads/coord_database.html. Accessed on 27 Nov 2015
16. Roth, A., Goldmann, F.: Funktionsprinzip und Konstruktionsweisen von Sandwich-Strukturen. Lightweight-Design. **2**, 36–40. Vieweg+Teubner, Wiesbaden (2009)
17. Lang, U.: Schaumkern als wirtschaftliche Alternative für Flugzeugbauteile. Lightweight-Design. **3**, 36–39. Springer Vieweg, Wiesbaden (2014)
18. Sportgesetz der FIA, Anhang J, Art. 258A "Technical Regulations for Sports Cars" (2006)
19. Katz, J.: New Directions in Race Car Aerodynamics, 2. Aufl. Bentley Publishers, Cambridge (2006)
20. Mühlmeier, M.: Virtual Design of a World Rally Car (Nov. 2006). Vortrag auf der Race.Tech, München (2006)
21. N.N.: Formula Renault 2000 Manual, Renault Sport Promotion Sportive (2001)
22. Piola, G.: Formel 1. Copress, München (2001)
23. Paefgen, F.-J., Gush, B.: Der Bentley Speed 8 für das 24-Stunden Rennen in Le Mans 2003. ATZ Heft. **4**, 280–289 (2004)
24. McBeath, S.: Competition Car Preparation, 1. Aufl. Haynes, Sparkford (1999)
25. Ulrich, W.: Audi der Sieger von Le Mans, Vortrag der ÖVK Vortragsreihe. Mai, Wien (2004)

Frame

6

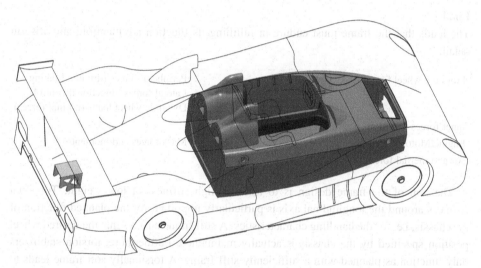

The frame or chassis literally forms the backbone of a racing vehicle. This is not (or no longer) the case with road vehicles with their self-supporting bodywork. In racing vehicles, the design method of connecting key assemblies as directly as possible to each other is much more pronounced, as there is no need to provide space for luggage, loads or passengers.

6.1   Requirements

Even if the frame is not the most important component of a racing vehicle, it must nevertheless fulfil numerous functions in the sense of a competitive vehicle and, in addition, fulfil requirements of the regulations.

Function
The following tasks are performed by the frame:

- Connection of the wheel suspensions
- Transfer of the initiated operating forces
- Guarantee of the required crash behaviour
- Accommodation of engine, drive train and auxiliary components
- Accommodation of the cockpit or accommodation of the driver.

Loads
The loads that the frame must endure in fulfilling its function are manifold and arise in detail:

From the wheel forces:	Peripheral forces (driving, braking)
	Lateral forces (directional control)
	Dynamic wheel loads (normal forces)
Inner forces and moments	
Engine/Motor vibrations:	Mass forces and moments
Air and inertial forces	

The aim of the frame design is to achieve high stiffness at low weight. Torsional stiffness around the longitudinal axis is particularly important for the planned function of the chassis, i.e. for the handling characteristics. A stiff frame ensures that the desired wheel position specified by the chassis is actually maintained. Furthermore, torsion stabilizers only function as planned with a sufficiently stiff frame. A torsionally soft frame leads to wheel load compensation of opposite wheels or is twisted by the stabilizer.

The bending stiffness is less decisive compared to the torsional stiffness. After all, bending around the transverse axis does not contribute to wheel load displacement (more precisely axle load displacement).

In the overall design process of the vehicle, the frame itself is not the most important subsystem. The driver's position, the chassis, the powertrain, etc. are much more crucial and are designed beforehand. The frame "merely" has to be adapted to the mounting points specified by it and not vice versa. In the case of integrated monocoque frames, the conditions are naturally completely different, because a large part of the vehicle is defined by the "frame".

6.2 Types of Construction

There are several basic types of frame in racing car construction. Whereby a development over the years is noticeable, starting from the carrier frame over ladder frame to tubular space frame. It was later replaced in the highest racing classes by the box frame. The box panels were first sheet metal panels later honeycomb core panels. Currently, a monocoque in fibre-reinforced plastic construction represents the pinnacle of development. However, there are also mixed designs that cleverly combine the advantages of individual design methods. For example, a tubular space frame construction can be bolted to a box frame, which forms the front end and the cockpit, to accommodate the engine with gearbox if this structure alone is not suitable for absorbing all the forces.

Basic considerations can be made irrespective of the type of frame, even if the findings cannot be implemented equally effectively for all types of construction. The load should be applied over a large area, which avoids stress peaks and thick-walled components. Forces acting in a nearly point-like manner, such as those that can arise from chassis connections, should be introduced at points where they are also transmitted directly (i.e. without causing bending or torsion). Such points are nodes in the case of tubular space frames or bulkheads in the case of planar frames.

6.2.1 Tubular Space Frame

Tubular space frames consist primarily of (sectional) tubes that form a spatial structure along the lines of a truss. The strength of the frame results from the arrangement of the material components relatively far away from the components that the frame accommodates. Such a frame thus represents a lightweight design of bar elements.

Tubular space frames are very old in terms of their use and have long since been replaced by other designs in some racing series. Because of their properties, however, they are still of interest to some manufacturers and racing series today. For example, many sports prototypes are based on a steel lattice tube frame, which is clad with fibreglass or CFRP bodywork parts. Likewise, this frame lends itself to touring cars, with the roll cage integrated directly into the frame. Many raid cars get their strength from a tubular space frame under their outer skin, which mimics the shape of a production car body. In the North American NASCAR series (which is much more popular in the US than Formula 1), all cars run with such a frame.

Advantages

- good stiffness/weight ratio
- weight-saving design with simultaneous potential for high bending and torsional stiffness
- simple and inexpensive production

- easy repair after accidents
- easy possibility of subsequent modifications in certain areas; e.g. engine/motor installation
- good accessibility to internal components
- rollover structure can be easily integrated.

Disadvantages

- partly complex preparation of the bar ends before joining, residual stresses due to tolerances and welded joints (local heat input)
- bodywork or cladding required in any case
- protection of the driver from flying parts (stones, etc.) by additional planking or similar required.

Design Principle

In an ideal tubular space frame, the bars only transmit tension/compression forces. This corresponds to the principle of direct load conduction and is therefore a measure to reduce weight while maintaining high stiffness. In fact, the bars can also transmit bending and torsional moments because they are usually rigidly connected at the nodes rather than articulated. An approximation to the ideal state is achieved by the specific application of forces. Forces may only be introduced at the nodes. At least three rods should meet at the connecting nodes, this results in the stiffest structure in space. Rods should not be connected with other rods between their connecting points (bending!). Thus, when designing the frame, one will aim for triangular structures (triangulation) and prevent bars in parallelogram arrangement by inserting (at least) one diagonal brace, Fig. 6.1. For triangulation, it should also be noted that large interior angles are much better than acute ones.

Another way of increasing rigidity is to fit structural panels: e.g. bonded or riveted aluminium sheets (sheet thickness 0.9–1.2 mm), rivet spacing approx. 50 mm; plywood, fibreboard (especially on the vehicle underbody where contact with the road is possible), plastic laminate panels.

As an example of a complete frame, Fig. 6.2 shows the tubular space frame of a single-seater. The frame connects the suspension of the front wheels as well as the steering gear with the engine connection and the front connection of the wishbones of the rear axle. The rest of the rear axle suspension carries the transmission. In addition, the rollover structures (bars at the front at steering wheel level and behind the driver) are integrated into the frame.

Components

Tubular space frames consist primarily of bars, components whose cross-section is small compared to their length. The first consideration for the bars is their cross-sectional shape. For a lightweight frame, the areas of the cross-sections must be as small as possible for the desired stiffness and strength. The decisive factor here is the distribution of individual surface areas in the cross-sectional plane. In addition, the type of stress (bending, tension,

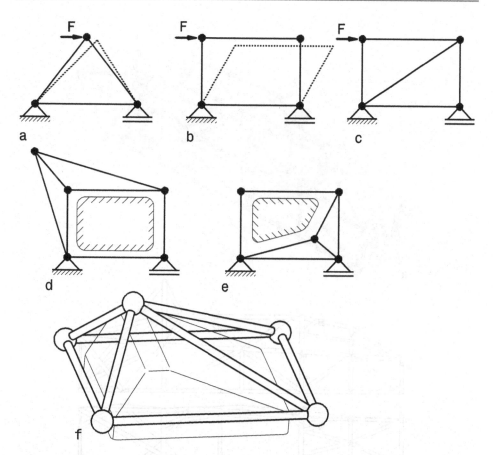

Fig. 6.1 Principle of the tubular space frame. All bars are hinged together in the nodes. (**a**) Plane triangular structure is rigid because rods only transmit compressive and tensile forces. (**b**) Plane rectangle cannot absorb force *F* at all because it is articulated, i.e. it still has one degree of freedom. (**c**) An additional strut creates two triangles out of the rectangle. This means that the rectangle is now able to balance the force *F*. (**d**) Stiffening by triangulation can also take place outside the rectangle, e.g. because components within the rectangle block the installation space. (**e**) Similar situation to (**d**), except that there is room for triangulation within the rectangle. (**f**) Spatial triangulation of a rectangle. Inside the rectangle, vehicle components prevent stiffening. Stiffening can still be achieved by using several triangles outside the plane of the rectangle

compression,...) influences the ideal distribution. The aim is therefore to arrange an existing surface as favourably as possible in the cross-section for the stresses that occur.

Figure 6.3 shows the most favourable cross-section shapes for individual loads. For pure tensile forces, the bar with solid cross-section is the most favourable shape. For pure compressive forces, the failure mode buckling must be taken into account. Therefore, the circular ring cross-section is much more suitable for this type of loading, especially for

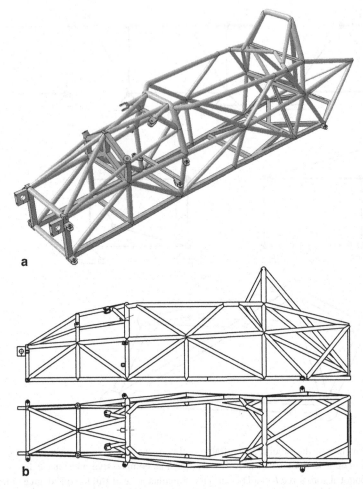

Fig. 6.2 Tubular space frame of a monoposto (Formula Ford). a Axonometric representation. b Floor plan and elevation. The frame has 57 nodes connected by 129 bars. The bars are tubes with circular and rectangular cross-section

limit-load loading. If an element is to transmit torsional moments, the closed pipe cross-section is the first choice.

Bending moments do not occur in the ideal truss, but they do in the real tubular space frame. Therefore, bending is also a criterion in the choice of cross-sectional shapes of bars. The bending stiffness of a member depends on the material and the geometry of its cross-section:

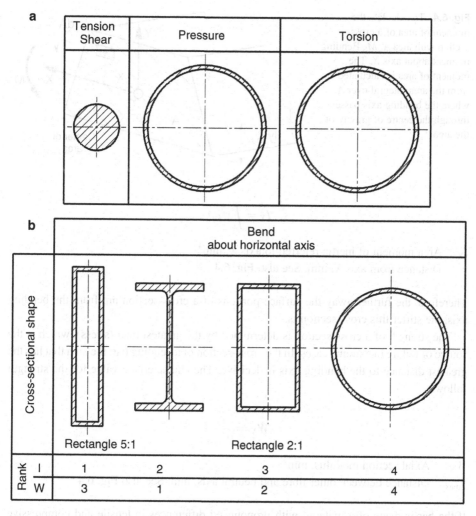

Fig. 6.3 Favorable cross-sectional shapes of bar-shaped components depending on the required component behavior. (**a**) tension/compression. Torsion, (**b**) Bending. To (**a**): The solid rod or a rope is the ideal element for tensile forces. Due to the risk of buckling, tubes are more suitable for compressive forces. To (**b**): The cross-sections all have the same area. If one considers stiffness (I) and strength (W), there are differences in the ranking, according to [1]

$$\textit{Bending stiffness} \propto E \cdot I$$

E Modulus of elasticity, N/mm^2
I Area moment of inertia, mm^4

The axial moment of inertia of a surface A with respect to the bending axis X is calculated as:

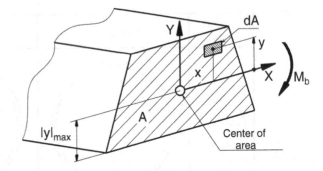

Fig. 6.4 To calculate the moment of area of a cross-section with area A. M_b Bending moment about axis X. The moment of area is calculated from the area integral over A, where the bending axis passes through the centre of gravity of the area

$$I_X = \int_A y^2 dA$$

I_X Area moment of inertia referred to axis X, mm^4

y Distance from axis X, mm. See also Fig. 6.4

Therefore, the further away the surface portions of a cross-section are from the bending axis, the stiffer this cross-section is.

The strength of a cross-section is determined by the greatest load (stress), which is the source of failure in extreme cases. In the cross-section of a bending bar, the area that has the greatest distance to the bending axis is decisive. The characteristic value for the strength follows:

$$W_X = \frac{I_X}{|y|_{max}}$$

W_X Axial section modulus, mm^3

y_{max} Distance between outer fibre and neutral axis, mm. See also Fig. 6.4

If the bar is made of a material with pronounced differences in tensile and compressive strength, both edge fibres (i.e. $+y$ and $-y$) are considered.

If we now look at some standard cross-sections in terms of their bending behaviour, Fig. 6.3b, the tall, narrow rectangle is shown to be the stiffest cross-section ahead of the I-section, which in turn offers the highest strength. The circular cross-section has the practical advantage of having no "preferred direction" of bending axis. If the axis of bending moments changes during operation or cannot be detected, the circular cross-section is preferred at this point. In this comparison, all cross-sections have the same area. Rods with these cross-sections are thus – when using the same material – all of the same weight.

Further criteria for the choice of a bar cross-section result from the manufacturing process. Rectangular sections are easier to cut and join than round tubes. As a result, they also offer better opportunities to rivet on cladding or stiffening plates. Seamless drawn (shaped) tubes offer greater strength, but are also more expensive than their welded counterparts. Conversely, welded (shaped) tubes are more cost-effective with strength disadvantages [2].

In the interests of optimum material utilisation, the cross-section of a bending bar should not be constant along its length if the bending moment changes along the length of the bar. Conified tubes have a greater wall thickness at their ends than in the middle. This makes them easier to weld and braze, while helping to keep the mass low.

Materials

Some basic physical considerations are helpful when dealing with the topic of stiffness. The stiffness, i.e. the deformation resistance of a component to loads, in the linear-elastic range depends only on the modulus of elasticity E and cross-sectional dimensions such as area or area moments of inertia. Assuming that the stiffness of a steel frame is to be increased, simply changing to a higher strength steel is pointless because it has the same modulus of elasticity as the steel with lower strength. In this case, an improvement is achieved by increasing the cross-section or by using a different material. A higher strength material shifts the load level upwards, above which plastic (permanent) deformation occurs. For this reason, high-strength steels are used for racing cars that have a long service life, even though they are more expensive and more difficult to process. It is important to remember that production sports cars, for example, are raced by hobby drivers for over 20 years.

If steel is replaced by aluminium in order to reduce the weight of the frame, then – depending on the type of stress – cross-sectional parameters such as area or moments per unit area must be increased in the ratio of the E moduli $E_{Steel}/E_{Aluminium}$.

Reynolds 531 (trade name for Mn-Mo steel with $R_m = 700$ to 850 N/mm^2), 25CrMo4 (DIN EN 10083), S355J2G3 (DIN EN 10025), S275J2G3 (DIN EN 10025). The excellent weldable material 1.7734 (15CDV6) is used for raid rally vehicles [3]. It is used in two heat treatment states (T4 and T5) and achieves tensile strengths of $R_m = 700$ and 1180 N/mm^2.

Because of their deformation capacity with high strength, (laser-welded) austenitic steels are also proposed for vehicle frames in [4]: X2CrNiN 18-7 and X2CrNiMoN 22-5-3 (EN 10088).

Wrought aluminium alloys, e.g. AlMg3 F23, AlMgSi1 F20.

Dimensions

Dimensions for steel pipes: $\varnothing 13 \times 0.9$ to 1.6 mm wall thickness, $\varnothing 26 \times 0.9$ to 1.2 mm. Largest pipes up to $\varnothing 40 \times 2$.

Sectional tubes are used in dimensions ranging from $20 \times 20 \times 1$ mm to $28 \times 28 \times 1.5$ mm.

Joints

Suitable connection methods for the rod-shaped components of a tubular space frame are:

- Brazing
- Glue
- Welding

For brazing, fillet brazing and socket brazing can be used.

Bonding with sockets enables the connection of different materials that cannot be welded or soldered and facilitates the joining process.

Welding takes less time than soldering and is easier to automate. In principle, all common processes (electrode rod arc, metal inert gas MIG, tungsten inert gas TIG, gas fusion) can be used. Electrode rod welding, however, hardly provides satisfactory results with small wall thicknesses (about less than 2 mm). Depending on the material, number and arrangement of the welds, heat treatment of the frame may be necessary to reduce the residual stresses. Welds are weak points in a construction and should therefore be placed in less highly stressed zones. Seam accumulations should be avoided. In the case of ribs or gusset plates, a recess is therefore made at the joint to other seams. By appropriate design of the connecting parts, welds can also be advantageously extended or the type of stress (e.g. from tension to shear) can be changed, Fig. 6.5.

In weld seams which are inclined to the axial direction of the pipes, not only does no pure tensile stress occur, but the weld seam is also longer than the pipe circumference (a, c). Slot welds (b) can also be combined with other welds. Recommendations for dimensions are given in the picture. In the case of cross-section transitions, the welded joint can remain in the pipe area with the same diameter as well as wall thickness by adapting the connection (d), thus avoiding the position in the highly stressed transition area.

Before welding or brazing, the pipe ends must be machined. In the case of circular pipes, it is advisable to mill off one end with a roller cutter of the appropriate diameter, Fig. 6.6a. In the case of pipe joints of pipes with the same diameter, the preparation can be carried out exactly with two plane cuts each, Fig. 6.6b. In this context, there are advantages for square sections: Generally, one plane cut on the butt tube is sufficient, the contact points are easy to weld and the weld seams are more accessible than with round tubes, Fig. 6.6c. A disadvantage of square sections is a position-dependent section modulus.

But also when connecting slimmer pipes to a thicker one, the welding preparation of the ends can be done by simple plane cuts, Fig. 6.7.

$$h = \frac{D^2}{2} - \sqrt{\frac{D^2}{4} - r^2}$$

$$\alpha_1 = \arctan\left(\frac{h \cdot \sin\Theta}{r + h \cdot \cos\Theta}\right)$$

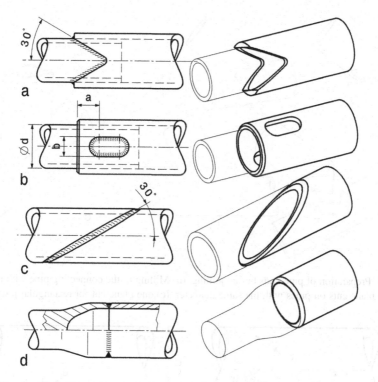

Fig. 6.5 Favourable position of weld seams for tube junctions. (**a**) Pipe joint with weld seam under thrust (*fishmouth joint*). (**b**) Slot weld (*rosette*). Width *b* approx. *d*/4, min. edge distance *a* = *d*/2. (**c**) Diagonal butt weld. (**d**) V weld in the pipe area of a transition piece

$$\alpha_2 = \arctan\left(\frac{h \cdot \sin \Theta}{r - h \cdot \cos \Theta}\right)$$

$$c = \sqrt{r^2 - (r-t)^2}$$

$$h = \frac{D^2}{2} - \sqrt{\frac{D^2}{4} - (r-t)^2}$$

$$\alpha_1 = \arctan\left(\frac{h \cdot \sin \Theta}{r + h \cdot \cos \Theta - c \cdot \sin \Theta}\right)$$

$$\alpha_2 = \arctan\left(\frac{h \cdot \sin \Theta}{r - h \cdot \cos \Theta - c \cdot \sin \Theta}\right)$$

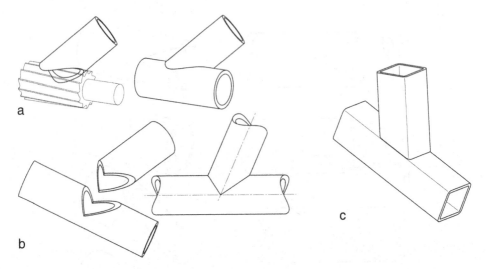

Fig. 6.6 Preparation of pipe ends before joining. (**a**) Milling of the connecting pipe with roll cutter. (**b**) Two plane cuts on pipes with the same diameter. (**c**) one plane cut for rectangular profile

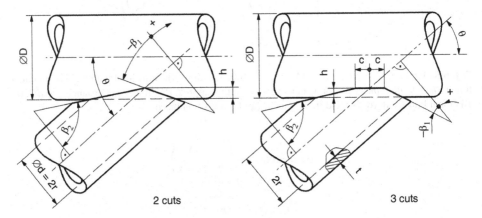

Fig. 6.7 Plane sections for filler rods, after [5]. The preparation for joining the filler rod consists of two or three plane cuts. α_1 and α_2 Auxiliary angles for calculating the cutting angles β_1 and β_2

$$\beta_1 = 90° - \Theta + \alpha_1$$

$$\beta_2 = -90° + \Theta + \alpha_2$$

Plane cuts of pipe ends can also be used to join two rods into a node, Fig. 6.8.

Complex cutting curves can also be generated at the end of the tube using a laser cutter, provided that the CAD data of the frame is available. In this case, the manufacturer only

Fig. 6.8 Nodes before welding together, according to [6]. The strut ends have been manufactured with two plane cuts. A reinforcing plate can also be inserted between the two struts

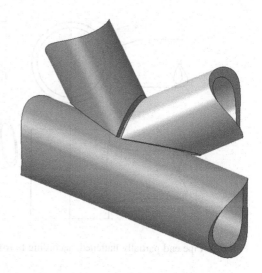

needs STEP files[1] (~. stp) of the desired pipe sections and delivers semi-finished products with both ends cut in the correct position and labeled with the part number. With the aid of a welding jig, these parts can be joined directly to form the frame.

In the case of round cross-sections, the end can be flattened on both sides before welding, Fig. 6.9. This saves mechanical processing and also widens the joining zone in one direction beyond the original diameter.

If members have to be connected axially in such a way that the connection can be released again, methods such as those shown in Fig. 6.10 can be used.

The head plate connection (a) can be used at a mitre cut as well as at a straight joint. The two head plates are screwed together.

Roll bar supports, engine mounts, chassis components and other elements may be arranged in such a way that they need to be removable and sometimes angle adjustable. Forked ends are advantageous for such struts. Some possibilities for attaching forked ends to section tubes are shown in Fig. 6.11.

For smaller dimensions a weld-in end (a) can be used. This clevis end is welded to the hollow section end. For larger loads, the weld should be arranged as in Fig. 6.11d. The connection between fork and tube can be additionally reinforced with sheet metal strips, Fig. 6.11e.

A pipe clamp provides a further adjustment option by sliding it along the pipe, Fig. 6.12.

If lugs are required at the end of a tube, for example for single-shear screw connections or to accommodate a fork, they can be designed in different ways. Figure 6.13 shows a selection of connections. A Y-shaped piece can be bent and welded from two sheets of metal and welded to the suitably cut section tube (a). The welds are stressed more favourably in this way than when butt-welding a T-piece. A flat plate can also be welded

---

[1] Standardized data exchange format for 3D data (Standard for the Exchange of Product Data).

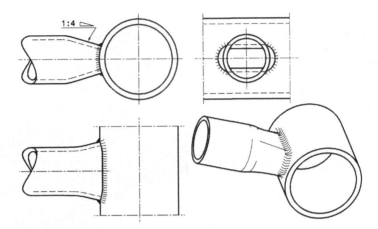

Fig. 6.9 Pipe end partially flattened, according to [6]. A joint with little preliminary machining

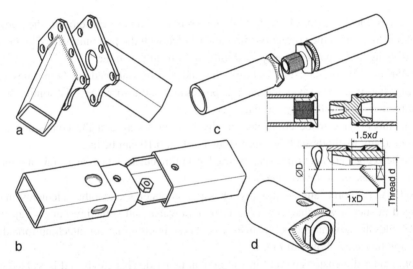

Fig. 6.10 Detachable axial connections of bars. (**a**) Head plate connection [6]. (**b**) Connection with sheet metal strip [6]. (**c**) Screw connection. (**d**) Threaded insert for high stresses

into a corresponding slot in a pipe (b). By forming a tube, there is a further possibility of producing a simple lug. The lug can be pressed to one side (c) or formed centrally to the pipe axis.

Further screw connections with ends of rods, Fig. 6.14.

A screw cup (a) at the end of a tube provides a simple way of setting the tube at almost any angle to the screw axis. Hexagon socket or hexagon head screws can be used for screwing. For the latter, the inner diameter of the cup must of course be large enough to fit the socket for tightening the screw.

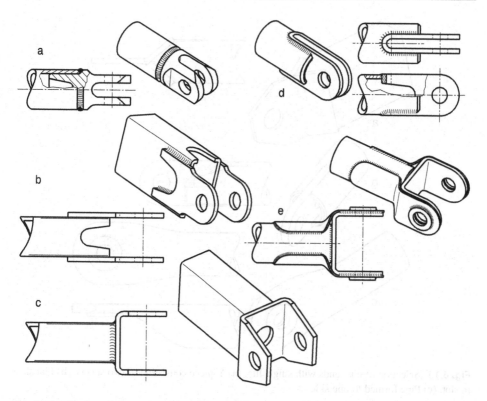

Fig. 6.11 Clevis ends on struts. (**a**) weld-in boss type tube end. (**b**) two tabs. (**c**) sheet angle. (**d**) sheet bent u-shaped. (**e**) wide yoke with reinforcement

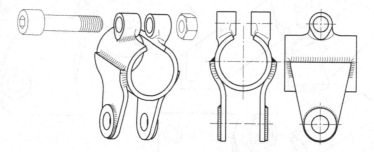

Fig. 6.12 Tubular clamp, according to [7].

Other ways of connecting tubes are also prescribed by the FIA for the design of roll cages, see also Sect. 3.3 *Protective Devices*.

According to the basic idea of a rigid tubular space frame, usually more than two tubes will be brought together in one node. It is important that the centre lines of all struts meet on the centre line of the main tube, Fig. 6.15.

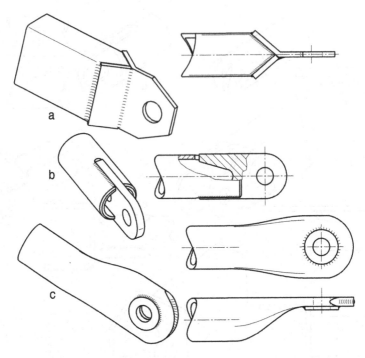

Fig. 6.13 Selection of pipe ends with single lug. (**a**) Y-piece consisting of two sheets. (**b**) Flat sheet in slot. (**c**) Pipe formed to one side

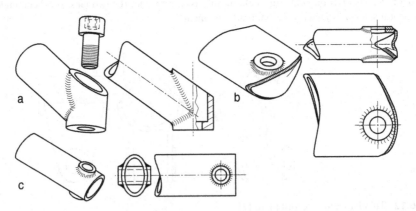

Fig. 6.14 Selection of pipe ends with sleeve. (**a**) Tube end with screw cup. (**b**) End of an elliptical profile. (**c**) End of an airfoil

Elbow Joints

If a frame is constructed strictly according to the truss principle, there are no bends in the tubes, they are all straight. A bend specifies a preferred direction of buckling of a strut and this must be avoided at all costs. Nevertheless, bends are used and they are used when the

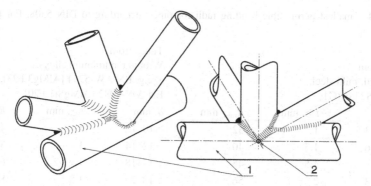

Fig. 6.15 Frame nodes in general. **1** Main pipe. **2** Common intersection of all centre lines

Fig. 6.16 Pipe bending. (**a**) Details of Table 6.1. (**b**) Pipe cross-section in the bend with strong ovalisation, as should be avoided, e.g. by pipe filling or support mandrel

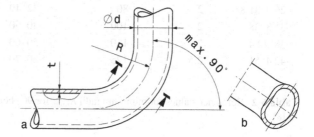

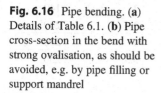

tube has a known loading direction and when the tube has to cover a change of direction in one piece. When the direction of load is known, the bend will oppose the load, actually increasing the load carrying capacity of the tube. Roll bars must be one piece in most racing classes. In any case, it is advisable not to choose the bending radius too small, because otherwise the tube cross-section will be ovalized too much and the strength limit of the tube material will be exceeded in the tensile zone. Another point concerning the strength is that the forming process can reduce the wall thickness by up to 20% [8]. Reference values for minimum bending radii can be taken from Table 6.1.

The values apply to cold bending with bending device without support mandrel or by hand.

Stiffeners

Stiffeners help to reduce the bending stress on frame tubes caused by stiff nodes, or they increase the stiffness of a node. The deflection of a unilaterally restrained member increases with the third power of its length, Fig. 6.17a. If its free length is reduced by a stiffener, e.g. to half, its bending is thereby reduced to one eighth, and its stiffness has thus increased eightfold. Thus, even smaller stiffeners can measurably increase the stiffness of a frame. In the case of bars that are only subjected to tension or compression, the stiffness of the tube does not increase, but their buckling length is reduced by the stiffener. With very high loads, this is an advantage for the frame because it fails later as a result.

Table 6.1 Smallest permissible bending radii for pipes, according to DIN 5508. For terms, see Fig. 6.16

Tubes from Unalloyed. mild steel, z. E. G. S235JRG1			Tubes from Wrought aluminium alloys, z. e.g. EN AW-5754 [AlMg3 F23], EN AW-6082 [AlMgSi1 F20]		
d, mm	$t \geq$, mm	R_{min}, mm	d, mm	$t \geq$, mm	R_{min}, mm
>10.2–13.5	1	32	>10–12	1	40
>13.5–16	1.5	40	>12–14	1	50
>16–19	1.5	45	>14–18	1	63
>19–21.3	2	56	>18–22	1.5	80
>21.3–25	2	63	>22–25	1.5	100
>25–26.9	2	70	>25–32	2	110
>26.9–31.8	2	80	>32–40	2	140
>31.8–38	2	100	>40–50	2.5	200
>38–42.4	2	110	>50–60	3	200
>42.4–51	2.5	125	>60–70	3.5	250

Table 6.2 Reference values for minimum radii $R_{i,min}$ for 90° bending of sheets and strips, according to [8]

$R_{i, min} = c \cdot t.$

Material	Steel[a] with R_m, N/mm^2			Aluminium			
	$\leq$390	390 to 490	490 to 640	Soft	Semi-hard	hard	Magnesium alloys
c	1–1.7	1.2–2.3	1.6–2.3	0.6–1	0.9–2	2–3	3–5

[a]Valid up to sheet thickness $t = 7$ mm

Stiffeners are also used to represent fixing eyes on a frame.

Stiffeners can consist of sheet metal plates which are welded onto the pipes to be connected in the joint area. By designing the plates accordingly, stiffness jumps in the frame can be avoided, Fig. 6.18.

For a pipe connection, seamed gusset plates that contact the pipes near their neutral fiber are much better than a simple gusset plate that is welded between the pipes like a stiffening rib, Fig. 6.19.

Stiffening ribs for a pipe bend should not be welded in the bend but on the outside of the neutral fibre of the pipe, Fig. 6.20. In the first case (a), there is an increase in stress at the transition from the pipe to the stiffening plate. This area at the inner bend is in any case more stressed than the wall in the area of the neutral fibre. In addition, sheets can tear or buckle if they are too high and thin. If a stiffener is welded along the neutral fibre (b), the stress distribution in the pipe is much more uniform. The plate should increase in height so that there is no jump in the stiffness curve of the welded structure. However, to ensure that

Fig. 6.17 Stiffening at a pipe connection. (**a**) Operating principle of a stiffener: left without, right with stiffener. (**b**) Round folded stiffener plate at a pipe connection. The effective height of the plate increases continuously and thus avoids a jump in stiffness in the frame tube

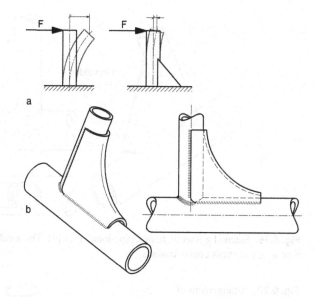

Fig. 6.18 Gusset plate for frame stiffening on a motorcycle frame. The gusset plate stiffens the connection of the frame tubes to the steering head, which accommodates the two fork bearings. To avoid jumps in stiffness, the stiffening plate has a continuous transition to the tubes

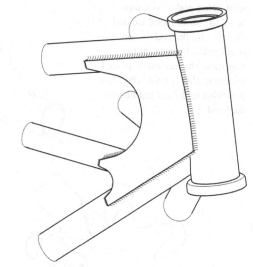

the two ends can also be welded well, it is essential to avoid a sharp taper in the sheet contour.

If several pipes meet in a node, a common reinforcing plate can be placed over all pipes, Fig. 6.21.

If pipes are to be stiffened at a greater distance from the joint, filler rods can be inserted, Fig. 6.22.

Reinforcements can also be placed inside a thin-walled main pipe to which pipes with a much smaller diameter are welded. A transverse tube welded into a corresponding hole in

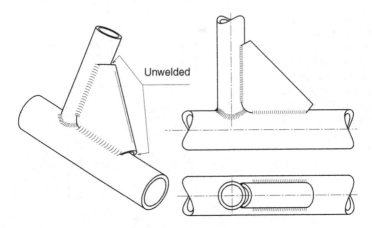

Fig. 6.19 Seamed gusset plate of a pipe joint, after [9]. The weld seam should not be led around the pipe to avoid stress concentrations

Fig. 6.20 Arrangement of reinforcing plates on a pipe bend. (**a**) Sheet metal welded directly into the arch produces stress peaks (arrows) at the transition. (**b**) Favorably placed stiffening sheet metal. The plate is welded to the neutral fibre of the bend

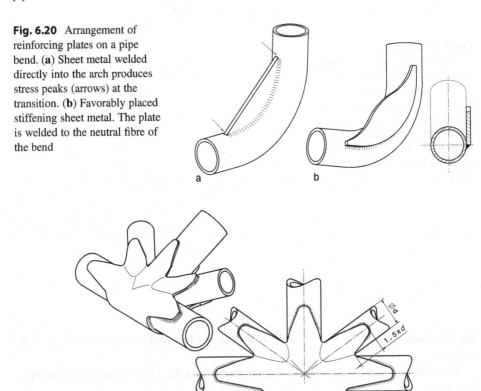

Fig. 6.21 Common reinforcing plate at a pipe node. To avoid jumps in stiffness in the individual pipes, the sheet ends run in a finger shape

Fig. 6.22 Stiffening with a filler rod

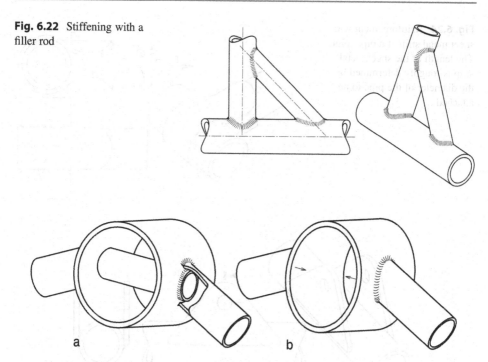

Fig. 6.23 Pipe joint of two thin pipes on a pipe with a larger diameter. (**a**) By inserting a cross pipe, the local stress on the thin-walled main pipe is reduced. (**b**) Without a cross pipe, the main pipe becomes oval

the main tube prevents oval deformation of the main tube due to the longitudinal forces of the two struts, Fig. 6.23.

At right-angled pipe joints with pronounced tensile stress, a sheet metal strip can additionally be attached as reinforcement to relieve the weld seam, Fig. 6.24.

Fixing lugs, as required for struts, chassis parts, heat exchangers, etc., must be attached to the tubes in the same way as stiffening ribs, so that no notches and thus weak points are created in the frame tube, Fig. 6.25.

Surface Treatment [13]
The frame requires surface treatment for corrosion protection and visual enhancement. There are several methods for this.

Galvanising
Although this technique promises an attractive finish, you have to expect high costs for first-class quality. In addition, this process highlights visual blemishes such as lumpy welds and scratches even more conspicuously.

Sometimes acid residues from the electroplating bath are trapped in the tubes – either through joints that are not fully welded or if vent holes have not been provided. Such vents can in turn cause stress concentrations and thus premature fatigue fractures. Trapped acid

Fig. 6.24 Reinforcement with sheet metal strip at a pipe joint. The length of the sheet metal strip is roughly determined by the diameter of the pipe to be attached

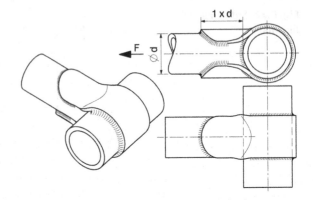

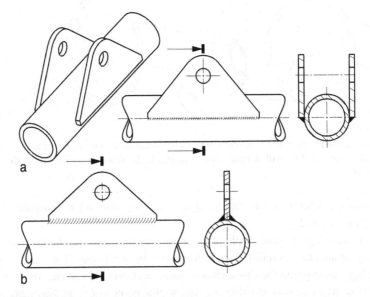

Fig. 6.25 Attaching a mounting bracket to a straight pipe. The plate should be welded to the neutral fiber of the bend. (**a**) favourable arrangement with two lugs on the neutral fibre of the pipe. (**b**) unfavourably placed lug

residues eventually create corrosion from within and can seep through at the welds; the result is rust spots and ugly stains.

Brittleness due to hydrogen exposure is another hazard of electroplating. During treatment in the bath, hydrogen is precipitated, which penetrates the surface structure of the steel and can cause material defects. In this respect, nickel plating is preferable to chrome plating. In many racing formulas, for example, the parts of the wheel suspension may not (no longer) be chromed.

Despite the disadvantages described, it must be said that motorcycle suspensions were sold nickel-plated for years, to all appearances without problems.

Painting

New types of paint are constantly appearing on the market, so you should always follow the manufacturer's processing instructions. Nitro paints may be good enough for tanks and other add-on parts, but they cannot meet the demands of frame painting. There, traditional stove enameling achieves the best all-around finish. The more modern electrostatic spraying methods with epoxy powder have found their supporters, but this type of paint – although more resistant to damage – is also more difficult to touch up.

Plastic Coating

This process, which is relatively new for frames, produces an excellent surface quality when used for initial painting. It covers blemishes, but is less practical in the long run because scratches cannot be polished out and repaired as easily as from painted surfaces. In addition, damage that extends to the metal can allow moisture to penetrate large areas and create large rust spots or blisters. With painted surfaces, corrosion remains limited to the damaged area and is easily repairable.

Anodising

Although some aluminium alloys are quite corrosion resistant, road salt in winter leads to visible corrosive attack (white coating). Anodizing is a protective process that creates a tough oxide film on the surface of the workpiece by immersion in an acid bath. This oxide film can even be coloured – grey, gold, red, blue or black – contributing not only to corrosion protection but also to more attractive appearance. Some alloys benefit from this process more than others; rolled material responds better than cast. However, the base material should have a high basic toughness due to unavoidable embrittlement during anodizing. The surface should be smooth without gaps (welds continuous!) and cracks.

Finally, regarding the tubular space frame, an executed example of a production sports car, Fig. 6.26.

6.2.2 Box Frame (Sheet Metal Monocoque)

Design Principle

The frame consists of box-shaped, hollow structures made of thin sheets and/or plates as well as solid single spars or frames, Fig. 6.27. The walls transfer all forces via tension/compression and shear. Panels with a sandwich structure are advantageous. These are capable of transmitting bending moments at a relatively low weight. In contrast to the tubular space frame, no additional cladding is required in some cases (apart from aesthetic and aerodynamic requirements). Reinforcements and spars are used to transfer concentrated individual forces.

Fig. 6.26 Tubular space frame for a production sports car. The frame can be seen from diagonally above. The engine area is on the right

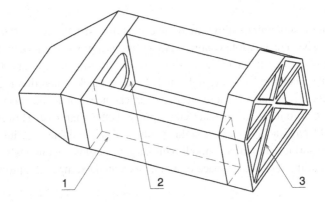

Fig. 6.27 Principle of a box frame. The frame consists of box-like, thin-walled structures. **1** single box. **2** bulkhead. **3** transverse bulkhead

Advantages

- Embodiment of the lightweight design principle, namely only tensile and compressive forces in the components and large distance between the edge fibres
- Embodiment of the lightweight design principle that each part fulfils several functions: The bodywork is load-bearing
- No or at least partially no bodywork required
- Crash elements easy to implement and easy to integrate all around (front, rear, side). Easy to repair because individual sections between the bulkheads can be replaced without having to renew the entire structure.

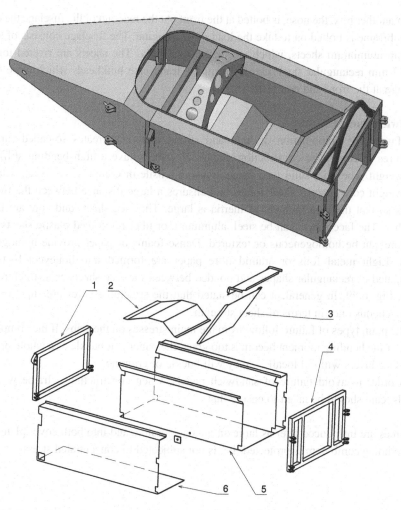

Fig. 6.28 Box frame of a single-seater (Monopin, 1100 cm^3 -class of the British hill-climbing championship). top: frame complete, below: Individual parts of the fuselage. **1** Front transverse bulkhead. **2** Sheet metal moulding under the driver's knee. **3** Bulkhead to engine = backrest. **4** Rear transverse bulkhead. **5** Cockpit hull with right-hand side panel. **6** Side panel left. The frame consists of folded sheet metal parts that are riveted together. Transverse frames made of sectional tubes accommodate landing gear parts and the engine

Disadvantages

- High production effort
- Introduction of point loads problematic
- Access to internal parts difficult.

An example of a box frame made from sheet metal parts is shown in Fig. 6.28. The vehicle from the 1100 cm^3 -class of the British hill-climbing championship consists of a fuselage to which another box, the nose, is bolted at the front. At the rear end of the fuselage the engine or a subframe is bolted on to take the load off the engine. The fuselage consists of several 0.7 mm aluminium sheets, which are bent accordingly. The sheets are riveted together. 20 × 2 mm rectangular steel sections form the transverse bulkheads which close off the fuselage at the front and rear [10].

Sandwich Panels

The favourable combination of light and strong materials creates so-called sandwich structures in composite construction. Sandwich panels have a high bending stiffness at low weight. The thin, solid face sheets transmit tensile or compressive forces, while the lightweight core absorbs shear forces and ensures a large distance between the two face sheets so that the area moment of inertia is large. The face sheets and core are bonded together. The face sheets can be steel, aluminum, or fiber-reinforced plastic sheets, while the core can be homogeneous or textured. Dense foams or paper provide homogeneous cores. Light metal foils or aramid fibre paper are formed into honeycomb, tubular, corrugated or rectangular shapes and bonded between the face sheets as a structured core layer, Fig. 6.29. In general, it can be stated that the structured cores are superior to the homogeneous ones in terms of sheet strength.

The main types of failure follow from the main stresses on the slabs. If the compressive force or the bending moment becomes too great, the panel will buckle as a whole or one or both face layers will fail locally – they will buckle or crumple.

In order to avoid failure of sandwich panels, when constructing a frame with such panels, care should be taken to ensure that

- forces are introduced over as large an area as possible and into both cover plates
- the honeycomb core is protected, i.e. is not subjected to concentrated stress.

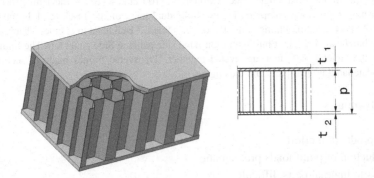

Fig. 6.29 Honeycomb-cored panel (aluminium). The cover panels have a typical thickness t_1 or t_2 of 0.6–1.5 mm. The panel thickness p is between 6 and 25 mm

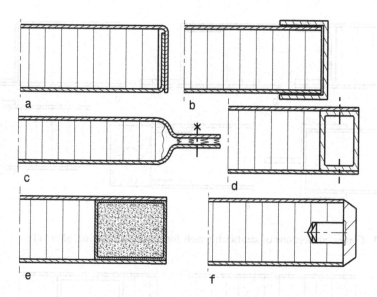

Fig. 6.30 Edge protection for panels. (**a**) Cover panel folded to size. (**b**) Edge closure with glued U-profile. The panel edge can be additionally filled with adhesive before applying the profile. (**c**) Edge area pressed. Only for panels up to 10 mm thick and Al honeycomb core. The edge can also be used for screw or rivet connection. (**d**) Profile tube glued in. This pipe can also be used for screw or rivet joints. (**e**) Thin-walled profile pipe filled with foam. (**f**) End strip glued on

Due to the special structure of sandwich panels, the edges, corner joints and centre joints deserve further attention, as well as the connection with struts at the edge and in the middle of panels.

When designing the panel edges, the guiding principle of connecting both face layers is also helpful, Fig. 6.30.

Starting from edge terminations suitable for force flow, suitable shapes of edges can be found for the connection with profile tubes or similar, Fig. 6.31.

If struts or supports need to be fitted within panels, or if panels need to be connected with transverse frames, suggestions such as those in Fig. 6.32 can be used.

If sandwich panels are extended or sections are repaired, centre joints must be implemented. Figure 6.33 shows some examples.

In order to create a spatial box frame from flat sandwich panels, panels must be joined at an angle. There are several possibilities for this, taking into account the idea of joining both face sheets. Figures 6.34, 6.35, and 6.36 give some ideas.

Not all forms of blind rivets are suitable for structural joints, but preference should be given to those which retain the mandrel during setting, Fig. 6.35.

Panels whose face sheets are made of metal can also be elegantly bent to any angle, Fig. 6.36. Before bending, a strip must be removed from the inner face sheet. The width of

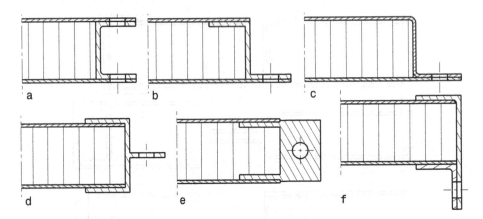

Fig. 6.31 Corner connections of sandwich panels for bearing and facing, after [11]

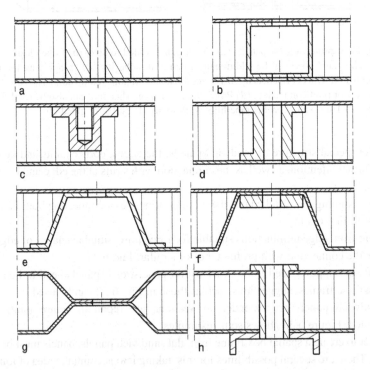

Fig. 6.32 Centre connections of sandwich panels for attaching struts or columns and for connecting to bulkheads, after [11]. Shown are some suggestions for force application within panel faces. The face sheets are bonded together and external forces are largely kept away from the core

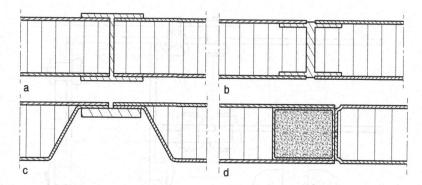

Fig. 6.33 Center joints of sandwich panels, after [11]. These shocks can also be used for repair purposes

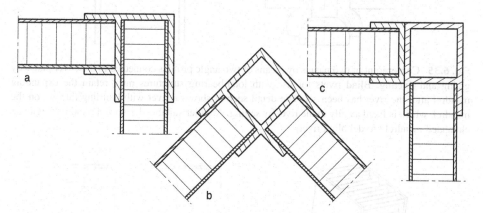

Fig. 6.34 Corner joints of sandwich panels with extruded profiles, after [11]

this strip is exactly equal to the arc length of the opposite face sheet when the inside radius is 0 and it is thus:

$$b_{Strip} = \frac{2\pi \cdot \beta \cdot p}{360^\circ}$$

b_{Strip} Width of the strip to be removed, mm
β Folding angle
p Panel thickness, mm

Bolted Connections
Some connections of a frame must be easily detachable, for example to access internal parts during repair and maintenance. In such cases, bolted connections are used. The problem

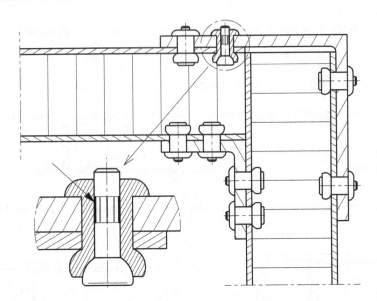

Fig. 6.35 Corner joint of two panels by means of two angle profiles connected to the cover plates by gluing and riveting. Blind rivets for use with load-bearing structures must retain the expanding mandrel after the rivet has been set. The detail section shows a rivet with knurling (arrow) on the mandrel, which is fixed axially in the rivet by setting. Another suitable design is a blind rivet with an elongated mandrel (Avdel blind rivet)

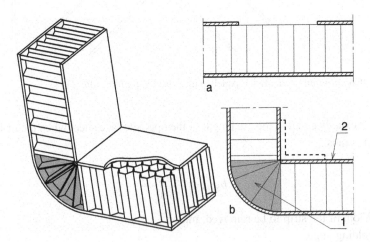

Fig. 6.36 Bending a panel. (**a**) Condition before bending: A strip of the top cover layer has been removed. (**b**) Panel folded down. **1** Filling compound. **2** inner cover plate, from which previously cut out a strip. On the inside, an angle profile (dashed) can be glued on in case of high strain

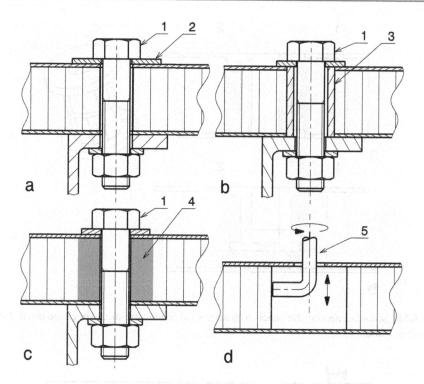

Fig. 6.37 Screw connections for panels. (**a**) Screw (1) with large washer (2) (body washer). Tightening torque limited by panel pressure resistance. (**b**) Screw with sleeve (3). Length of sleeve = panel thickness − 0.2 mm. (**c**) Screw with core filler (4). (**d**) Core removal tool (5)

with the design of bolted connections is, as always with sandwich structures, that the local forces should act as evenly as possible and that they must be introduced into both face layers. Attention must also be paid to relieving the load on the core. Figure 6.37 shows some possibilities. They are not all equivalent. On the contrary, variants a and b are only suitable for medium loads. If large forces have to be transmitted, the more labour-intensive variant c must be used. In this case, the core or core structure is first removed with a simple tool (figure part d) and then hardening core compound is introduced into this cavity. After hardening, the actual screw hole can be drilled.

Fig. 6.37 (**a, b**) for medium load, (**c**) for high stresses

Inserts are suitable for the direct screw connection of elements within a panel surface, Figs. 6.38 and 6.39. Such inserts are designed with and without thread. Blind hole threads can also be represented in this way. This type of screw connection is suitable for medium to high loads.

Detachable covers and cladding parts can be screwed to existing structures via lugs and angle profiles. Rivet nuts that are set in the manner of a blind rivet are suitable for this, i.e. access from one side is completely sufficient, Fig. 6.40.

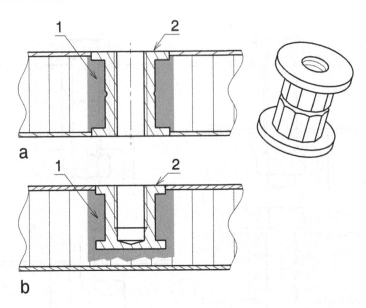

Fig. 6.38 Screw connections for panels with inserts. (**a**) continuous insert. (**b**) closed insert. **1** Insert adhesive. **2** Insert

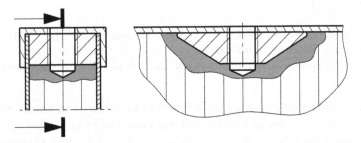

Fig. 6.39 Reinforcing plate under edge strip of a slab. The reinforcement plate was glued in place before an edge profile was attached and the thread was drilled afterwards

Fig. 6.40 Rivet nut. This fastener, suitable for thin-walled structures, is available in thread sizes from M3 to M12

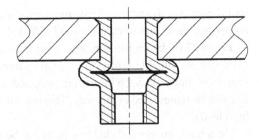

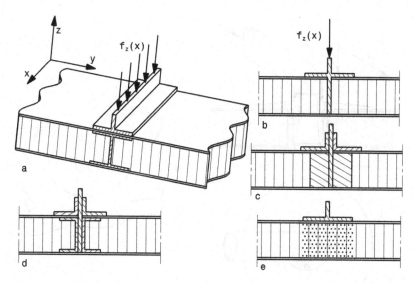

Fig. 6.41 Force application in sandwich panels, after [11]. $f_z(x)$ related shear force (line load). The load is applied over a large area and in such a way that the core is not damaged. (**a–d**) Core locally reinforced by battens and profiles. (**e**) Structured core, locally foamed

Force Application

The application of force at points in lightweight structures is a delicate problem. It is best to distribute the force over a distance (Fig. 6.41) or, better, a surface (Fig. 6.42).

Figure 6.42 provides an example of solving the problem of local introduction of a single force. This is done by an inclined support in a sandwich plate. By designing the connection using two plates with screw threads, the arrangement is detachable. If the connection must be non-detachable, the two plates can have any shape and can be glued and riveted to the cover layers of the plate. In this way, for example, bearing journals for bell cranks or abutments for spring/damper units can be riveted to plates.

Structures for larger bending forces (e.g. side impact protection of the cockpit) can be represented by a shear stiff combination of two panels, basically like the sandwich panels themselves, where two tensile stiff cover layers are kept at a large distance by a core (Fig. 6.43).

Bulkheads

The box-like structure is stabilized at suitable sections of transverse bulkheads. Furthermore, local forces can be introduced into the frame. Such sections are behind the driver, in front of the steering wheel, connection planes of steering gear and wishbones. In addition, the engine can also be bolted directly to a bulkhead, if it is suitable for this.

Bulkhead walls are milled from the solid, cast or built up from individual parts (sectional tubes, sheets, etc.) and integrated into the rest of the load-bearing structure by riveting, bonding or welding, Fig. 6.44.

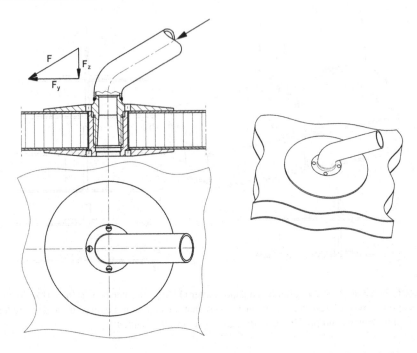

Fig. 6.42 Connection of a column to a sandwich panel, after [11]. The inclined tubular support ideally dissipates shear forces (F_y) and compressive forces (F_z). The screw-in element distributes the support force as a surface load in both cover skins and the core is kept free of compressive forces. This connection is designed to be detachable

Fig. 6.43 Bonding two panels with PU foam. **1** Outer wall of the frame. **2** Two-component foam (e.g. Ewidur). **3** Cover plate. **4** Spacer plate. The outer wall of the frame and the cockpit side wall form the side impact protection by bonding them with a hardening foam. Additional Z-shaped sheet metal strips (4) are glued between the panels. A sheet metal (3), which is only applied after foaming, forms the top closure

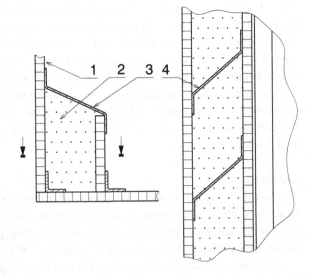

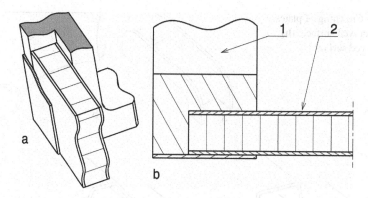

Fig. 6.44 Connection of a panel as a side wall to a transverse frame of a box frame. (**a**) Oblique section. (**b**) Sectional view. **1** transverse bulkhead. **2** panel. Due to the milling in the frame, the outer contour of the frame does not show any difference in level. An additionally glued cover layer connects the bodywork of the panel with the frame

Adhesives

The following adhesives can be used as structural adhesives when processing sandwich panels: SikaFast® -5211, a two-component acrylic-based adhesive or 3 M Scotch-Weld® Adhesive 9323.

A filler suitable for honeycomb core panels is: 3 M™Scotch-Weld™ Filler EC 3439, an epoxy resin.

Repair

A repair can be carried out relatively easily on a local basis. Entire plates can be replaced or only partial areas can be "patched". A damaged panel is removed by drilling out the rivet. If an additional bond is present, this bond is loosened using a hot air blower. After cleaning the connection points, a new plate can be inserted as in the original manufacture.

Stiffening of Sheets

For box frames, not only sandwich panels but also thin sheets are used. Although these transfer tension and shear in a weight-saving manner, they have low dimensional rigidity and are very susceptible to failure under other types of load. Therefore, stiffeners are required, especially for large-area parts, to increase the load-bearing capacity and safety against failure due to instability (buckling, crumpling, buckling,. . .). In terms of light-weight design, it is ideal if this is done without the need for additional material and/or with the gain of additional functions. The most common methods are:

- Crowning (shell formation)
- Beading
- Ribs
- Edge stiffeners
- Swipes.

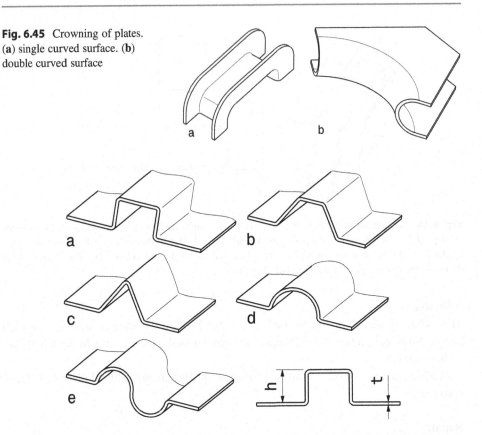

Fig. 6.45 Crowning of plates. (a) single curved surface. (b) double curved surface

Fig. 6.46 Bead shapes. (a) Rectangular bead. (b) Trapezoidal bead. (c) Triangular bead. (d) Semicircular bead. (e) Example of a shaped bead. Bead depth $h_{max} \leq 5$ to $6\ t$

Crowning

Crowning creates a spatial shell structure from a flat plate, which has a much higher moment of inertia per unit area for the same wall thickness and thus offers greater resistance to bending than the flat plate, Fig. 6.45.

Beading

Large sheet metal surfaces can be stiffened by beads. In this case, the sheet is embossed linearly out of its plane and thus the area moment of inertia is increased disproportionately transverse to the direction of the beading (quadratic parallel-axes theorem component of the moment of inertia). The contour of the coining or rolling tool determines the shape of the bead, Fig. 6.46.

When increasing the area moment of inertia, it is particularly important to achieve a large distance between the material cross-sections and the bending axis. This is why rectangular beads are stiffer than semicircular or triangular beads, Fig. 6.47. This also means that sharp-edged beads are stiffer than those with large rounding radii. However, the

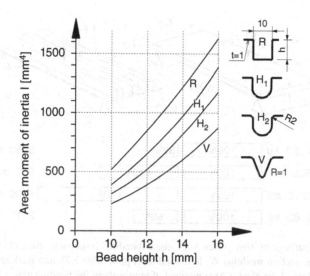

Fig. 6.47 Influence of the bead shape on the area moment of inertia, after [11]. A sheet with a thickness of 1 mm and a bead width of 10 mm can be beaded to the height h in different ways. The area moment of inertia is greatest with the sharp-edged rectangular shape because the specified width × height area is utilized as best as possible. *R* Rectangular bead, H_1 Half-round bead, sharp-edged, H_2 Half-round bead, rounded out, *V* Triangular bead

smallest bending radius that can be produced is dictated by the strength of the sheet and, depending on the sheet material, is of the order of the thickness of the sheet.

However, beads must not be designed too high, because otherwise the side walls of the bead itself will become too soft again. A guide value for the greatest bead depth is 5–6 times the sheet thickness.

A comparison between plates with the same basic dimensions (length × width) and the same section modulus, clearly demonstrates the weight and stiffness advantage of a beaded plate, Fig. 6.48.

The solid sheet steel plate (left) deflects almost twice as far as the thin sheet steel plate (right). The solid plate requires about three times the mass for the same section modulus.

Beading also significantly reduces the noise and buckling sensitivity of sheet metal panels.

Some guidelines for the strength-related arrangement and design of beads are summarized in Fig. 6.49.

Ribs

Ribs increase stiffness in the same way as beads, by lifting material cross-sections out of the sheet plane and thus disproportionately increasing the area moment of inertia. The difference with beads is simply that ribs are additionally applied to the sheet surface, whereas beads are pressed directly into the sheet. Ribs are therefore built up in a differential design and the material and manufacturing costs increase accordingly compared to

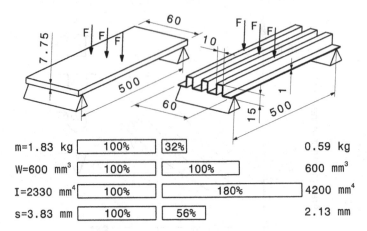

Fig. 6.48 Comparison of two plates with equal bending resistance, after [11]. The basis for comparison is the section modulus W. left: solid steel plate with 7.75 mm thickness, right: beaded 1 mm plate. *m* Mass of the plate, *I* Area moment of inertia about the bending axis, *s* Deflection under load *F*

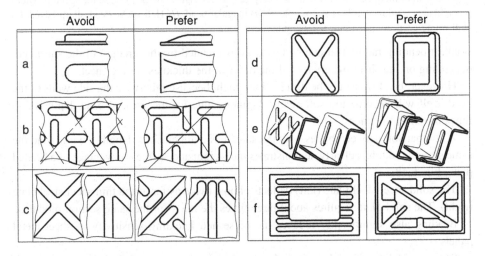

Fig. 6.49 Design guidelines for beads. **a**: Beads should not end with sharp edges, but should decrease in height towards the edge of the sheet. **b**: In the case of sheet metal fields, individual bead sections are usually embossed. Care must be taken to ensure that these individual beads do not form a straight-line pattern. This will result in line-like areas (dashed lines in figure b, left) where the moment of inertia of the sheet is determined only by its thickness. The sheet will therefore bend preferentially along these lines. **c**: If several beads run together, nodes with stress peaks are formed. Such nodes should be broken up. **d**: Long diagonal beads on flat plates have a lower stiffness-increasing effect than circumferential beads with mandrel-like corner areas. **e**: In the case of long U- and Z-shaped plate spars, there is a risk of buckling despite the beads if the beads only act linearly, i.e. increase the moment of inertia about one axis only. Beads that also stiffen the profile edges are more favorable. **f**: On sheets subjected to dynamic loads, regular bead patterns lead to fatigue fractures at the bead edges. In this case, resolved beading arrangements are preferable

Fig. 6.50 Examples of ribs. These ribs are made from straight battens

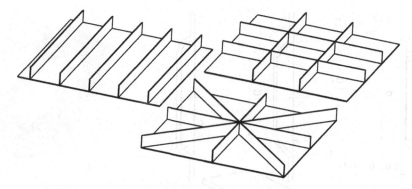

Fig. 6.51 Examples of arrangements of ribs. Due to the intersecting arrangement of the ribs, the stiffness of the plate is independent of direction

beading. For the joining technique, gluing, riveting, welding or soldering can be considered. In the case of cast parts, however, ribs are usually cast on.

However, ribs do not have to be parallel as in Fig. 6.50, but can also run diagonally or crosswise, which increases stiffness not only in one direction, Fig. 6.51.

If ribs do not extend from one force application point to the next, but are only applied locally, they must be designed to extend in height so that there is no jump in stiffness at the end of the rib.

Edge Reinforcement

Free sheet metal edges should be reinforced for two reasons. Firstly, the low stiffness of flat sheet metal leads to instability (buckling, dents) and, secondly, free edges with their knife-like shape pose a risk of injury to people and cables. Some possibilities for effectively stiffening sheet edges by folding or rolling are summarised in Fig. 6.52.

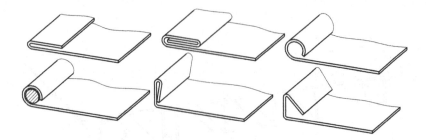

Fig. 6.52 Edge stiffeners of plates. The edge stiffeners not only lift the surface moment of inertia but also form usable edges

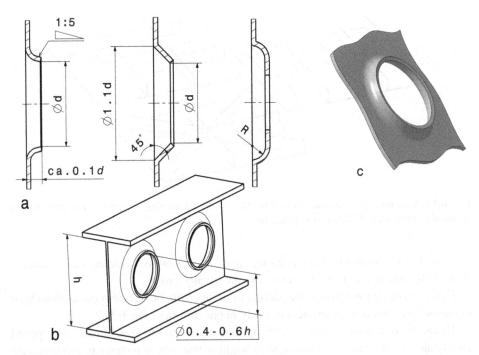

Fig. 6.53 Swaged holes. (**a**) Some types of drafts with indications of proven dimensions. (**b**) Beam with perforated web. (**c**) Axonometric representation

Swaged Holes

While swaged holes are not a general type of stiffener, they are used where large areas are to be lightened by removing material. This is the case with webs of tall beams and large, flat sheet metal panels. If only simple holes were made in these areas, the stiffness would decrease. The goal in designing a pull-through must therefore be that the wall with a pull-through has a higher moment of inertia per unit area than the unperforated wall.

The webs of beams can also be stiffened with pull-throughs (Fig. 6.53b). Although for the web the shear stress is decisive for failure and a simple hole inevitably increases the

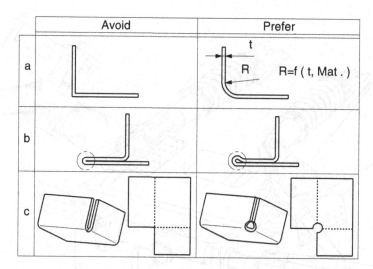

Avoid	Prefer
a	
b	
c	

Fig. 6.54 Design information for sheet metal parts. Measures to reduce residual stress by rounding out the edges of the fold before bending. (**a**) If sheets are folded with too sharp an edge, there is a risk of breakage at the edge due to overstretching in the tensile zone, which is why curves are much more favourable and also easier to produce. The minimum rounding radius depends on the sheet thickness and the material, see Table 6.2. (**b**) Even when sheet metal is folded, compression of the rounding leads to a reduction in strength in the area. This can be remedied by an eyelet-shaped rounding. (**c**) If several sheet metal parts meet at one edge during folding, the formation of a bead in the compression zone leads to a gaping of the edges and to unfavorable stress conditions in the corner. Circular recesses (radius $\geq 2\ t$) at the corners of flat sheet metal blanks create favorable conditions for manufacturing and strength.

stress, the shear stress drops with higher webs and pull-throughs [11]. Thus, beams can indeed be lightened without loss of strength.

Design Information for Sheet Metal Parts

In the case of bends in sheet metal parts, stress peaks can be avoided by appropriate design of the fold edges, Fig. 6.54.

Examples of aluminium honeycomb core panels in racing cars:

The production sports car DJ Firecat has a vehicle weight of 420 kg with an overall length of 3870 mm. Its frame is made of panels with a thickness of 25 mm with 0.6 mm cover plates for the monocoque and 1.2 mm for the engine compartment [12].

The box frame made of aluminium and carbon fibre plates with steel cross members of the Renault RE20 Formula 1 car (1979) weighed 46 kg [13].

Figures 6.55, 6.56, 6.57, and 6.58 show details of completed box frames.

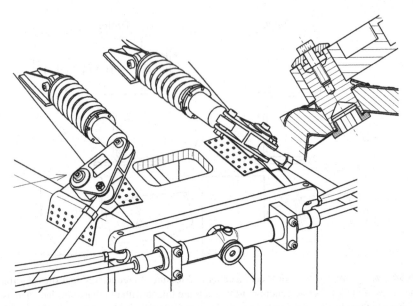

Fig. 6.55 Front end of a monocoque made of honeycomb core panels with attachment of the front suspension bell cranks. The bulkhead at the front, which supports the steering gear and holds the front arms of the wishbones, is milled from a solid aluminium plate. The mounting of the bell cranks is done according to the principle of Fig. 6.42. See also the sectional view of the axle journal for the right bell crank (arrow)

Fig. 6.56 Large-scale connection of a handlebar to the frame (Lotus 49 R6, 1970). The suspension of the right front wheel can be seen. On this vehicle, the upper wishbone also transmits bending moments to the inner body spring. Therefore, its bearing is dimensioned accordingly strong and riveted to the frame over a large area by a wedge-shaped plate

Fig. 6.57 Box frame of a monoposto (Opel Lotus)

Fig. 6.58 Box frame of a production sports car (Osella). The front part of the bodywork is removed, therefore the frame is easily visible. The box structure is made of thin sheet metal. The engine is bolted to the rear wall of the cockpit and additionally connected on the gearbox side with struts to the rollover structure (This can be seen in the Racing Car Technology Manual Vol. 3 *Powertrain* Chap. 5 in Fig. 5.122). At the front of the nose is the crash structure. The gear lever is on the right outside the cockpit. The driver reaches through the oval opening in the side panel to shift gears. A lifting eye can be seen at the front right

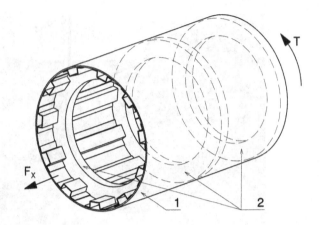

Fig. 6.59 Shell construction schematic. The frame consists of a shell that can transmit normal and shear forces. The bodywork (cladding) is co-supporting. **1** Bodywork (with longitudinal stringers). **2** (Transverse) bulkheads resp. partition walls

6.2.3 Composite Monocoques

The monocoque construction is currently the most integrated form of frame. The relatively thin-walled structure, built in shell form, transmits all forces and moments. A monocoque comprises the cockpit and connection points to neighbouring assemblies in one piece. It thus also represents the entire load-bearing structure of the vehicle and has a mass of only about 45 kg for a single-seater.

With great design freedom, wall thicknesses and fibre directions can be varied and open or closed frame structures can be realised. The weight saving compared to comparable aluminium designs can be up to 40%, with steel this value can reach 60% [14].

Principle
A monocoque consists of "one" shell, which absorbs all loads, Fig. 6.59. The bodywork is also integrated into the overall structure in a load-bearing manner and is usually constructed as a sandwich structure. Transverse ribs determine the cross-sectional shape and stiffen the structure primarily at points where forces have to be introduced at specific points, for chassis connections or for engine mounting.

As always, there are not only advantages to this design method, but also counter-arguments that allow other frame types to exist. In the highest leagues of European motorsport, however, this design dominates, unless the regulations expressly forbid it.

Advantages

- highest integration of functions in the frame possible
- at least partially no bodywork required
- great design freedom
- high reproducibility
- low weight.

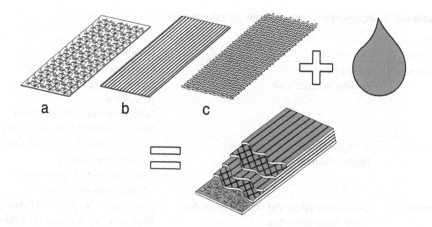

Fig. 6.60 Layout of thin-walled fibre-composites laminates. From (different) fibre mats (**a, b, c**) plus resin with hardener a layer composite is formed. (**a**) Mat. (**b**) roving. (**c**) Woven fabric

Disadvantages

- Complex production, which can only be carried out by a few specialist companies
- Repairs difficult
- Relatively low temperature resistance
- No subsequent concept changes possible due to integral design
- Screw-on points must be specified in the concept and cannot be changed subsequently
- Expensive.

Fibre Composite Laminated Material

Monocoques and many other components (bodywork, wings, air control elements, brackets,...) are built up from fibre composites in layers (laminates). A composite consists of several different materials that are combined with each other in a material-bonding manner. Basically, the structure of a layered composite is as follows, Fig. 6.60.

Several layers of thin, flexible fiber mats are embedded with a filler (matrix) in a layer composite and thus obtain the desired dimensional stability. The mats can be produced by processing the fibers in different ways. Short fibres pressed together form mats (a). If the fibres of the mat are aligned parallel in one direction (unidirectional), it is called a roving. Finally, the individual fibres can be woven into a fabric (e.g. by twill weave like a textile fabric). The fibres can be made of glass, Kevlar®, aramid and carbon. The plastic matrix is usually formed by epoxy resins, polyethers and also carbon (→ CFC, carbon-reinforced carbon). The epoxy resins used for CFRP can usually withstand temperatures of approx. 130 °C. Above this temperature, they soften and the composite loses its strength. Above this temperature, they soften and the composite loses its rigidity. If higher temperatures have to be endured (e.g. in the vicinity of the engine, exhaust system, brakes), resins for continuous temperatures above 180 °C are also available [15]. Due to continuous

Table 6.3 Comparison of common resins for laminates, [17]

Resin system	Advantages	Disadvantages
Polyester	Easy to use Lowest material costs	Comparatively poor strength Strong styrene emission during processing Strong shrinkage during curing Short processing times (open time)
Vinyl ester	Very good chemical resistance Higher strength than polyester	Annealing required for high strength High styrene content More expensive than polyester Curing shrinkage
Epoxi	Good mechanical and thermal properties Good water resistance Very long processing times possible Heat resistant up to 140 °C (wet) or 220 °C (dry) Low shrinkage during curing	More expensive than vinyl ester High accuracy required for mixing/dosing Critical for skin contact

development, maximum values of up to 400 °C are now possible. These resins are cured at lower temperatures and reach the stated values due to the higher service temperatures [16]. A common resin for monocoques is Cytec 950-1.

In general, the following properties will be considered when choosing a matrix material: Adhesiveness, mechanical properties (shear/tensile strength, Young's modulus, elongation at break, shrinkage on curing), resistance to micro-fractures, fatigue behaviour and behaviour on exposure to moisture. An overview of the most common resins is given in Table 6.3.

The rough sequence of manufacturing steps for CFRP components, which are carried out by completely different companies, is shown in Fig. 6.61. First, the fibres have to be produced. They form the starting basis for fabrics. Resins must be available to impregnate the fabrics. Moulds can be made directly from the solid or they can be made by moulding a full-scale model. For the final production of structural parts from composites, cores are also needed, which must be pre-machined at the joints and in the area of inserts.

Principle Fibre Reinforced Polymers
The advantage of fibers is that their orientation can be adapted to the stress curve in the sense of lightweight design. In this way, stiff, organic load-bearing structures can be created with comparatively little material input – even if this still involves some time-consuming manual work. This is demonstrated by an illustrative example, Fig. 6.62. A spatial shear field girder is replaced by a frame element made of fibre composite. The frame element integrally combines the individual elements of the shear panel girder in a shell construction. Fiber orientations can be clearly assigned to these individual elements of the

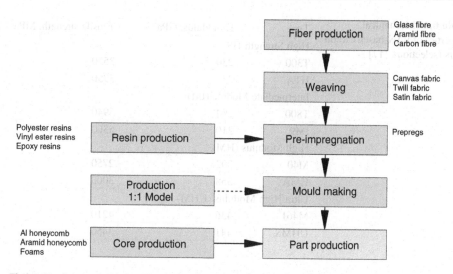

Fig. 6.61 Sequence of manufacture of CFRP components (schematic)

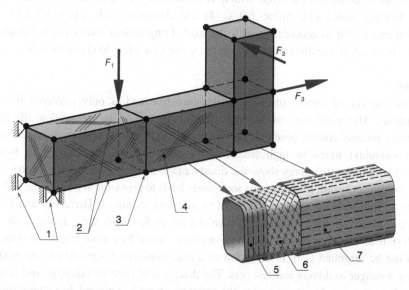

Fig. 6.62 Principle of a fibre-reinforced support structure, after [18]. A frame element (right) made of fiber composite material is derived from a shear field girder (left), which is loaded by individual forces F_1, F_2 and F_3. Due to the spatial arrangement of the forces, bending and torsional moments are generated in the structure in addition to longitudinal and transverse forces. 1 Support, 2 Flanges, 3 Nodes, 4 Structural panel, 5 Ply with fibres oriented for force application, 6 Ply for shear and torsion, 7 Ply for tension/compression and bending tension/compression and bending

Table 6.4 Mechanical properties of PAN-based carbon fibers (selection), [17]

Type	E-modulus, GPa	Tensile strength, MPa
High Strength HS		
T300	230	3530
HTA	238	3950
Intermediate Module (IM)		
T800	294	5940
IM9	310	5300
High Modulus (HM)		
M40	392	2750
HMA	358	3000
Ultra High Modulus (UHM)		
M46J	436	4210
UHMS	441	3450

shear field girder, essentially nodes, chords and shear fields. The load introduction (5) is carried out by unidirectional layers with fibre orientation analogous to the bars between the load bearing nodes (3). Shear fields (4) are designed with layers of ±45° fibre reinforcements (6) in accordance with the load. Longitudinal forces are transmitted by plies with axial (0°) reinforcements (7) in the sense of a direct load conduction.

Fibres

The production of carbon fibres is energy-intensive and is only mastered by a few companies. The most common starting material (precursor) is polyacrylonitrile (PAN). Different process control (combustion conditions, oxidation temperature) results in stiff (high modulus) fibres or high tensile fibres with numerous gradations in between, Table 6.4. Highly stiff fibres show the disadvantage of impact sensitivity. In the case of monocoques and crash elements, this sometimes leads to problems in passing the FIA crash test or to irreparable damage in the event of mechanical impact. Therefore, tensile fibers with comparatively lower modulus of elasticity are preferred for these applications.

Despite all the strength and weight advantages offered by carbon fibers, the downsides must not be forgotten [19]. In the event of a fire, respirable fragments can be produced, posing a danger to drivers and rescuers. The detection of (internal) damage and fatigue is difficult, as is repair. In extreme cases, the entire part must be replaced. For long-term use, a separate damage monitoring system is required. For series production vehicles, there are additional aspects such as poor performance in life cycle analysis and increased requirements for rescue personnel when recovering vehicle occupants.

Fabric

In textile production, numerous ways of interlacing (weaving) threads have been developed. This know-how is used for semi-finished products of fibre-reinforced parts. In a loom, threads are stretched parallel longitudinally in the direction of production (*warp*) and

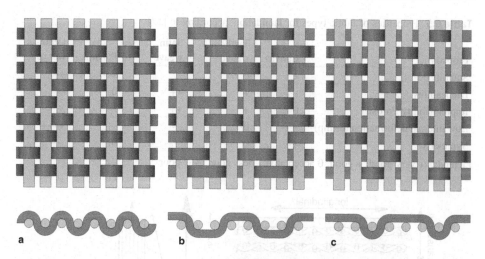

Fig. 6.63 Common weave styles Weave styles. (**a**) Plain weave, (**b**) Twill weave, (**c**) Satin weave warp threads grey, weft threads brown

a thread with the "shuttle" is weaved across the warp threads (*weft*). In the process, the warp threads are crossed alternately in a specific sequence. This results in the so-called weave. The best-known weaves are illustrated in Fig. 6.63. The simplest weave is the plain weave (plain weave, a). Warp and weft threads cross each other regularly. The weave is symmetrical and dense. It is abrasion resistant, can be pushed well but hardly laid in three-dimensional curved shapes: Drapeability is thus poor. The strong deflection of the threads (crimp) also weakens them. Crimping creates additional compressive and bending stresses in the thread. A straight thread can bear the greatest tensile stress. Unidirectional (UD) fibres in a UD scrim are used in highly stressed areas. In twill weave (Fig. 6.63b), a weft thread runs alternately over two or more warp threads. This reduces the crimp and increases the strength of the fabric. However, this is at the expense of the pushability. On the other hand, the drapability is better than with the linen weave. It is therefore suitable for complex shapes. In addition, a twill weave is easier to impregnate with resin than the denser plain weave. The fabric shows the typical herringbone pattern. The atlas weave (Fig. 6.63c) is an asymmetrical variation of the twill weave. The weft thread is passed alternately over three or more warp threads before coming to rest again under a warp thread. The number of strength-reducing crossing points is even lower than with twill weave, but the slippage suffers. The flat fabric can be soaked well and impresses with its excellent draping properties.

Other types of fabric that can be used for laminates are the Panama fabric and the leno weave. Finally, Table 6.5 provides an overview of the properties of fabrics that are decisive for laminating.

A layered composite has different properties to the individual materials and is anisotropic, i.e. the properties are directional, Fig. 6.64.

Table 6.5 Properties of fabric types for lamination, according to [17]

Property	Canvas	Twill weave	Satin weave	Panama	Leno weave
Stability	+	0	-	–	++
Drapeability	–	+	++	0	– –
Impregnability Porosity	0	+	++	–	– –
Low internal deformation (crimp)	–	0	++	–	–/++
Symmetry	++	0	– –	0	– –

Legend: ++ very good, + good, 0 acceptable, – less good, – – poor

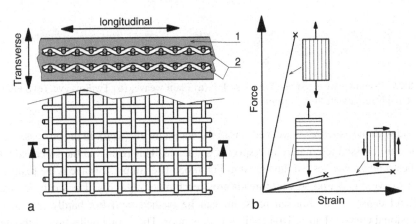

Fig. 6.64 Anisotropy of a layer composite. (**a**) Structure of the layers. 1 matrix. 2 fabric layers. (**b**) Strain behaviour of parallel fibre-reinforced composites: highly rigid in fibre direction, compliant transverse to fibre direction and compliant, non-linear in shear loading

The unidirectional parallel fiber directions most commonly used in high-performance structures exhibit pronounced, different stiffnesses and strains in the fiber direction and transversely. In the fiber direction the fibers determine the tensile strength and transversely only the matrix. In the fiber direction, the composites are highly rigid with a low elongation at break. Transversely to the fiber orientation, these materials exhibit much lower stiffness, but greater elongation at break. When fracture occurs, the layers are locally separated again (delaminated) by failure of the matrix. In both of the above directions, the force-strain relationships are nearly linear. The situation is different for shear stress. In both main directions, the bond behaves strongly non-linear.

Because of this directional dependence of the material properties, it is crucial to place the fibres in the component in such a way that they are mainly subjected to tensile stress. This is not always possible, so several layers with different fibre configurations are usually combined, Fig. 6.65.

Symmetrical ply structure keeps the distortion small during curing under temperature supply (a). The layer structure in the picture is 45°-fabric, 90°-fabric, 90°-fabric and 45°-fabric. If smaller forces are introduced locally at certain points, for example by a

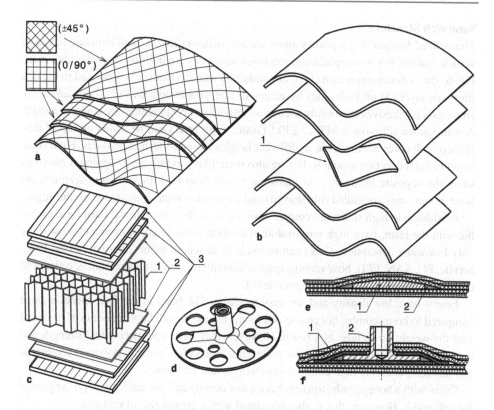

Fig. 6.65 Examples of the layer structure of a fibre composite structure. (**a**) Layers are laid symmetrically to reduce component distortion during curing: 45° – 90° – 90° – 45°. (**b**) Local reinforcement (1). (**c**) Sandwich structure: 1 Honeycomb core. 2 Honeycomb adhesive. 3 fibre layers. (**d**) Force introduction element. (**e**) Laminated plate (1, Hardpoint) with modelled ramp (2). (**f**) Force introduction element (2) laminated with reinforcing plaster (1)

fastening for the bodywork, reinforcing layers are placed locally between them (b). For larger forces, such as for bolting brackets, metal inserts are used, so-called force introduction elements (d). These can consist of a simple plate that is drilled out after curing (e) or of elements that already carry a spigot with an internal or external thread (d). For more details, see Figs. 6.79, 6.80, and 6.81. Even if aluminium offers a weight advantage, these elements should be made of stainless steel or titanium alloys. With other metals, there is a risk of electrochemical corrosion when combined with carbon fibre mats [21]. A ramp of thickened resin or structural adhesive is modeled at the transition points from the insert to the plies, this mitigates this jump in stiffness, Fig. 6.65e (2). Reinforcements for bolted joints can also be represented by laminating prefabricated thick CFRP sheets (approx. 10–15 mm).

Sandwich Structure

Thin-walled laminated composites alone are not sufficient for the shell structure of racing vehicle frames. For this application, sandwich structures are built up from the fibre mats, Fig. 6.65c. A honeycomb core (1) of aluminium or aramid (Nomex®) is bonded to the first fibre mat layer (3) on both sides by means of honeycomb adhesive (2) (*core adhesive*). Honeycomb adhesives are usually used as adhesive films because they are easier to handle. A well-known adhesive is MTA® 240.[2] Further fibre layers can be added on top of this. Honeycomb cores are available in different heights as semi-finished products. High cores result in higher section modulus, but are also more likely to fail due to tipping. Low cores show the opposite behaviour. At transition points from high to low cross-sections, the honeycomb cores are milled off (shaved) and a transition ramp is modelled on the edge.

In addition to high shear and compressive stiffness, the core material must be compatible with the resin, have high temperature resistance in the intended area and must show only low water absorption. Cores can be made of structural foams (thermoplastics: PVC, acrylic, PU, SAN, PEI), honeycombs (paper, aluminium, aramid), wood (balsa, cedar) and other sheet materials (Coretex®, Spheretex®).

Foams have low density and are easy to shape. The bond to the face sheets is large compared to honeycombs, but coarse-cell foams tend to absorb large amounts of resin and lose the weight advantage as a result. However, they have low mechanical strength and tend to outgas. A foam known by the trade name Rohacell is made of polymethacrylimide and has comparatively high compressive strength and stiffness.

Cores with a honeycomb structure have a low density and are easier to drape the greater the cell width. However, this is also associated with a greater risk of collapse.

Aluminium honeycomb cores have a high shear strength, good impact resistance and good drapeability. On the other hand, they are sensitive to corrosion (humidity and salt). The formation of a corrosive local element is further enhanced by the electrical conductivity of the carbon fibers. An insulating layer, e.g. of glass fibers, is recommended in any case.

Cores made of aramid honeycomb are characterized by good drapability, but are mostly used for non-load-bearing areas. They have a lower density than foam cores, but are considerably more expensive.

The height of the cores and thus the thickness of the sandwich depend on the area of application. Figure 6.66 shows an example of the locations of different cores on a monocoque. The number of layers is also not the same at every point and varies – as does the fibre orientation – depending on the expected stress at the point.

The monocoque of a Formula Renault car consists of an aluminium honeycomb core 20 mm thick. The two cover layers of CFRP are each a total of 1.5 mm thick [22]. In the hand lay-up process of laminates, the weight fraction of the (actually load bearing) fibers is about 30–35% of the total weight [16]. So the rest is the resin forming the matrix, whose

---

[2] See https://www.cytec.com/sites/default/files/datasheets/MTA240.pdf.

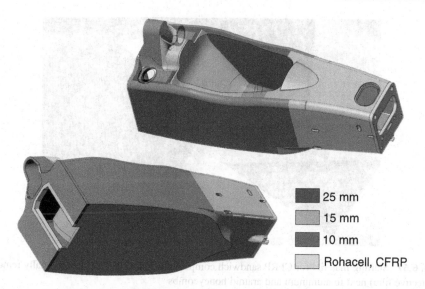

25 mm
15 mm
10 mm
Rohacell, CFRP

Fig. 6.66 Core occupancy on a monocoque by honeycomb height. A monocoque is shown translucent in two views so that the essential cores are visible. The color distinction is made according to the height of the honeycombs. The frames at the front and rear are partly made of foam cores and partly of solid CFRP sheets (yellow). On the one hand, the side panels are the main areas that have to bear the loads in the event of a side impact and, on the other hand, their stiffness is particularly important so that the torsional stiffness in the open cockpit area does not collapse too much compared to the closed cross-sections. They therefore have the highest cores

strength is much lower. For racing vehicles, the plies are therefore built up with so-called prepregs. Prepregs are mats that have already been *pre-impregnated* with resin in the desired ratio and can be cured under pressure and temperature in a pressure chamber (autoclave) without any further additives. The prepregs are stored frozen at −15 to −18 °C so that the resin remains in the fabric. Because no resin is added manually with this technique, the weight percentage of the fibres in the finished composite increases to 65% and more. This increases the strength and reduces the weight compared to structures produced by hand lay-up. The designation of prepregs is determined by their components and properties. An example: GG 200 T (T800) – DT120 – 42

GG	Internal company code
200	Basis weight, g/m^2
T	Type of fabric, here twill (twill weave), see also Fig. 6.63
(T800)	Fibre designation, see Table 6.4
DT120	Resin system, here: 120 °C curing
42	Resin content in % of the mass per unit area

Fig. 6.67 Starting material for CFRP sandwich components. Prepregs (left, with partially removed protective film) next to aluminum and aramid honeycombs

In racing, carbon fibre prepregs are primarily used for frames. The composite material thus produced is a *carbon fibre reinforced plastic* (*CFRP*). The starting semi-finished products for a monocoque are shown in Fig. 6.67.

CFRP structures also prove advantageous in absorbing crash energies. Drivers have already been able to survive accidents above 200 km/h without major injuries. Approximately 12 monocoques per car were used in Formula 1 per year as long as the rules did not limit the count.

Moulds

The molds for manufacturing the laminates are milled from aluminum, especially for smaller components. A disadvantage is the strongly differing thermal expansion behaviour of aluminium and CFRP, which worsens the dimensional stability, especially for large components. After the autoclave process, the molds should be heated in a tempering furnace and the components removed from the mold while still warm. Graphite moulds are an alternative. This material is easy to machine and exhibits similar thermal expansion behavior to CFRP, although it has lower strength. For large, dimensionally accurate parts, the mold itself is made by hand from CFRP. This provides a stable mold with the same thermal expansion as the part. This method is used for monocoques and is described in the following section. To make the CFRP mould, a positive mould is first required from which the negative CFRP mould can be moulded. The positive mold is milled, for example, from Necuron® 651.[3]

---

[3] Polyurethane-based plastic. For data sheet and further information see: www.necumer.de.

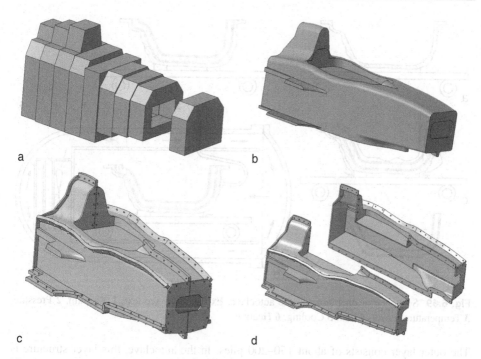

Fig. 6.68 Creation of negative moulds

Manufacture

The manufacturing process of a monocoque starts with the creation of a positive model (*buck*), similar to a sand casting. This reproduces the outer shape of the later part with all its nuances, which is why high demands are placed on the surface quality of this model, Fig. 6.68b. The model is usually milled from a block of plastic or wood fibre boards glued together based on the CAD data of the component, Fig. 6.68a. From this model, manual processes (hand lay-up, resin infusion, etc.) are used to create the negative moulds (*female mould*). This can be a single mould or two mould halves can be removed, with which two half-shells are first created, which then have to be bonded together to form the final monocoque. Depending on the complexity of the model, several parts are required for each mould half, which are bolted together using flanges, Fig. 6.68c. The negative mold must be removable without destroying the model, i.e. it must not have any undercuts. In the case of a modern monocoque, six parts are required for two mould halves (d).

The actual production process of a chassis begins with these moulds, Fig. 6.69. The individual layers are inserted into the mould or mould halves (a), which are coated with release agent (b). The required prepregs are usually cut by CNC using laser cutters or by hand. In any case, the individual plies must have exactly the fiber direction and sequence intended by the design, as specified in the *ply book*. The first ply, which is placed in the mold, deserves special care because it forms the visible surface of the finished component.

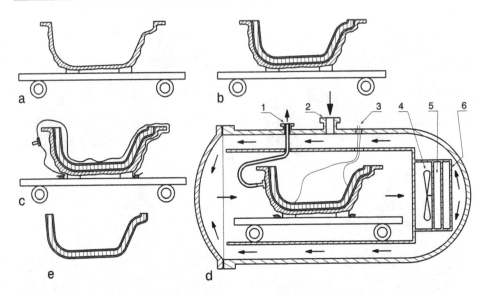

Fig. 6.69 Sequence of operations in the autoclave. Explanations see text. **1** Vacuum. **2** Pressure. **3** Temperature sensor. **4** Fan. **5** Cooling. **6** Heating

The outer layer consists of about 150–200 plies. In the autoclave, this layer structure is pressed together to form a composite. The production of a monocoque involves a great deal of manual work, which makes the quality of the result highly dependent on the person carrying out the work, and at least eight weeks must be allowed when planning for the production of a raw monocoque in two-shift operation.

The basic structure of a monocoque is a sandwich structure, i.e. the honeycomb cores are placed on the outer layers and inserts are inserted at the intended screw connection points. The structure is then covered with the inner layers.

Treatment in the autoclave takes place at several stages of the production process and is basically always as follows. The layer structure in the mould is covered with a porous release film and covered with an absorbent fabric. A vacuum film is placed over all this and sealed to the base plate (Fig. 6.69c). The vacuum bag thus formed is evacuated. This promotes degassing of trapped air in the layered structure and atmospheric pressure compresses the layers. Excess resin is absorbed by the suction fabric. Such a lay-up is shown in detail in Fig. 6.70. This lay-up is placed in the autoclave (Fig. 6.69d). An autoclave is a vessel in which pressure, vacuum and temperature can be set and changed separately. Sensors in the layer assembly monitor and document the progress of these variables. A typical autoclave cycle for the production of CFRP components is shown in Fig. 6.71.

In the autoclave, in addition to the atmospheric pressure, overpressure acts on the component and mold and the compression of the fiber layers is further increased. The maximum pressures vary: the outer layer alone is compressed at 7 bar, the entire sandwich structure at only about 3 bar, so that the honeycomb core does not collapse. Both the

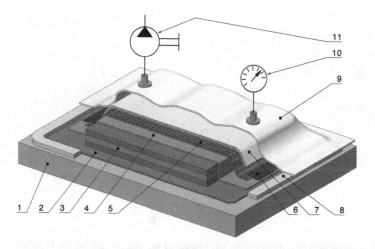

Fig. 6.70 Detail of the setup for the vacuum process in cross-section. 1 mold resp. bed plate, 2 PTFE release layer, 3 laminate, 4 peel ply, 5 porous release film, 6 flash tape, 7 absorber mesh (breather/bleed mat), 8 sealant tape, 9 vacuum bag, 10 vacuum gauge, 11 vacuum pump

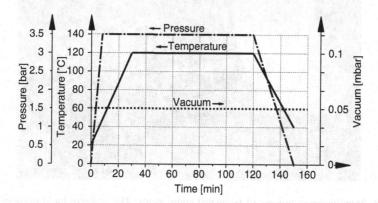

Fig. 6.71 Typical autoclave cycle for CFRP parts, after [23]. The predefined ramps of pressure, temperature and vacuum are controlled by computer in the autoclave and documented by measuring sensors

component and the mold are subjected to the same pressure. Therefore, the mold only has to be inherently rigid but not pressure-resistant.

After the last hardening cycle in the autoclave, the finished component can be removed from the mould (Fig. 6.69e). If the chassis consists of two parts, these are then bonded together, Fig. 6.72a. There are also manufacturers who bond and cure both halves of the mould in the autoclave at the same time, which further increases the strength of the entire frame. For some manufacturers, however, bonding represents such a great fundamental risk that they prefer a one-piece construction of the chassis. They consciously accept the increased production effort this entails. Now the monocoque can be machined. Mounting

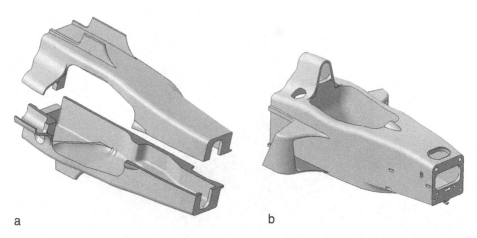

a   b

Fig. 6.72 Completion of a monocoque. (**a**) Gluing of the chassis halves. (**b**) Finished monocoque

Fig. 6.73 Manufacturing process of laminated components. The negative mold made of CFRP is laminated in a wood/plastic positive mold (left). After curing, this can be removed from the mother mold (center) and in turn serves again as a negative mold for laminating the final workpiece (right). In the picture you can see the "beautiful" (smooth) surfaces of the parts

and bolting points for chassis and engine are drilled or milled at the points prepared with inserts and reinforcement layers (Fig. 6.72b).

The most important stages in the manufacture of thin-walled components are illustrated in Fig. 6.73 using the example of a seat shell.

Monocoques may also receive separate frames to stiffen the shell structure, as described in Fig. 6.74.

Bulkheads may also be directly incorporated into the design of the overall structure to form part of the rollover structure. Figure 6.75 shows such a solution with ring-shaped bulkheads for a production sports car. The central bulkhead (2) thus forms a roll bar and the A-pillar at the same time.

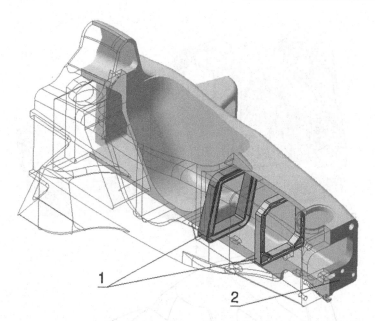

Fig. 6.74 Arrangement of frames in a monocoque. Two bulkheads (1) and one bulkhead (2) stiffen the front area of the cockpit. The frames are made of aluminium or CFRP laminate. The wheel forces are introduced into the frames via the wishbone mountings

Finally, some details of manufacturing and design rules for components made of fibre-reinforced plastics follow.

Ideally, a layer of one piece of fabric can be laid. This avoids interruptions of the load flow in the fibers. However, this will not be possible for more complex shapes. Nevertheless, one will strive to require as few individual pieces of fabric as possible. So that the strength of the laminate is not affected too much, fabric layers are overlapped by at least 15 mm at joints. If several butt joints meet at one point, they should overlap by at least 20 mm and be staggered, Fig. 6.76. In the case of plies of twill weave and 45° fibre orientation, many manufacturers attach importance to the herringbone pattern of the first ply being symmetrical about the pitch.

A possible, complete assembly of the main layers and cores for a sandwich structure in a two-part mould is shown in cross-section in Fig. 6.77. Such a cross-section occurs, for example, in the nose area of a monocoque for a formula car and presupposes for production that there is access from at least one side for covering the closed mould. In the case under consideration, the mold can be loaded through the cockpit opening and from the front through the nose connection (closing bulkhead).

At abutting points of cores – at edges or in small bends which a honeycomb core cannot follow without deformation – a suitably machined core of structural foam is placed, Fig. 6.78.

Some generally applicable design rules for fibre-reinforced components are summarised in Fig. 6.79.

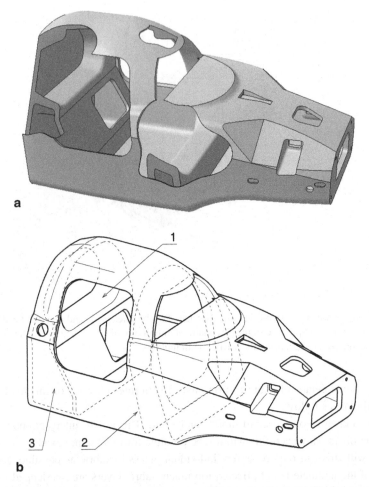

a

b

Fig. 6.75 CFRP monocoque of a sports prototype with closed cockpit. (**a**) Shaded representation, partially sectioned. (**b**) Axonometric representation. Two annular bulkheads form the rollover structure. **1** Bulkhead aft. **2** Bulkhead forward. **3** Fuel tank area

In principle, the introduction of concentrated loads (e.g. bolted connections) in a sandwich structure can be carried out in the same way as shown for panels in Sect. 2.2. In contrast to this semi-finished product, however, local force application elements can also be installed directly in a sandwich structure produced using the lay-up method, Fig. 6.80. In [20], these elements were investigated numerically, whereby a force was applied upwards in the axial direction of the metal sleeve. Variant (a) turned out to be unfavourable. Strong excess stresses occur in the contact area between the cover layer and the side of the sleeve. Version (c) has the advantage that the screw connection does not overhang the underside. The most favourable variant is (b). The large-surface dome transfers the load to the upper cover layer without any jumps in stiffness. A washer on the lower cover layer ensures the advantageous distribution of the concentrated force.

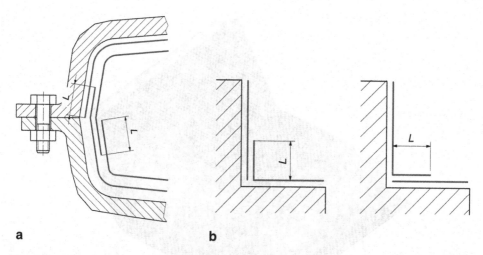

a   **b**

Fig. 6.76 Design of overlaps of the first layers. (**a**) In a two-part mould, several joints of fabric layers meet. The first layer (red) is particularly important because it is the visible layer on the finished part. The following layer (green) is inserted according to the sketch. The overlaps should be min. $L = 20$ mm and must be staggered. (**b**) Overlap at angle joints

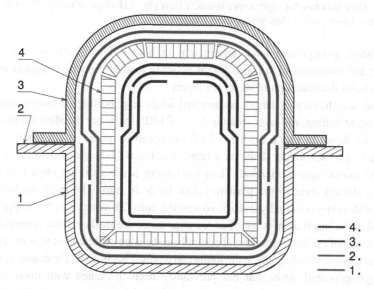

Fig. 6.77 Cross-section through a complete layer structure (schematic). 1 lower mould half, 2 parting plane, 3 upper mould half, 4 honeycomb core. The individual main layers are color-coded according to the order in which they are inserted. In addition, there are reinforcing layers in the cockpit area as well as reinforcing plasters for inserts and screw joints

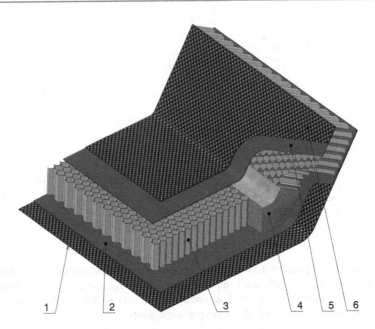

Fig. 6.78 Core structure for tight corner bends. 1 First ply, 2 Honeycomb core adhesive, 3 honeycomb core, 4 Foam core, 5 Honeycomb core adhesive, 6 top ply

In general, a large-area load introduction in both(!) face sheets is recommended. Likewise, the avoidance of stiffness jumps between load introduction element and cover layers prevents delamination of the cover layers.

For the introduction of large concentrated loads (e.g. bolting of the restraining belt mount, engine bolting, roll-over structures, . . .) CFRP inserts are installed in the sandwich structure as shown in Fig. 6.81. The CFRP insert can be water jet cut from a plate and reworked to a frustoconical shape on a lathe. The insert (3) is glued onto the first layers (1) during monocoque fabrication. Then reinforcing patches (5) are placed on top. The following plasters overlap the plasters below by 5–10 mm. The entire surface is then covered with honeycomb adhesive (6) so that the core (2), which has been appropriately machined in the insert area, is later bonded over the entire surface in the autoclave. In the processed area of the honeycombs and basically at the edge of the core, the honeycombs are reinforced to prevent collapsing at the joints and edges. This is done, for example, by filling the honeycombs with resin that has previously been thickened with micro-balloons[4]. Before the top layers (8) are applied, the area around the hole is covered with reinforcing plasters (7). Top layers and reinforcing patches are made of the same fabric (e.g. 300 T (T800)), but differ in fiber orientation.

---

[4] Also called micro-spheres: Tiny hollow spheres, usually made of glass. Diameters range from a few μm to 5 mm, depending on the application.

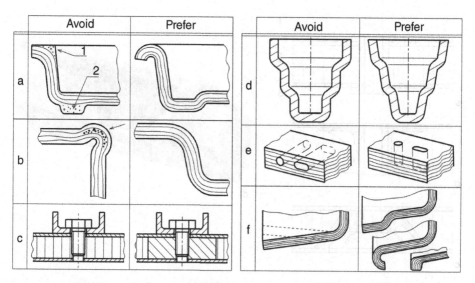

	Avoid	Prefer		Avoid	Prefer
a			d		
b			e		
c			f		

Fig. 6.79 Design rules for fibre-reinforced plastics. (**a**) The strength is determined by the fibres in the direction of the thread. The resin is merely a binding material and is brittle in relation to the fibres when bent. Resin accumulation due to cross-sectional jumps (1) or tight radii and indentations (2), which the fiber mats do not follow due to their inherent stiffness (so-called *bridging*), must therefore be avoided. (**b**) At edges and corners, the fiber mat can break during lamination if the radius is too small and the bending angle too large. Large radii and smooth transitions prevent this. (**c**) If the components are to have high bending stiffness, sandwich structures must be built over the tensile stiff layers with spacing, lightweight cores. Inserts made of metal, wood or CFRP sheets must be laminated into the structure at the points where screw connections are made in the finished component. See also Figs. 6.32, 6.37, and 6.38 in the Sandwich Plates section and below. (**d**) Demoulding slopes allow non-destructive release of the component from the mould. The slopes should be at least between 1:25 and 1:100, depending on the depth of the component and the manufacturing process. (**e**) Holes and milled-out areas should be drilled transversely to the layers, otherwise there is a risk of delamination. f: Large-area component edges should be stiffened. This is done with appropriate shaping or local reinforcements

Frames made of metal or CFRP plates as well as roll bars can also be laminated directly, Fig. 6.82. The metal tube (5) of the roll bar is sandblasted and covered with honeycomb adhesive (6) before installation. The transition between the tube and the honeycomb cores is filled with appropriately machined foam cores (4). The honeycomb cores are stabilized in the machined area with micro-balloon thickened resin (7) and the remaining cavities are filled with swelling resins (e.g. Amspand)[5] during the autoclave cycle.

Photos from the production of a monocoque will illustrate some of the steps and details described above, Figs. 6.83 and 6.84.

In the following, Figs. 6.85, 6.86, and 6.87 show some examples of executed CFRP monocoques.

---

[5] See https://www.tencatecomposites.com/product-explorer/products/tq9N/Amspand-ES72A-2.

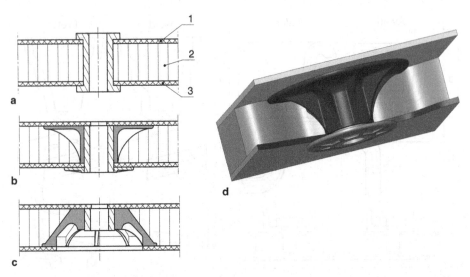

Fig. 6.80 Load introduction in sandwich construction, after [20]. Three different concepts are presented that allow bolting in a sandwich structure The sandwich structure consists of: 1 upper cover layer(s), 2 (foam) core, 3 lower cover layer(s). The load introduction elements made of long-fibre reinforced plastic are shown shaded grey in the section. Production begins with the top cover layer, followed by application of the load introduction element, insertion of the core and application of the cover layers. The area around the load introduction elements is finally filled with foam. (**a**) metal sleeve, (**b**) metal sleeve in cone-shaped load introduction element with cover plate, (**c**) metal sleeve in funnel-shaped element, (**d**) axonometric representation of (**b**) (core and lower cover layer partially removed for visibility reasons)

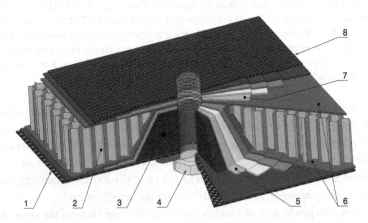

Fig. 6.81 Sandwich structure with insert for load bolting. 1 First ply, 2 Honeycomb core, 3 Insert, 4 Bolt with washer, 5 First reinforcement patch, 6 core adhesive, 7 Fourth reinforcement patch, 8 Topplies

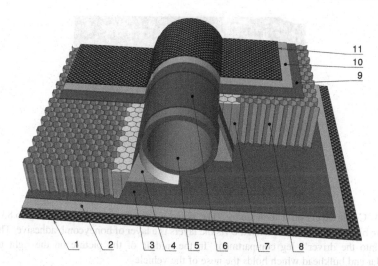

Fig. 6.82 Laminated pipe in sandwich structure. A metal tube is laminated between two honeycomb cores of different heights of a sandwich structure. In the case of a monocoque, for example, this is the area of the front rollover structure, which separates the cockpit area from the driver's leg compartment. 1 First ply, 2 Second ply, 3 Core adhesive, 4 Foam core, 5 Roll-over hoop, 6 Core adhesive, 7 Resin fill, 8 Honeycomb core, 9 Honeycomb adhesive, 10 Third ply, 11 Top ply

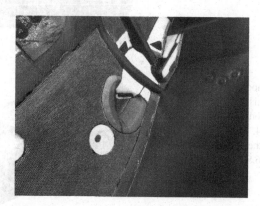

Fig. 6.83 Core loading on a monocoque. The picture was taken during production. The view is down through the cockpit opening. The brown aramid honeycombs form the majority of the (left) sidewall. These are interrupted by white and grey foam cores only at bolted joints to accommodate inserts. The top image area shows a tubular laminated roll bar and rectangular stiffener. The right image area shows the bottom of the monocoque. The next steps will be the filling of the core edges and the covering with honeycomb glue

Fig. 6.84 Cover layers on a monocoque. The photo shows the monocoque of Fig. 6.83 after the cores have been covered. Underneath the visible layers is a layer of honeycomb adhesive. The view is forward into the driver's leg compartment. In the middle of the picture on the right there is a rectangular end bulkhead which holds the nose of the vehicle

Fig. 6.85 CFRP monocoque (Dallara). Such monocoques are used in the IRL

Fig. 6.86 CFRP monocoque (Formula Renault). This monocoque has a structure with formers as in Fig. 6.74. The engine connection at the rear side is shown in Fig. 6.87

Fig. 6.87 Engine connection of a CFRP monocoque (Formula Renault). The picture shows the rear view of the monocoque of Fig. 6.86. The engine is bolted to the rear wall of the monocoque by means of cross members in the cylinder head area and directly in the oil pan area via the laminated inserts (This can be seen in the Racing Car Technology Manual Vol. 3 *Powertrain*, Chap. 1 Fig. 1.157). Individual openings for shift linkage, fuel lines, electrical lines and throttle cable can also be seen. In addition, the glue joint between the upper and lower part of the chassis can be seen

6.3 Strength

The frame must have a high degree of strength. It literally holds the vehicle together and forms the survival cell for the driver. Not least for this reason, the vehicle body is subjected to various tests prescribed by the individual regulations. For more details, see Sect. 3.5 Tests.

In addition to strength, low deformation under load is also important. Of particular interest in a frame is the torsional stiffness between the axles, or more precisely between the points of application of the individual vertical forces of the wheels. The frame should be as stiff as possible in relation to a torsional moment about the longitudinal axis, so that the tyres take up their position in relation to the road surface as planned by the chassis designer and so that torsional stabilisers can shift wheel loads in the desired way and not distort the frame. Successful racing cars have been reported to have a frame stiffness of only four times the total suspension roll stiffness [24]. This ratio apparently forms a good compromise between mass (too high stiffness means unnecessary weight) and responsiveness (a frame that is too soft leads to sluggish behaviour) of the car. The bending stiffness of frames is considered to be that about the horizontal axis (Y-axis), as that which is decisive for the bending due to the dead weight and wheel loads. Table 6.6 provides a comparison of stiffnesses.

Table 6.6 Stiffness of components or frames of different vehicles

Bending stiffness [25]	N/mm	Torsional stiffness, partly [25]	kN m/°
Motorcycle rim	1000	Motorcycle frame engine removed	0.5–0.7
Telescopic fork motorcycle	100	Motorcycle frame engine installed	1–2
Swingarm motorcycle	400	Car body sedan [26]	12–24
Motorcycle frame engine removed to Z-axis	70–350	Car body hatchback	10–14
Motorcycle frame engine mounted to Z-axis	200–450	Car body station wagon	10–14
		Tubular space frame Formula Student	1–3
		Monocoque Formula Student	2–3
		Roadster [27]	5.6–14.5
		Formula 3, Monocoque	8
		Minivan body	10–12
		Formula Renault Monocoque	10–12
		NASCAR	17
		Toyota racing car [28]	38
		Touring Car [29]	40
		Formula 1, Monocoque	up to 70
		Formula 1, axle to axle [30]	6.5
		LMP 1, Monocoque	100

The torsional stiffness can be determined by holding the wheel carriers of one axle by means of a device and a wheel carrier of the other axle only snaps onto a ball cap, i.e. it can rotate freely when the chassis is twisted.[6] A weight G is now attached to the other wheel carrier (upright) of this axle for loading. The wheel carrier lowers by the value Δz due to the load. The torsional stiffness of the chassis is then:

$$c_{ts} = \frac{M_{ts}}{\beta} = \frac{Gb}{1000 \arctan\left(\frac{\Delta z}{b}\right)}$$

c_{ts} Torsional stiffness, N m/°
M_{ts} Torsional moment, N m
β Torsion angle, °
G Weight, N
b Track width of loaded axle, mm
Δz Lowering, mm

---

[6]For more details, see Racing Car Technology Manual Vol. 5 Data Analysis, Tuning and Development, Chap. 6 *Development*.

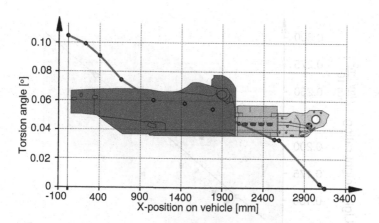

Fig. 6.88 Torsion angle over the vehicle position in longitudinal direction of a Formula 1 car (BMW-Sauber) [31]. The engine (2.4 l V8) is a fully supporting element between the monocoque and the transmission

The torsional stiffness curve of a Fomel 1 car along the longitudinal direction of the vehicle is shown in Fig. 6.88. The bolted monocoque engine-transmission assembly is loaded at the nose with a torsional moment and the torsional angles are recorded at specific points. As expected, the most torsionally stiff area is the protruding area around the cockpit. Towards the slimmer nose the structure becomes similarly "soft" as towards the rear end. The engine-to-monocoque and gearbox-to-engine connection areas do not present any pronounced weak points.

However, simply comparing the stiffness values of different vehicles is not very meaningful. If the mass and size used are also taken into account, a more objective value is obtained. This is done, for example, with the lightweight quality L:

$$L = \frac{c_{ts}A}{m}$$

L Lightweight quality, $(N\,m/°m^2)/(kg\,10^3)$
c_{ts} Torsional stiffness of the frame, $N\,m/°$
A Contact area of the frame, m^2
m Mass of the frame, kg

Figure 6.89 shows the trend in lightweight quality over recent years for passenger car bodies-in-white.

The influence of the torsional stiffness of a vehicle frame on the driving behaviour comes to bear during cornering. A simple substitute model can be used to illustrate the relationships, Fig. 6.90.

When cornering, the centrifugal force twists the frame and this torsional moment M_e is supported by the spring moments $M_{Ro,f}$ and $M_{Ro,r}$ of the two axles. The body is articulated

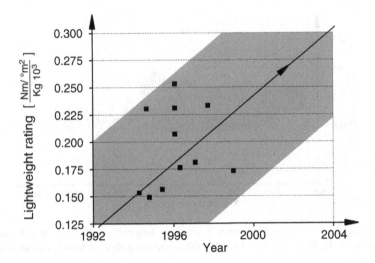

Fig. 6.89 Lightweight quality of car body shells, according to [26]. The diagram shows the trend with scatter band

at the roll centres Ro_f and Ro_r. Via these, the centrifugal force coming from the tyres is introduced into the frame. And vice versa, the corresponding load on the tyres and axles is applied via these forces in the roll centres. This moment M_e to be absorbed by the springs (body springs and stabilizers) (therefore also called the elastic portion of the wheel load displacement) is

$$M_e = m_{Bo} a_y h' + m_{Bo} g \sin(\varphi_t) \approx m_{Bo} a_y h' + m_{Bo} g \varphi_t \qquad (6.1)$$

m_{Bo} Body mass (sprung mass), kg
a_y Transverse (lateral) acceleration, m/s²
h' Distance between centre of gravity Bo and roll axis, m
φ_t Total roll angle, rad

If the torsion spring pairs acting on both sides of the "lever" are combined to form a total torsion spring rate c_f and c_r respectively, the roll angle can be expressed from known quantities. The two total spring rates act in parallel in relation to the lever and the following applies

$$M_e = (c_f + c_r)\varphi_t \qquad (6.2)$$

$$\Rightarrow \varphi_t = \frac{M_e}{c_f + c_r} = \frac{m_{Bo} a_y h'}{c_f + c_r - m_{Bo} g h'} \qquad (6.3)$$

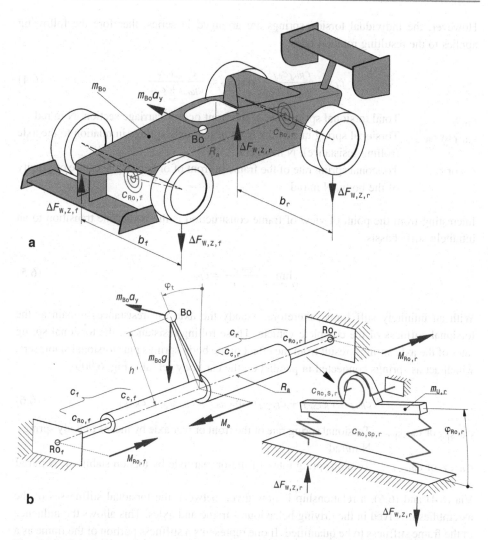

Fig. 6.90 Substitute model for consideration of the torsional stiffness of a vehicle (**a**) Overview representation, (**b**) Replacement model. In this model, only the swaying of the body with mass m_{Bo} about the roll axis R_a is considered at first. The frame (index c) and the axles (index Ro) are represented by torsion springs which are held in place at the outer ends (by the road surface). At the body centre of gravity Bo, the centrifugal force and the weight of the body act during cornering. This torsional moment M_e (elastic part of the wheel load displacement) is balanced by the spring moments of the front and rear axles $M_{Ro,f}$ and $M_{Ro,r}$. These spring moments are generated by body springs and torsion stabilisers acting in parallel (detail shown for rear axle). Bo centre of gravity of the body, R_a roll axle, Ro_f or Ro_r roll centre of the front or rear axle, $m_{U,r}$ axle mass (unsprung mass) at the rear, $c_{Ro,Sp}$ axle-related torsional spring rate of the body springs, $c_{Ro,S}$ axle-related torsional spring rate of the torsion stabiliser, $\Delta F_{W,z,f}$ or $\Delta F_{W,z,r}$ wheel load change at the front or rear, respectively

However, the individual torsion springs are arranged in series, therefore the following applies to the resulting rotation rates

$$c_f = \frac{c_{Ro,f} c_{c,f}}{c_{Ro,f} + c_{c,f}} \quad \text{resp.} c_r = \frac{c_{Ro,r} c_{c,r}}{c_{Ro,r} + c_{c,r}} \tag{6.4}$$

c_f, c_r Total torsional spring rate of the front or rear carriage section, N m/rad

$c_{Ro,f}$ or $c_{Ro,r}$ Torsional spring rate of the front or rear axle springs in relation to the axle (rolling resistance), N m/wheel

$c_{c,f}$ or $c_{c,r}$ Torsional spring rate of the frame in front of or behind the centre of gravity of the body, N m/rad

Interesting from the point of view of frame construction is the borderline transition to an infinitely stiff chassis

$$\lim_{c_c \to \infty} \frac{c_{Ro} c_c}{c_{Ro} + c_c} = c_{Ro} \tag{6.5}$$

With an infinitely stiff frame, therefore, exactly the axle roll resistances remain as the torsional stiffness of the complete vehicle. These rolling resistances, the torsional spring rates of the axle springs, result from the effect of the body springs plus torsional stabilisers, which act as springs connected in parallel on the car body (cf. also Fig. 6.90b):

$$c_{Ro,f} = c_{Ro,Sp,f} + c_{Ro,S,f} \text{resp.} c_{Ro,r} = c_{Ro,Sp,r} + c_{Ro,S,r} \tag{6.6}$$

$c_{Ro,Sp,f}$ or $c_{Ro,Sp,r}$ Torsional spring rate of the front or rear axle by means of body springs, N m/rad

$c_{Ro,S,f}$ or $c_{Ro,S,r}$ Torsion spring rate of front or rear axle by torsion stabiliser, N m/rad

Via (6.4) and (6.6), a relationship is now given between the torsional stiffnesses of the assemblies involved in the driving behaviour – frame and axles. This allows the influence of the frame stiffness to be quantified. If one represents a stiffness portion of the frame as a multiple of the rolling resistance of an axle, this influence becomes evident:

$$c = \frac{c_{Ro} c_c}{c_{Ro} + c_c} = \frac{c_{Ro} k c_{Ro}}{c_{Ro} + k c_{Ro}} = c_{Ro} \frac{k}{1+k} \Rightarrow \frac{c}{c_{Ro}} = \frac{k}{1+k} 100\% \tag{6.7}$$

If $k = 1$ (frame part and axle have the same rotation rate), the total rotation rate c of the considered carriage part reaches 50% of the rolling resistance c_{Ro} of the connected axle. At $k = 5$ (frame part 5 times stiffer) 83% and at $k = 10$ 91% are reached. The measured torsional stiffness c_{ts} of this (for the sake of a simple overview) symmetrical frame (see (6.13)) amounts to

$$c_{c,f} = c_{c,r} = c_c, c_{ts} = \frac{c_c c_c}{c_c + c_c} = \frac{c_c}{2} = \frac{k}{2} c_{Ro} \tag{6.8}$$

Thus, if you want a frame whose front and rear parts are 10 times more torsionally stiff than the two axles ($k = 10$), the measured torsional stiffness c_{ts} of the frame must be 5 times the rolling resistance c_{Ro} of one axle.

The total moment balance for the body establishes the relationship for the driving behaviour via the torsional spring behaviour (6.2):

$$M_e = M_{Ro,f} + M_{Ro,r} \tag{6.9}$$

$$M_{Ro,f} = c_f \varphi_t, M_{Ro,r} = c_r \varphi_t \tag{6.10}$$

$M_{Ro,f}$ or $M_{Ro,r}$ Spring torque of front or rear axle, N m.

The spring moments are generated by body springs and stabilizers during roll and, in addition to other influences, cause the wheel load to shift:[7]

$$\Delta F_{W,Z,f} b_f = M_{Ro,f} + M_{g,f} + M_{U,f} \text{ bzw.} \Delta F_{W,Z,r} b_r = M_{Ro,r} + M_{g,r} + M_{U,r} \tag{6.11}$$

$\Delta F_{W,z,f}$ or $\Delta F_{W,z,r}$ Wheel load change at the front or rear, N
b_f or b_r Track width front or rear, m
$M_{g,f}$ or $M_{g,r}$ Geometric portion of the roll moment, N m. Arises due to the centrifugal force at the centre of roll at the front or rear
$M_{U,f}$ or $M_{U,r}$ Share of unsprung masses (axle masses) in roll moment, N m.

From (6.2) and (6.9) it can be seen that via the total stiffnesses c_f and c_r respectively, the proportion of the spring moments of the respective axle is specified. If the remaining moments in (6.11) remain unchanged, the proportion of the spring moments M_{Ro} of an axle can be increased or reduced by increasing or decreasing its torsional spring rate c_{Ro} and thus the driving behaviour can be influenced by increasing or reducing the wheel load displacement at this axle. Greater wheel load displacement results in greater slip angle requirement for that axle and vice versa. The influence on the driving behaviour can be quantified by the proportion of the front torsional stiffness to the total stiffness[8]

$$\Phi_{Ro,f} = \frac{c_f}{c_f + c_r} 100\% \tag{6.12}$$

---

[7]For more details see series Racing Car Technology Manual vol. 4 *Chassis*, Sect. 2.2 Wheel *Suspension*.

[8]This is an important tuning key performance indicator (KPI), see also Racing Car Technology Manual Vol. 5 *Data Analysis, Tuning and Development*, Chap. 5 *Tuning*.

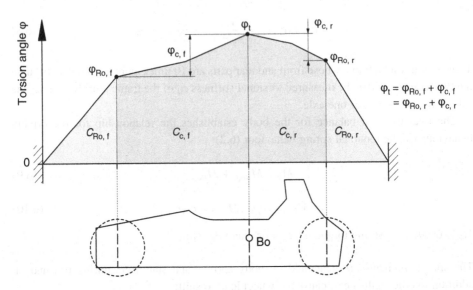

Fig. 6.91 Division of the torsion angle. The vehicle is loaded and twisted by centrifugal force at the centre of gravity Bo as shown in Fig. 6.90. The wheels are intended to be fixed, but the axles are intended to be torsionally soft. The frame cross sections at axle height are twisted by angles $\varphi_{c,f}$ and $\varphi_{c,r}$ respectively. The axles at the front $\varphi_{Ro,f}$ and at the rear $\varphi_{Ro,r}$ contribute to the total roll angle φ_t.

For balanced, tending to understeer handling, $\Phi_{Ro,f}$ should be equal to the proportion of the front axle load to the total weight plus 0.5%. If, for example, the frame is torsionally soft in the front area ($c_{c,f}$ small), this will have to be compensated for balanced driving behaviour by increasing $c_{Ro,f}$ (6.4), i.e. a stiff stabiliser will have to be provided on the front axle (large $c_{Ro,S,f}$).

A note is now appropriate because the measurement of the torsional stiffness has formed the introduction to this section. The measured total torsional stiffness of the frame c_{ts} results from the interaction of the two partial stiffnesses $c_{c,f}$ and $c_{c,r}$ in series(!), because when measuring the frame is not loaded "centrally" as in Fig. 6.90b, but the test moment is introduced at one end and transmitted over the entire frame:

$$c_{ts} = \frac{c_{c,f}c_{c,r}}{c_{c,f} + c_{c,r}} \tag{6.13}$$

c_{ts} Total torsional stiffness of the frame from one axle support to the other, N m/rad

The section stiffnesses of the carriage and the transmitted torsional moment can be used to determine the proportional torsional angles, Fig. 6.91:

$$\varphi_{c,f} = \frac{M_e}{c_{c,f}} \text{ resp.} \varphi_{c,r} = \frac{M_e}{c_{c,r}} \tag{6.14}$$

$\varphi_{c,f}$ resp. $\varphi_{c,r}$ Angle of twist of the frame cross-section at the height of the front or rear axle, rad.

6.4   Add-on Parts

6.4.1   Brackets and Mounts

The frame not only connects the two axles together, but also carries all the add-on parts. These include fuel and oil tanks, wings, battery, control units, parts of the exhaust system, fire extinguishers, etc. Brackets are usually welded, riveted or bolted to the frame to secure these components. Brackets and mounts may be simple parts, but if they fail, it can still cause the car to fail in the race. So consoles, like all other areas of the vehicle, should perform their function safely with as little mass as possible. When designing consoles, it is important to bear in mind that they will have to bear several times the weight of the add-on part due to acceleration (e.g. 1.5 g in the transverse direction) and impact, and that the load will usually be oscillatory. The general rule for guards is that they must be able to withstand decelerations of 10 g. In the case of engine mounts and gearbox mounts, it must be borne in mind that, due to dynamic effects, the maximum moment to be absorbed will be far higher than, for example, the maximum engine moment or the largest tire moment that can be transmitted by calculation. A realistic assumption is: $M_{dyn} = 2M_{stat}$.

The principle of direct load conduction can be used to advantage in the design of supports, see Sect. 2.5 *General Principles of Embodiment Design*.

In Fig. 6.92, some notes are given for welded plates.

If the strength of light metal alloys is impaired by the heat input during welding or if different materials meet, only riveting and/or bonding can be considered for joining. The rivet diameter d (Fig. 6.93) is based on the smallest sum t_{min} of the sheet thicknesses that are loaded in the same direction by the hole pressure and is:

$$\text{for single shear connections} : d = 2 \cdot t_{min} + 2$$

$$\text{for double shear connections} : d = t_{min} + 2$$

d Rivet diameter, mm
t_{min} Sum of the relevant plate thicknesses, mm

In the case of lugs that are riveted to walls, unequal force distribution on the rivets occurs due to unequal component expansion. The first rivet in the direction of force experiences higher stress than those behind it. Outer rivets at the ends transmit greater forces than those in the middle of the sheet. Therefore, no more than 5 to 6 rivets should be arranged one behind the other in the direction of force and if more rivets are needed for reasons of load-bearing capacity, they must be staggered, Fig. 6.94.

Rivets should not be subjected directly to a tensile load, but should be stressed in shear. The solution of Fig. 6.95a should therefore be avoided and the variant Fig. 6.95b should be preferred, especially for large dynamic forces.

In general, holders should be designed in such a way that the forces are introduced over as large a surface area as possible and that no jumps in stiffness occur, Fig. 6.96.

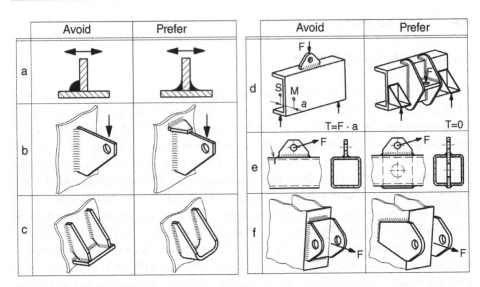

Fig. 6.92 Brackets and holders. Design information for welded brackets. To (**d**) S centre of gravity of the profile cross-section. M shear centre of the profile. (**a**) The seam root of welds should not be located in the tensile zone. The most secure welded joint, especially under dynamic loading, is the butt weld. In the case of dynamically loaded single-sided fillet welds, the redirection of the force flow has a particularly strength-reducing effect. Double-sided seams designed as fillet welds are much better. (**b**) The load-bearing capacity can be increased by extending the seam in the tension zone. A T-cross-section reduces the stress in the tension zone of the bracket loaded on one side. (**c**) In general, welds should be used sparingly. Distortion and residual stresses thus remain low. Brackets can thus be bevelled from a strip and only then welded on. This is easier and provides the same result with fewer welds as a structure made of individual sheets. (**d**) Although open profiles are avoided in frame design because of their unfavourable torsional stiffness, if a bracket is nevertheless to be attached, the shear centre M of the profile must be taken into account. If the force acts at the centre of gravity S of the cross-section, a torsional moment T *is* generated. This additional stress is avoided if the force acts at the shear centre M. (**e**) If a link plate is welded to a hollow section, clever force application must be ensured. Inserting the strap through the section tube prevents cracking (arrow) in the tension zone and also does not only load the flexible wall of the tube. (**f**) If brackets are welded to a wall, the fillet welds should not only be loaded in tension. This is achieved by designing individual brackets with longer seams which transfer the tensile forces of the bracket to the wall primarily via shear

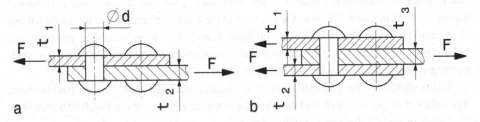

Fig. 6.93 Rivet terms. (**a**) single shear connection. $t_{min} = t_1$. (**b**) double shear joint. $t_{min} = t_1 + t_2$ or $t_{min} = t_3$ depending on the size of the actual sheet thicknesses

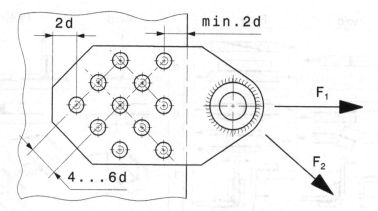

Fig. 6.94 Rivet arrangement for lugs. The recommendations for edge and rivet distances apply to light metal design

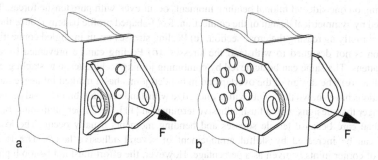

Fig. 6.95 Brackets riveted to wall. (**a**) Rivets subjected to tensile stress: Avoid. (**b**) Rivet arrangement to be preferred

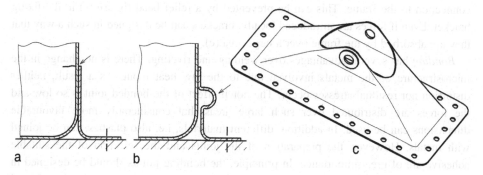

Fig. 6.96 Strength-compliant design of brackets. (**a**) Vessel mounting with stiffness step. (**b**) Vessel mounting with relief bead. (**c**) Bracket with large-area force application

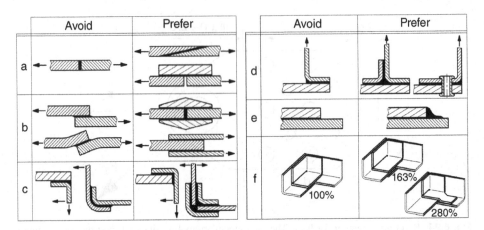

Fig. 6.97 Design information for bonded joints. (**a**) The joint surfaces must be sufficiently dimensioned. The joints can be scarfed or additional battens can be used. (**b**) In the case of connections overlapping on one side, additional bending moments occur even with pure tensile forces. These can be avoided by symmetrical design of the connection. Roof-shaped profile battens reduce the stiffness jump by abruptly increasing the cross-section. (**c**) Peeling stresses occur in angled connections if the connection is not designed to withstand the stresses. (**d**) Peeling can be prevented by additional angled battens. The same can be achieved by combining joining techniques, e.g. stapling or riveting and bonding. (**e**) The design of the overlap end in bonded joints has a marked influence on strength. Investigations have shown that joints with a fillet edge withstand certain vibration loads about a factor of ten longer than designs with a flush adhesive termination [11]. (**f**) Corner joints must be designed with special care because tensile stresses and bending moments usually occur. The load-bearing capacity can be increased by skilful arrangement of several adhesive layers. The load-bearing capacity of corner joints is given as a percentage. However, the effort must not be too high, because otherwise welding can be more favourable in terms of weight

Rigid container fastenings lead to cracking (a, arrow) under dynamic load and direct connection to the frame. This can be prevented by a relief bead (b, arrow) in the fixing bracket. Even if forces are introduced locally, brackets can be designed in such a way that they are absorbed by the frame over a large area (c).

Bonding has several advantages over welding and riveting. There is no change in the microstructure of the metals involved due to the low heat input. As a result, neither distortion nor residual stresses occur. The notch effect of the bonded joint is so low and the forces are distributed over such large areas that considerably more favourable dimensions can be used. In addition, different materials, i.e. also plastics, can be joined with metals. However, the preparation of the joints and the selection of the suitable adhesive are of great importance. In principle, the bonding points should be designed in such a way that the adhesive is subjected to shear and compressive stress by the external loads. Tensile and peel stresses must be avoided in the interests of long-term durability.

Some basic design considerations for bonded joints are summarized in Fig. 6.97.

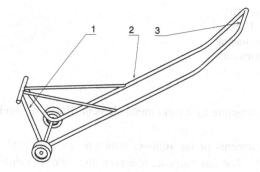

Fig. 6.98 Quick jack for the rear of the vehicle. This device enables one person alone to lift a trolley and at the same time jack it up stably by folding the jack down to the ground. For this purpose, the lifting fork (1) makes an acute angle with the stirrup (2). The transverse handle (3) is bent out of the plane of the handle so that it can be easily grasped even when the jack is lying on the ground

Fig. 6.99 Mounting for the quick lift on a Formula 3000 car rear end. Attached to the narrow supporting structure are two forked tabs that engage the quick lift. You can also see the diffuser after the gearbox

6.4.2 Towing Systems

For the rapid recovery and removal from the track of vehicles that have been involved in an accident or have "broken down", eyelets and receptacles are used that must be easily recognised by the track marshals. They are in signal colours (yellow, orange or red) and marked with stickers. The supports must be able to bear the proportional weight of the vehicle.

A lifting device as shown in Fig. 6.98 is used to lift vehicles, e.g. by marshals, and quickly pull them off the track or to lift the vehicle for changing wheels.

Figure 6.99 shows what the receptacle for such a quick lift looks like.

Towing Lugs

. Towing eyes (Fig. 6.100) are fitted to touring and production sports cars at the front and rear ends of the vehicle. They must have a minimum internal diameter of between 80 and 100 mm. The sheet thickness must be at least 5 mm for steel. The eyelets must be connected

Fig. 6.100 Towing eye. The
eyelet is screwed to the front and
rear end of the frame of touring
cars and production sports cars

directly to the main structure by a rigid metal element. Ropes or similar are therefore not permitted.

In some cases, elements of the rollover structure are also used to recover vehicles involved in accidents. For this purpose, however, the rollover element must be suitable for simple attachment with lifting gear. This is the case, for example, with hoop-shaped elements. Furthermore, the manufacturer of the vehicle must provide the race control with his consent to such use of the rollover structure in writing.

References

1. Kunz, J., Lukic, D.: Biegesteifigkeit und Biegefestigkeit in der beanspruchungsgerechten Auslegung. Konstruktion. **7/8**(Juli/August), 65–67. Springer, Berlin (2006)
2. McBeath, S.: Competition Car Preparation, 1. Aufl. Haynes, Sparkford (1999)
3. Schweighart, T.: Das optimale Dakar-Rallye-Fahrwerk. Diplomarbeit an der FH JOANNEUM, Graz (2006)
4. Cunat, P.-J.: A New Structural Material for Passenger Cars: Stainless Steel. AutoTechnol. **3**, 40 ff. Vieweg, Wiesbaden (2002)
5. Dutta, D., et al.: Herstellung, Zusammenbau und Aufstellung von Hohlprofilkonstruktionen, aus der Reihe "Konstruieren mit Stahlprofilen", Bd. 7. TÜV Rheinland, Köln (1998)
6. Wardenier, J., et al.: Anwendung von Hohlprofilen im Maschinenbau, aus der Reihe "Konstruieren mit Stahlprofilen", Bd. 6. TÜV Rheinland, Köln (1996)
7. Smith, C.: Prepare to Win, 1. Aufl. Aero Publishers, Inc, Fallbrook (1975)
8. Hintzen, H., et al.: Konstruieren und Gestalten, 3. Aufl. Vieweg, Braunschweig (1989)
9. Foale, T., Willoughby, V.: Motorradfahrwerk, Theoretische Grundlagen, Entwurf, Ausführung, 1. Aufl. Motorbuch Verlag, Stuttgart (1988)
10. Staniforth, A.: Race and Rally Car Source Book, 4. Aufl. Haynes Publishing, Sparkford (2001)
11. Klein, B.: Leichtbau-Konstruktion, 5. Aufl. Vieweg, Wiesbaden (2001)
12. McBeath: Bericht über DJ Firecat. Racecar Eng. **6**(Juli), 54 ff. (2000)
13. Incandela, S.: The Anatomy & Development of the Formula One Racing Car from 1975, 2. Aufl. Haynes, Sparkford (1984)
14. Strambi, G.: Assembly Technology for Carbon Fibre Body Structures. AutoTechnol. **4**(Aug.), 56 ff. (2006)
15. Hölscher, M.: Carrera GT. Der neue Hochleistungssportwagen von Porsche. Vortrag der ÖVK Vortragsreihe, Wien (2003)
16. Wright, P.: Formula 1 Technology, 1. Aufl. SAE, Warrendale (2001)
17. N.N.: CTM Compositleitfaden. CTM GmbH Composite Technologie & Material, Schleswig (2005)
18. Hufenbach, W., Helms, O.: Zum methodischen Konstruieren von Leichtbaustrukturen aus kohlenstofffaserverstärkten Kunststoffen. Konstruktion **62**(Okt), 69–74 (2010)

19. Köllner, Ch.: Welche Nachteile haben Carbonfasern? www.SpringerProfessional.de/ Verbundwerkstoffe/Leichtbau. Accessed on 01.10.2018
20. Fliegener, S., Hohe, J.: Mehrstufiges numerisches Screening für neuartige Lasteinleitungskonzepte. Konstruktion **68**(11/12), IW 4–IW 6 (2016)
21. Schwarz, M., et al.: Zur Gestaltung von punktuellen Krafteinleitungen in Faserverbund-Strukturen. Konstruktion. **6**(Juni), 90–96. Springer, Berlin (2007)
22. N.N.: Formula Renault 2000 Manual. Renault Sport Promotion Sportive (2001)
23. Fuchs, C.: Herstellung von CFK-Leichtbaukomponenten in Autoklav- und Injektionstechnik. Vortrag auf der RaceTech, München (2004)
24. Roberts, N.: Think Fast, The Racers's Why-To Guide to Winning, 1. Aufl. Eigen, Charleston (2010)
25. Breuer B.: Motorräder, Vorlesungsskriptum. TH Darmstadt, Darmstadt (1985).
26. Braess, H.-H., Seiffert, U.: Vieweg Handbuch Kraftfahrzeugtechnik, 1. Aufl. Vieweg, Wiesbaden (2000)
27. Kleemann, W., Kolk, M.: Der neue BMW Z4 Roadster. ATZ. **6**, 550. Vieweg, Wiesbaden(2003)
28. Bericht über Reynard 2KQ in Racetech Heft Nr. 27, Dez. 1999/Jan. (2000)
29. Indra, F.: Grande complication, der Opel Calibra der ITC-Saison 1996. Automobil Revue. **50**, 50 (1996)
30. Wright, P.: Ferrari Formula 1. Under the Skin of the Championship-winning F1-2000, 1. Aufl. David Bull Publishing, Phoenix (2003)
31. Theissen, M., et al.: 10 Jahre BMW Formel-1-Motoren, Beitrag zum Wiener Motorensymposium. VDI Reihe 12 Nr. 716 Bd. 2, VDI, Düsseldorf (2010)

Bodywork

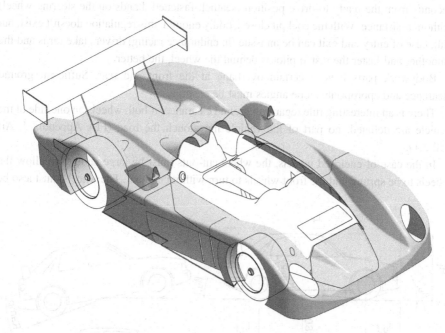

The outer skin (bodywork) has far more tasks than simply covering the innards. It contains many aerodynamic functions and shapes the overall appearance of the car to the observer like no other subsystem.

M. Trzesniowski, *Complete vehicle*,
https://doi.org/10.1007/978-3-658-39667-1_7

7.1   Terms

To begin with, a few terms should show that the outer shell determines the external appearance and its overall dimensions are covered by the law or regulations, Fig. 7.1.

7.2   Requirements

When designing the outer skin or the body, legal requirements and regulations must first be observed. For example, the position of lights, bumpers, crash elements, windows and doors or flaps are specified or restricted. In addition, the fire resistance of the materials used is important.

Despite all the aesthetics and aerodynamics, the bodywork should enable the driver to get in and out of the vehicle and not hinder him. In single-seaters with an open cockpit, for example, all regulations require that the driver must be able to exit the vehicle within five seconds from the ready-to-drive position (seatbelt fastened, hands on the steering wheel) without assistance. With the cockpit closed, oddly enough, this regulation doesn't exist, but still, ease of entry and exit can be an issue. In endurance racing, drivers take turns and the smoother and faster the rested pilot is behind the wheel, the better.

Bodywork parts have a certain overhang at the front and rear. Sufficient ground clearance and appropriate slope angles must be ensured.

There is an interesting rule regarding ground clearance: if both wheels on one side of the vehicle are deflated, no part of the vehicle may touch the road (FIA Appendix J, Art 252 2.1).

In the case of enclosed wheels, the wheel cut-outs must be large enough to allow the wheels to be sprung and the front wheels to turn without contact. The wheels must also be

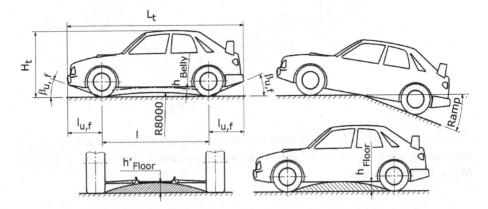

Fig. 7.1 Dimensions on the outer skin. L_t Overall vehicle length, H_t Overall height, l Wheelbase, h_{Belly} Belly clearance, $\beta_{u,f}$ and $\beta_{u,r}$ Front and rear overhang angles respectively, β_{Ramp} Ramp angle. $l_{u,f}$ and $l_{u,r}$ Front and rear overhang respectively, Ground clearance h_{Floor} and h'_{Floor}

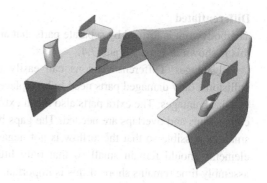

Fig. 7.2 Engine cover on a formula car. The cover includes the cladding of the air supply to the engine, the cladding of heat exchangers, cooling stacks, additional wings and air guide surfaces. The inside is partially covered with thermal protection film

sufficiently enclosed. Wheel house ventilation becomes important from an aerodynamic point of view: see Sects. 5.2 and 5.6.

Aerodynamic considerations and effects also come into play in numerous other places. The air resistance should be as low as possible for greater acceleration and higher top speed, while the downforce should be as high as possible. Due to the fact that the vehicle also builds up a sideslip angle and a wind can also act from the side, the reduction of crosswind sensitivity is remarkable, especially at high speeds. Targeted influencing of the flow or well-considered positioning of openings and deflectors prevents rapid windshield fouling and enables cooling air to be supplied to and removed from heat exchangers (see Sect. 5.6 *Heat Removal and Ventilation*) and brakes as well as combustion air to be supplied to the engine, Fig. 7.2. On many vehicles, reliable exhaust gas routing must be ensured in the underbody area.

Strength considerations are subordinate in the areas that do not have a load-bearing function, but the vibration and buckling behavior must be taken into account, especially for large-area parts. The engine cover of Formula 1 cars is deliberately designed to be light so that the car's centre of gravity is only slightly affected. However, the cover has to be replaced regularly (after a few races). In many cases, this is not a problem because the latest aerodynamic findings are incorporated into the new parts anyway.

Finally, aesthetics should not be forgotten: a sleek appearance is rewarded by sponsors and the public alike.

7.3 Design

The design of the bodywork is based on the data of the interior design and the unit arrangement and takes aerodynamic aspects into account. Basically, there are two possible designs:

- differentiated,
- integrated.

Differentiated

The bodywork is formed by separate parts that are slipped over the frame. An example of this is shown in Figs. 7.3 and 7.4.

Advantages: Different versions can easily be provided for different routes. After collisions, only damaged parts need to be replaced.

Disadvantages: The extra parts also mean extra mass. In addition, holders to the frame, connections and overlaps are needed. The gaps between the individual parts should be as small as possible so that the airflow is not negatively affected. The number of fastening elements should remain small so that only little additional weight is created and the assembly time remains short (if this is important in the race).

Integrated

With this design, the chassis itself already determines the external shape. An example is shown in Fig. 7.5.

Advantages: The mass is lower compared to the previous design. Interfaces including closures are eliminated.

Disadvantages: Subsequent modifications, for example to adapt the cooling or the downforce, are practically no longer possible on the existing chassis. Collision damage caused by an accident is much more costly to repair.

When designing the bodywork, the production must not be disregarded in any case. This already starts with the division into individual sections or parts. The parts should be manageable, which not only facilitates assembly and disassembly but also mould

Fig. 7.3 Bodywork parts of a production sports car (Osella PA 20 S). In the foreground (with the start number) is the part that is put over the cockpit. Behind it is the nose section, which encloses the front wheels on the vehicle in front of the cockpit section. Next to it is the engine cover

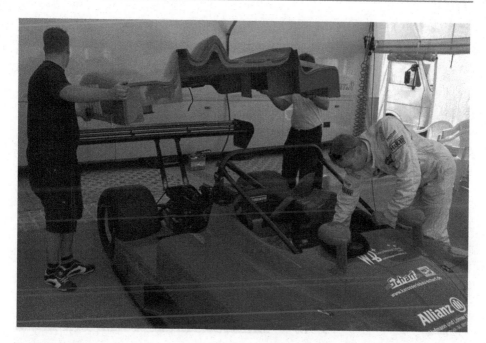

Fig. 7.4 Assembly of the engine cover of a production sports car. The engine cover is lowered over the engine compartment and connected to the other bodywork parts with eccentric locks. At the front end of the engine cover you can clearly see the beads to fix the position to the cockpit part in front of it

construction. In the event of an accident, the damage can also be limited in this way. The division lines, with which the bodywork is divided into individual sections, will – if the choice is free – be placed in such a way that by removing only one part, maintenance and adjustment areas become accessible. If two parts abut, an offset supports the connection of the neighbouring part, cf. also Fig. 7.12. Demoulding bevels (draft angles) are required in the preferred manufacturing processes of bodywork parts and monocoques both for direct moulding of negative moulds and for manufacture via a positive pattern.

The bodywork and fairing may represent additional mass on the racing vehicle, but this design allows the aerodynamic development and tuning to be track-dependent at relatively short notice. Not least for this reason, cover parts that have integrated additional functions, such as the engine cover, Fig. 7.2, can also be found on vehicles with monocoques.

The nose on formula cars not only supports the front wing, but is also the energy-absorbing element in a frontal crash (*impact attenuator*). The fastening must be designed in such a way that not only longitudinal forces but also large lateral forces can be absorbed. Nevertheless, the aim is to enable rapid installation. A common solution for this is shown in Fig. 7.7. The nose is connected to the monocoque by four pins. The four pins are bolted to the nose (figure part a). When assembling the nose, the pins are inserted into corresponding locating holes of the monocoque. The pins are secured by an eccentric lock (3, figure part c), as known from furniture construction. If the lock is rotated about its axis by approx.

Fig. 7.5 Nose of a formula car (Formula Renault). The nose is bolted directly to the monocoque. This outer shape is therefore given by the chassis itself. Only on the top side remains an area which must be accessible for assembly and setup (damper, spring, stabilizer) and which is still covered with an additional cover, Fig. 7.6

180°, its inner eccentric surface pulls the head of the pin (4) inwards and thus presses the contact surfaces of the lug (7) and the monocoque (5) together. For fine adjustment of the locking force, the pin can be turned in the joined state from the footwell of the cockpit using a pin wrench through a small hole (1). For disassembly and subsequent reassembly of the nose, only the four fasteners need to be turned from the outside. Transverse forces are introduced into the surrounding frames by a centring collar (6) or a corresponding sleeve. This type of fastening allows a continuous smooth surface and avoids recesses as they are necessary for tool access when bolting in the longitudinal direction of the nose.

Catches
Fairing parts that have to be removed for maintenance purposes are fastened with quick-release fasteners, e.g. Dzus fasteners, pins, swivel fasteners. Such fasteners basically consist of two parts, which are attached to the frame and to the bodywork, respectively, so that they cannot be lost. As an example of the many existing variants, Fig. 7.8 shows such a quick-release fastener. During assembly, the pin (1) is pressed in against the spring force and pretensioned and locked in the lower part of the lock (6) by a quarter turn. Disassembly is carried out in reverse order.

Fig. 7.6 Mounting the nose cover of a formula car. The maintenance side of the bow from Fig. 7.5 is closed with a single lid. The cover is fixed with quick release fasteners

Such quick-release fasteners can also be used to secure other parts to the vehicle. Depending on the weight and size of the parts to be fixed, there are numerous fasteners that have been optimised for their intended use. Figures 7.9, 7.10, 7.11, 7.12, and 7.13 show some examples.

Undertray

The undertray should be as level and smooth as possible. A slope of 1° to 2° is sufficient to generate a vacuum with a smooth underbody, Fig. 7.14a. The front of the vehicle is closer to the road than the rear. The car floor is therefore inclined and, in conjunction with the road surface, acts as a long diffuser which facilitates air flow (underbody rise, *rake*). With the ideal shape of the car bottom (Fig. 7.14b), the air is initially accelerated smoothly under the nose and the pressure decreases. At the rear, reduced wake reduces the drag of the car. In addition, less air flows over the top of the vehicle because a larger proportion takes the path below.

Under no circumstances should the car bottom out metallic in extreme driving conditions. This leads to instability due to the unloading of a wheel and the lack of lateral guidance. It has proven effective to provide defined wear points on the undertray. These can be wooden boards on the outer edge, which are rubbed off when the vehicle sits on

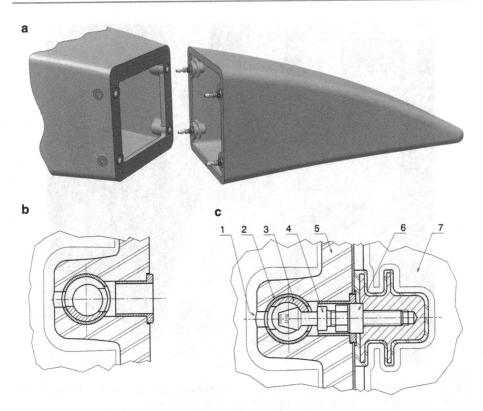

Fig. 7.7 Attachment of a nose (GP2 Series 2009, Dallara). (**a**) Axonometric view before assembly, (**b**) Detail section through a receptacle in the monocoque before assembly, (**c**) Detail section of a joined joint. 1 Access hole for allen key, 2 Steel sleeve, 3 Cam lock, 4 Fixing pin, 5 Monocoque, 6 Locking nut with centring collar, 7 Nose

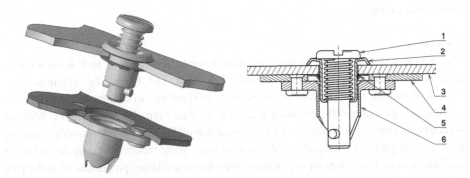

Fig. 7.8 Quick release quarter-turn fastener. 1 Locking pin, 2 Guide sleeve with spring, 3 Bodywork, 4 Frame, 5 Locking washer, 6 Lower part of lock. The shutter is closed via a quarter turn

Fig. 7.9 Quick release fastener on a bonnet (AE: hood) of a touring car. On the frame side there is a screw-in pin (Fig. 7.10) with a cross hole. On the bonnet is the counterpart, a pin with circlip, which is in a guide. The photo shows the open position. To close, the mandrel is inserted through the hole and the retaining ring is folded over the screw-in pin

Fig. 7.10 Screw-in spigot for quick-release fastener. Such a spigot is accompanied by a fastener as shown in Fig. 7.9. This spigot is screwed into the front end area of a touring trolley

them and thus give way, Fig. 7.15. On some vehicles, the flat underbody area is formed by a continuous plywood board.

For two-seater racing sports cars, i.e. those operated exclusively on the race track, the FIA prescribes a 20 mm thick continuous board, Fig. 7.16. The material is only restricted in its density. This must be between 1.3 and 1.45 g/cm^3. The board must be smooth without openings, except for those for fastening. The fastening elements (screws, rivets, …) must not protrude beyond the underside.

Similar regulations also exist for Formula 1 cars, Fig. 7.17. In these, however, the 10 mm thick floor board is examined for wear at certain points after the race. Teams generally use a board glued from beech veneer (trade name Jabroc) and for qualifying, where the sliding board may be thinner from the start, Tufnol, because this material has better wear properties [3].

Fig. 7.11 Quick release fastener on a touring car bonnet. A simple variant of the fastener of Fig. 7.9. A male plug, which is secured against loss with a nylon thread, is inserted through the pin on the frame side

Fig. 7.12 Eccentric catch (*over-centre catch*) of a production sports car (Norma N20). The latch is hooked and pretensioned on the upper part of the fuselage (removed on the left picture and not to be seen) when closing. The two beads on the butt surface serve to fix the position of the upper part by engaging in the corresponding recesses of the counterpart

Ground Clearance

Low ground clearance has the advantage that only a small amount of air flows underneath the vehicle, thus reducing lift. To keep this ground clearance within narrow limits, the road surface must be level and the suspension stiff. Regulations often stipulate that no part of the sprung mass may be lower than the underbody, with the exception of the sliding board [2].

For some vehicle categories, a minimum ground clearance is also prescribed, which is checked by sliding a block of defined height under the vehicle.

Fig. 7.13 Eccentric locks on the bodywork of a production sports car (Norma N20). The photo shows the joint at the left front of the vehicle. This is where the nose (with the wheel cutout and wheelhouse vent), the cockpit rim and the lower fuselage section meet. These three bodywork parts are joined together at this point with two eccentric fasteners

a   b

Fig. 7.14 Design of a smooth carriage floor to generate downforce, according to [1]. (**a**) Flat underbody. (**b**) Ideal shape of the carriage underside

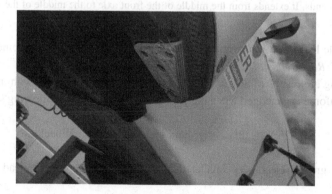

Fig. 7.15 Wear board on the underbody of a formula car (Reynard D94 F3000). The vehicle is jacked up and the view is directed from below to the right side box. A triangular shaped plywood board is attached to the outer area of the underbody. You can see beside the splitter under the fuselage also the side mirror and the front wishbones

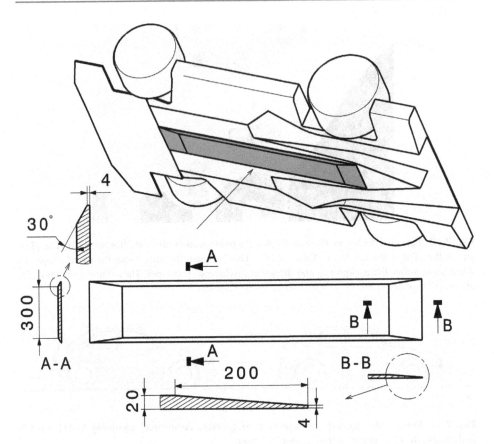

Fig. 7.16 Skid block for sports cars according to FIA [2]. The plate is attached to the underbody and bevelled at its ends. It extends from the middle of the front axle to the middle of the rear axle

A possible loss of air in a tyre must be taken into account in at least some regulations, see Sect. 7.2 *Requirements*.

Interesting are results of an investigation on conventional vehicles with uneven underbody: Downforce is generated when the ground clearance is in the following size range [1]:

$$0.125 \text{ Wheelbase} < \text{ground clearance} < 0.6 \times \text{Wheelbase}$$

Conversely, ground clearances smaller than 0.125·wheelbase actually lead to lift of the vehicle.

However, the average value of the ground clearance of such vehicles is outside this range, namely about 0.05·wheelbase. This illustrates why production vehicles generate lift with their uneven underbodies.

In racing cars the ground clearance is much lower, but the underbody is ideally flat, which is why these vehicles are also able to generate downforce through the underbody. A

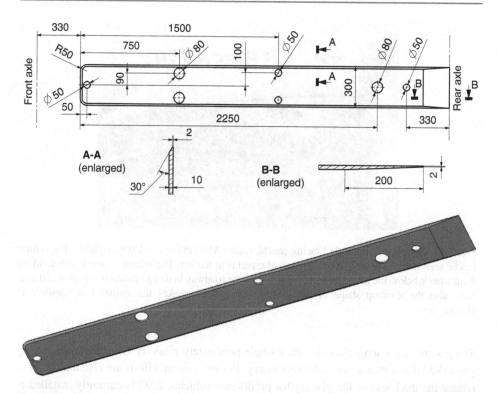

Fig. 7.17 Sliding board for Formula 1 cars, according to FIA. The board must be fitted symmetrically along the longitudinal axis of the vehicle. It ends exactly at the height of the middle of the rear wheels. The abrasion is measured after the race in the 50 and 80 mm holes

splitter (name!) divides the airflow so that one part flows under, the other next to the fuselage, Fig. 7.18.

Some rough numerical values follow below for orientation. Rough because the static ground clearance and thus inevitably the ground clearance is also changed in the course of tuning and because the ground clearances at the front and rear can be intentionally different. Production sports car Mercedes C291 (1991) approx. 46 mm [4], Mercedes C-Class (DTM '94) approx. 40 mm [4], Formula Ford front 45 mm and rear 70 mm, Formula 1 partly below 25 mm [5], Ferrari F1-2000 front from 14 to 20 mm and rear between 50 and 61 mm [6].

Openings

With the cockpit closed, openings for viewing areas (windscreen and side windows) must also be planned together with the seals required for their installation.

The windscreen (*AE: windshield*) is often the standard part in touring cars, or the regulations require road-approved glazing, e.g. in accordance with UN/ECE regulation ECE R 43. In many cases, an additional bracket is recommended for the standard solution.

Fig. 7.18 Splitter on a Formula 1 racing car (Mercedes AMG Petronas F1 W05 Hybrid). The vehicle can be seen in the side view. The direction of travel is to the left. The splitter is mounted behind the front wheels below the raised bow. The distance to the roadway in design position is approx. 30 mm. Note also the teardrop shape of the leading edge, which makes the splitter less sensitive to oblique flow

The glazing in passenger cars is either single-pane safety glass (ESG) or laminated safety glass (VSG) and in any case relatively heavy. For this reason, efforts are also underway to reduce the thickness of the glazing for production vehicles. ESG is currently installed at around 3.15 mm and VSG, which consists of two panes separated by a plastic film, at around 5–6 mm. Thickness reductions of a single pane to about 2.1–1.6 mm have been carried out in isolated cases. However, the pairing of 2.1/2.1 mm currently represents the lower limit for laminated safety glass for mechanical and aeroacoustic reasons [7].

Heated windshields are available for touring cars, which show their advantages in damp, cold weather. For endurance cars (GT1, GT2, Le Mans series), tear-off films made of plastic (e.g. Mylar) – similar to the films on the visors of safety helmets – are also used on the windscreens. During a pit stop, a mechanic can thus quickly remove all contamination from the entire windscreen by peeling off the top layer of film, Fig. 7.19.

Some production sports cars (e.g. LMP) are also permitted by the FIA to have a rigid plastic windscreen. These windscreens are made of polycarbonate with special coatings that improve optical properties and increase scratch resistance [8]. These windshields can also be designed to be electrically heated via microwires. Special manufacturing processes also allow complex shapes of the pane. Glazing made of this material also offers great resistance to impacting parts (stones, fragments of other vehicles in collisions, ...).

Polycarbonate windows are also used in the NASCAR series. In 2013, ballistic tests led to the replacement of the single-pane designs (monolith) that had been common up to then with a laminate design [9]. Among other tests, full Al beverage can and 0.39-kg steel cylinder were fired at the disc at 320 km/h. The steel cylinder penetrated the monolithic target but was stopped by the laminate target. The laminate consists of two 3-mm-thick

Fig. 7.19 Peel-off films on the windscreen of an LMP-1 vehicle. These films are stuck to the window of the Audi R18 e-tron quattro (2013). The area of the right-hand edge of the window can be seen. The mechanic can easily grasp the top film through the white tab and pull it off to the other side of the vehicle

polycarbonate discs joined by a 0.8-mm-thick urethane layer. The laminate thus has a total thickness of 6.8 mm.

Figure 7.20 shows how such a plastic washer can be attached. A mounting bracket (5) made of CFRP carries rivet nuts (3) and is itself bolted to the monocoque (7) or the cockpit bulkhead. The windscreen (1) has tapered countersunk holes into which countersunk screws (4) together with tapered washers (2) are inserted. A sealing cord (6) is inserted in the gusset between the bracket and the windscreen. The glass panes on series-production vehicles are held in the sheet metal cut-out by a profiled rubber or are glued directly to the body.

For side openings, all-plastic discs can be used in many racing series.

On rally cars, the FIA requires that side windows that are not made of laminated glass are covered with transparent anti-splintering film. The thickness of this film must not exceed 0.1 mm.

Materials

Single-pane safety glass: DIN 1249-12 or E EN 12150 Laminated safety glass: DIN 52337 or prEN 12600 Plastic panes are made of PC (polycarbonate, trade name e.g. Makrolon, Lexan) and are manufactured by injection compression moulding.

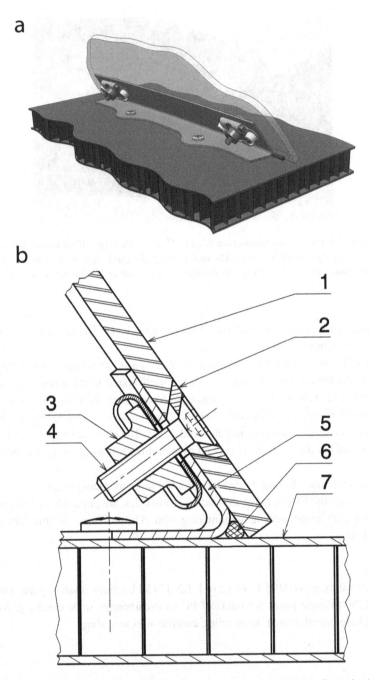

Fig. 7.20 Fastening of a plastic windshield. (**a**) Axonometric view (seen from the inside). (**b**) Section through a bolted joint. 1 Windscreen, 2 Bevel washer, 3 Rivet nut (floating version), 4 Countersunk screw M3.5, 5 Mounting angle, 6 Sealing cord, 7 Monocoque

Fig. 7.21 Structure of a deflector (Lola Zytec F3000). This deflector in front of the right rear wheel is damaged, but this is what makes its structure visible in the first place. Between the cover layers of CFRP mats, a plastic honeycomb core provides the stiffness-increasing distance between the outer structures

7.4 Materials

The following materials are used for the production of the large-surface bodywork parts:

Fibre-reinforced plastic laminates: Polyester or epoxy resins form the matrix. Reinforcement is provided by glass fibres (GRP), carbon fibres (CFRP) and Kevlar fibres (AFRP), Fig. 7.21.

The manufacturing process roughly consists of pattern making, mold making and molding. It is described in more detail in Sect. 6.2.3 *Composite Monocoques*.

However, metals such as aluminium or magnesium sheet are also used for parts.

References

1. McBeath, S.: Formel 1, Aerodynamik. Motorbuch, Stuttgart (2001)
2. Sportgesetz der FIA, Anhang J, Art. 258A "Technical Regulations for Sports Cars". Okt. (2006)
3. Glimmerveen, J.H.: Hands-On Race Car Engineer, 1. Aufl. SAE International, Warrendale (2004)
4. Ludvigsen, K.: Mercedes Benz Renn- und Sportwagen, 1. Aufl. Motorbuch, Stuttgart (1999)
5. Staniforth, A.: Competition Car Suspension, 3. Aufl. Haynes, Sparkford (1999)
6. Wright, P.: Ferrari Formula 1. Under the Skin of the Championship-winning F1-2000, 1. Aufl. David Bull Publishing, Phoenix (2003)
7. Braess/Seiffert: Vieweg Handbuch Kraftfahrzeugtechnik, 4. Aufl. Vieweg, Wiesbaden (2005)
8. http://www.isoclima.net/de_pc-secure.html. Accessed on 26.11.2016
9. http://articles.sae.org/11805/. Accessed on 20.03.2013

Comparison Series: Racing 8

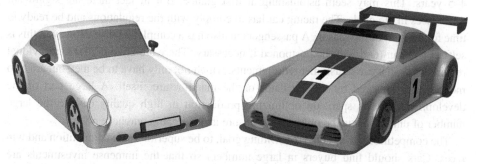

Comparisons between production and racing vehicles are made throughout the book to increase clarity, but here the similarities and differences are to be directly contrasted in a compact summary. This says a lot about both industries, their working methods and products in just a few pages.

8.1 Introduction

Developments in motorsport have always gone much faster from idea to implementation than series developments. This is also understandable. Once an idea has been born, it has to be turned into an advantage on the race track as quickly as possible, otherwise the opposition might take care of it. The results of the developments can be described in general terms as increased performance, increased efficiency, weight reduction, increased reliability and simplified handling. In other words, all achievements that are also quite

M. Trzesniowski, *Complete vehicle*,
https://doi.org/10.1007/978-3-658-39667-1_8

welcome in production vehicles. Numerous innovations in modern passenger cars have their origins in racing. The development results flowed from the sport into the production vehicles (disc brake, four-valve engines, direct injection, light alloy wheels, double clutch transmission, composite materials, . . .). In the meantime, however, the roles have changed to some extent. In the meantime, the race track serves the development departments of large companies as an unbureaucratic test laboratory for preliminary development. For example, when it comes to the practical testing of new materials or extreme designs of components. And if the development is successful, it is duly marketed. From the point of view of the car manufacturers, nothing has changed: Racing success brings sales success.

8.2   Development *Process*

The design and construction of an F1 car takes about 5 months. Depending on the team, around 300 engineers are employed. Thousands work on a production car for about 4–5 years. This may seem astonishing at first glance. But in fact there are significant differences in the goals. The racing car has to comply with the regulations and be ready in time for the start of the season. A passenger car also has a completion date. However, this is self-set and can (and will) be postponed if necessary. There are around 15,000 individual requirements in a car's customer requirements, which not only have to be met, but are also meticulously checked by the legislator or the manufacturer itself. A large part of the development time is spent on optimizing production in high quality despite the large number of units. This is the only way to ensure an affordable purchase price.

The competition car has ONE overriding goal, to be superior to the competition and win races. Cars should find buyers in large numbers so that the immense investments are worthwhile. They must therefore be adapted to the needs of the market and meet the (future) taste of the public.

A direct, reliable comparison between series and motorsport development is provided by companies that carry out both. As an example, consider the chronological sequence of engine development for Ferrari's production cars and Formula 1 cars, Fig. 8.1. While in production an engine matures from concept to producible unit in 42 months, the racing department goes through three full racing seasons. It not only designs, builds and develops new engines for each season, but also constantly improves them during the season in order to keep pace with its opponents or, ideally, to outdo them.

Even if the development processes are therefore different, one circumstance is (meanwhile) the same: the design of the vehicles takes place from the outside in. Based on market analyses, trend research results, strategic product specifications, etc., the design department specifies the size, external shape and appearance of the new car. The size, external shape and appearance of the new car. It is now the task of the design department to accommodate all the necessary and desired assemblies and components. In the case of racing cars, the process is exactly the same in that the aerodynamics department specifies the shape and

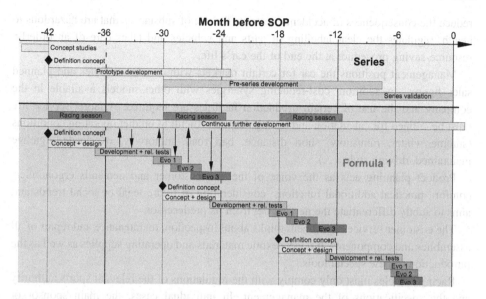

Fig. 8.1 Comparison of the flow charts of Ferrari's engine development, after [1]. Timing of engine development in series production (top) and in the Formula 1 team (bottom). The starting point is the start of production (SOP) of the series engine. Abbreviations: Gen.1: Generation 1, rel. tests: reliability tests, Evo 1: Evolutionary stage 1

thus the desired airflow around and through the car based on CFD calculations[1] and preliminary wind tunnel model tests, and all other development units must subordinate themselves to these specifications.

Often, in the course of a debate about profiting from a racing commitment, the question is asked as to which parts or assemblies have been incorporated from a racing vehicle into a production vehicle. The answer to this question does not have to be about components. Development tools and test methods are also advantageously adopted, as reported in [2], for example.

8.3 Development *Goals*

As mentioned, significant differences can be found in the development goals. Objectives are derived from requirements. In the case of passenger cars, these mainly come from the legislator (sometimes contradictorily from different states), from management, from product planning and from customer service.

Legislation limits emissions (exhaust, evaporation, noise), specifies maximum dimensions and maximum masses, lays down safety standards to prevent accidents and

---

[1] Computational Fluid Dynamics. Numerical flow simulations on a 2D or 3D computer model.

reduce the consequences of accidents, restricts the use of substances that are hazardous to health, regulates the clear labelling of parts and vehicles and takes care of an orderly, resource-saving procedure at the end of the car's life.

Management positions the car for certain markets with certain attributes and planned sales figures. In addition, cost-reducing synergies with other models available in the company and the use of certain production facilities are required. A target service life must be defined for calculation and testing despite difficult to predict operating conditions (summer/winter, rain/snow, short distance, bad road, motorway, overloaded, garage maintained, driving style, ...).

Product planning acts as the voice of the end customer and demands ergonomics, comfort, practical additional functions, consideration of future legal or social trends and aims to subtly differentiate the new model from its predecessor.

The customer service department thinks about inspection, maintenance and repair of all assemblies and components. It rejects exotic materials and operating supplies as well as the introduction of new special tools.

Racing vehicles must only comply with the regulations of the relevant sports authority and the specifications of the management. In individual cases, the main sponsor or purchaser will also make his demands and wishes known. The choice of material is only limited by the regulations and the time availability.

In general, it can be stated that the goals and methods of series and racing developments converge when the planned production quantity and the period of use of racing vehicles increase. Formula 3 cars are a good example of this. The cars are designed for a specific series but for unknown customers and are built for at least 3 years. Customer taste, cost and spare parts supply thus become significant issues in the design process.

8.4   Research and Development (R&D)

Continuous further developments are necessary – even if not for the same reasons – in both camps under consideration. The financial resources required for this must be generated. Twenty-five percent of the turnover of a racing team flows back into R&D expenditure. The automotive industry spends only 6% [3]. In this sense, a racing vehicle is never finished, but is subject to constant improvements, which are nourished by own research or ideas "inspired" by the opponent's solutions. Compared to the car industry, a competition vehicle is a permanent prototype.

Traditionally, racing has been extremely stingy when it comes to publishing research results or key findings. On the contrary, they even try to keep solutions they have found secret for as long as possible. After all, they promise an advantage over the opponents in the race. It is not uncommon for ingenious innovations to have decided entire world championships. Lotus, for example, was the lucky discoverer of winged cars using the ground effect. Williams developed an active suspension that made the aerodynamic effects work even more comprehensively. McLaren benefited from the fiber composite monocoque. Ferrari dominated races with a semi-automatic gearbox. Renault won the title not least thanks to mass dampers on both axles. The double diffuser was the clever exploitation of a loophole in the regulations that brought Brawn GP laurels. And that's just a chronological list of a few cases.

However, history also offers reverse cases, i.e. those in which a team was able to successfully use an alien innovation. The most famous example is the turbocharger, which Renault introduced and developed for years, but in the end it was Honda's turbo-charged engines that won multiple world championship titles.

The fear that one's own ideas can be implemented faster or even more successfully by others is therefore not unjustified. Secrecy, protection against industrial espionage and compartmentalization of the development departments are important strategic measures that shape the everyday lives of those involved. There are no major differences in mindset, regulations, contracts and delivery terms between racing and production development. The goals are basically the same: No one should know about groundbreaking creations before they go into the race or are offered in the new model, certainly not the direct opponents or the competition.

This omnipresent fear sometimes produces strange blossoms and makes cooperation with development service providers, suppliers and external manufacturers more difficult. Especially the change of drivers and employees is treated with great suspicion. One of the ways this is solved is that individuals only have access to very specific knowledge and do not have an overview of the big picture. Gone are the days when a handful of engineers designed and developed a complete car, complete with engine. The five-man engineering group of the Formula 1 teams of the 1980s has grown to over 200 engineers today. The developer of the brake caliper has no idea what his colleagues are doing to improve the steering. And drivers aren't privy to the technical details "under the hood" anyway. What's the point? They need to know which switch activates which function and concentrate on the track, literally.

Racing teams have to accept one major disadvantage in this context. Just as successes can be celebrated with great publicity, failures are also not hidden from the public. However, the press has never seen some of the mistakes made by car manufacturers, and for good reason.

8.5 Costs

Even if costs are at the top of the priority list for passenger cars, this does not mean that they do not play a role in motorsport. On the contrary: in numerous racing series, costs became an insurmountable financial problem in the course of the "arms race" between the participants and led to the end of this series of events. Even the top leagues in motorsport are by no means immune to this phenomenon. It is therefore with good reason that rule writers also look at the cost side when regularly adjusting the rule book. Particularly cost-driven racing series are GT3 and GT4, for example. GT3 cars are expected to be able to race for about 6000 km without having to replace parts. In GT4, this figure is as high as 10,000 km or one year. Production cars drive around 250,000 km in 10–15 years (some wear parts such as tyres, brake pads, brake discs, exhaust systems, timing belts, etc. are of course replaced in the process).

However, costs are also becoming a problem in professional racing due to increasing demands, more complex development and internationalisation of the series. History bears witness to several prominent examples where precisely such a development has forced changes in regulations or even meant the end of the series.

Customers' expectations are rising. What was a sensation in the luxury class yesterday is tacitly expected in the small car segment today. Cars must be low-maintenance with long service intervals. Long gone are the days when the valve clearance had to be adjusted after 5000 km or, in winter, a radiator grille with hinged elements was used to partially seal off the cooling air intake so that the engines warmed up (faster). Even punctures are hardly accepted and the average customer wants to be able to drive to the next garage without getting his hands dirty. Sensors that can continuously report the internal pressure of the tyres to the on-board computer while driving have proven their worth both in series production and in racing. Admittedly, pressure from the legislator was required for widespread use in passenger cars. For the race engineer, the current values of tire pressure and temperature are valuable information that can be used to determine the tire behavior and thus the driving performance of the vehicle. Over the course of the race, the development of the tires crystallizes and measures can be derived to influence traction and durability. Sensors located directly on the tire inner liner instead of at the usual rim position turned out to be ideal. They indicate the temperature close to the tread and provide an approximation of the actual internal pressure. The values of the pressure transducers on the rim have to be compensated with its temperature. Because the temperature of the rim rises much faster than that of the air in the tire during hard braking, overcompensation sometimes occurs and as a result annoying false alarms occur as the isochore-compensated cold pressure falls below the warning threshold [4].

In series vehicle production, the aim is to have the same (spare) parts for different versions of a type for reasons of cost and logistics. This can lead to parts being undersized for the strongest variant and oversized for the weakest variant. Rally teams are familiar with a similar dilemma, as they are used to maintaining their vehicles during competitions (under the most adverse conditions) and getting them back on the road again after damage. The fewer parts that have to be taken along, the better. As a concrete example, consider the wheel carriers. They are designed in such a way that one version can be fitted to all four wheels. The same applies to special tools. Incidentally, this is also a request made by customer service to the series designers. It should be possible to adjust and mount spare parts without special tools. During series assembly in the factory, robots carry out many work steps and assembly locations must be correspondingly accessible and assembly parts must be correspondingly clearly designed. This is where racing designs differ. The desired uncompromising fulfilment of function takes precedence. In any case, assembly is done by hand. Parts that work together – such as bearing journals and bearing shells, pistons and bushings, meshing gears, axial washers and housings, inner seal and inner channel contours – are measured, classified and meticulously mated beforehand. Not only dimensional tolerances, but also mass and surface tolerances are taken into account. In Formula 1, even screws with thread pitch matched to the elongation[2] are used.

8.6 Environmental Protection

In one area in particular, which is used by many sides as the main argument against motorsport, fundamental changes have taken place in recent decades. Thanks to environmental management, motorsport, like environmental technology, is regarded as a pacemaker for technical progress. This applies in particular to elementary areas such as energy efficiency, avoidance of pollutants, choice of materials and handling of hazardous substances [5]. Based on the ADAC Environmental Plan 2000+, environmental strategies are successfully implemented in 4 central fields of action: organization, technology, infrastructure, and research and science [5]. Lightweight design, tyre technology, occupant protection, alternative drives, increased efficiency and wear minimisation can be listed under the area of "technology" which is of interest here [5]. "Green racing" or "clean racing" are the buzzwords under which manufacturers, teams and sponsors visibly (and hopefully pioneeringly) pursue sustainable racing.

In the case of series-produced vehicles, the idea of environmental protection has been imposed much earlier. With mass motorization in the middle of the last century, vehicles not only became affordable for everyone, but also became a burden due to the large number of vehicles. Traffic regulations had to be set up, traffic guidance systems became necessary and ultimately exhaust and noise emissions had to be restricted. However, this does not only concern the operation of the vehicles themselves, but also their production and disposal. In the case of racing vehicles, exhaust emissions are not (yet) the focus of the regulation writers. And for good reason. The magnitudes of the effects compared to series production vehicles are completely different: At major DMSB events,[3] around 95% of the total vehicle kilometres travelled are attributable to spectators travelling to and from the event, who thus contribute around 93% of CO_2 emissions [5]. The following estimate illustrates even more clearly the impact of millions of vehicles compared to a few hundred. The total fuel consumption in German motor sports is less than 3% of the evaporative losses during parking and refuelling of passenger car road traffic [5]. To this should be added that the total vehicle evaporative emission of vehicles with gasoline engines at

---

[2] In common bolted joints, due to the elongation of the bolt and tolerances of the bolt and nut threads during pretensioning, the first thread generally carries approx. 25–35% of the load.

[3] Deutscher *Motorsport* Bund e. V., umbrella organization responsible for motor sports in Germany. Exercises national sporting authority for automobile and motorcycle sport.

standstill is regulated by law and checked in the so-called SHED[4] test. According to the US typing regulations, hydrocarbon emissions are also limited and monitored during the refuelling process.[5]

CO_2 emissions are directly (and linearly) related to fuel consumption with hydrocarbon fuel. A legal restriction on the fuel consumption of production vehicles and thus on climate-damaging CO_2 emissions therefore makes sense. In the case of racing cars, another consideration has led the teams themselves to use liquid energy storage as sparingly as possible. The less fuel a car needs for the targeted distance, the smaller and lighter the fuel tank can be. As a result, the car has lower driving resistance (mass, cross-sectional area) and benefits from higher performance and greater range. Conversely, higher efficiency means lower losses. These become noticeable through the need for heat dissipation and wear. Increased efficiency pays off directly through smaller heat exchangers as well as shrunken cooling air ducts and less coolant required, which in turn reduces drag and helps reduce mass. Reduced wear lowers the need for lubricating oil and its volume, which must neutralize or suspend abrasive particles with special additives (detergents and dispersants). In addition, the component with the wearing surface can be made thinner. It is no longer necessary to maintain so much wear volume so that the residual load-bearing capacity of the affected component wall remains large enough.

8.7   Technology

8.7.1 Frame and Body

Early on in the history of development, the paths of production and racing vehicles diverged when it came to frame construction. A passenger car is supposed to accommodate passengers and luggage, protect them from wind and weather ... A racing vehicle is, to put it exaggeratedly, an engine on wheels that can be manoeuvred on board by one person, or in exceptional cases two people. Initially, the ladder frame was the standard for all vehicles, but for racing cars it was replaced by tubular space frames, box frames and ultimately spatial shell structures made of fibre-reinforced composites. In passenger cars, these solutions are only found in very small numbers in sports cars, i.e. a preliminary stage to the pure-bred, purpose-oriented racing car. The self-supporting steel body has become established for passenger cars. Strictly speaking also a shell structure. This is no different for apparently production-based vehicles such as rally and touring cars (DTM, NASCAR). They may have a similar shape, which is crucial for recognition value, but under the outer skin there is sometimes completely different technology. Naturally, the difference between the road car and its track counterpart is much smaller for GT3 and GT4 cars. As is so often

---

[4] Sealed Housing for Evaporative Determination.
[5] ORVR: On-Board Refuelling *Vapour* Recovery, *vapour recovery* system during refuelling.

the case, the degree of specialisation to become a race car is, after all, a question of money. The more cost-effective a racing series is to turn out for the participants, the smaller the extent of the permitted or undertaken conversions may be. This is where the important role of the regulations comes into play, to maintain the alignment of the racing series and equal opportunities.

The numerous tasks of the bodywork or body also include the ventilation of vehicle areas (passenger cell or cockpit, engine compartment, brakes, ...) and thus, in the broadest sense, the aerodynamic behavior. While air resistance, noise generation, soiling of windows and lights are the main issues in passenger cars, downforce dictates the development of high-performance vehicles in particular. The extreme in this respect is Formula 1, which even subordinates chassis designs to this issue. For passenger cars, it is sufficient if the shape does not generate too much lift; sports cars should at least achieve slight downforce. Racing car designers also have an easier time of it; they don't have to take into account fields of vision, pedestrian protection, parking pile-ups or tyre covers, but can place wings, chimneys, deflector plates, splinters, spoilers, vortex generators, etc. on the vehicle according to purely technical aspects, subject to the regulations.

8.7.2 Engine

The main differences in the internal combustion engine result from the list of requirements. Passenger car engines should be easy to start in all seasons, regardless of previous operation, operate quietly, have low pollutant emissions, be fuel-efficient to run over a wide speed and load range, and function as intended for a long time with long service intervals. A racing engine must also last, but in the crassest ideal case only until the finish line. By then, however, it should have converted maximum power or torque from the energy provided in the fuel. The noise released in the process is perceived favourably, at least by the public, and is not perceived as a noise nuisance.[6]

If there is no need to compromise on suitability for everyday use, component designs can be precisely aligned with the desired objective. In extreme cases, this goes so far that racing engines may only be transported in a certain position and with pressurized valve cups. That they must be preheated before they can be started because the bearing and piston clearances are only suitable at operating temperature. That separate spark plugs are screwed in for the warm-up phase.

The engine is a heavy subsystem and should therefore be installed as low as possible in the vehicle. The flywheel diameter becomes the determining factor in this context. Not least for this reason, racing engines have small flywheels or none at all. High idle speeds and low elasticity only interfere with passenger cars.

---

[6]This is a phenomenon that can also be observed at major musical events.

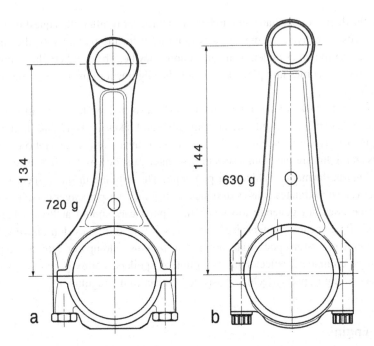

Fig. 8.2 Comparison of connecting rods of a 3.0-l petrol engine, according to [6]. A racing engine was derived from an in-line six-cylinder. With the height of the crankcase remaining the same, the longer connecting rod of the racing engine resulted in a shortened compression height of the piston. Both connecting rods were forged, but the material of the racing connecting rod was of higher quality. This was therefore also lighter in spite of the larger gauge. (**a**) Series connecting rod, (**b**) Racing connecting rod

Again, a direct comparison should make us aware of the differences that characterise this chapter. A DTM engine is derived from a series engine. In the example considered, Fig. 8.2, the racing connecting rod is longer and more flexurally rigid in the transverse direction, yet its total mass is lower. However, it must be mentioned that the material of the racing connecting rod is of higher quality.

Car engines have to cope with different fuel qualities, which, thanks to electronic control and corresponding sensors, involves fewer compromises nowadays than it did a few decades ago. Nevertheless, racing engines have an advantage here because they only have to be designed for a specific fuel and this fuel is also delivered with a much narrower tolerance of its composition.

Exhaust systems have the task of removing the combustion gases in a targeted manner, lowering the noise emission and reducing the proportion of harmful components. At the same time, the design of the pipe/reservoir system enables targeted tuning of the gas-dynamic behaviour for specific speed ranges. The positive scavenging gradient from the intake to the exhaust side is supported and the residual gas quantity (performance!) or the scavenging losses (fuel consumption!) are kept small. In the production vehicle, a compromise is made in the direction of low emissions (noise, pollutants). In the case of a racing vehicle, the focus is on performance. A power-demonstrating soundscape is even enjoyed by the public to a certain extent. Other uses of exhaust gas in racing vehicles are as an energy source in *Energy Recovery Systems* (ERS) and (at least in large-volume, high-revving engines) in supporting aerodynamic elements such as the rear diffuser and rear wing. In production vehicles, ERS will be used in hybrid powertrains. This allows the energy present in the fuel to be used more fully, or at least stored for later use.

Basically, the same applies to electric motors, even if in this case the many years of field experience are still lacking on both sides. Electric vehicles eke out a niche existence in road traffic and on the racetrack. Nevertheless, a lot has already been done in the comparatively short development time for racing vehicles and they often attract attention by setting a new record. The higher electrical currents caused by the greater power output in racing vehicles drive batteries with high internal resistance to the limits of their thermal load capacity and these thus become the bottleneck in the increase in power. And this is true in both directions, i.e. when driving and when (regenerative) braking. Other energy storage devices – such as supercaps – are more advantageous in this context. Improvements have also been made to current-carrying parts in the motor. For example, conductor cross-sections have been optimized (trapezoidal shape, etc.). Cooling methods, media and magnetic flux directions are varied with the aim of further increasing the power density. Wheel hub motors or motors close to the wheel make it possible to come remarkably close to the driving dynamics goal of allocating torque to individual wheels (torque vectoring).

8.7.3 Power Train

The differences in the powertrain are similar to those in the engines. Service life, operability and comfort characterize the development and results of series-produced vehicles. Low-noise, easy-to-change gears or transmission systems that automatically change gear ratios to save fuel are in demand for passenger cars.

Lightweight transmissions that transmit engine power to the wheels with low losses and are also a load-bearing part of the vehicle are preferred for racing vehicles. Damping elements, synchronising devices and additional masses on the outer gearshift to support the gearshift movement are not to be found on a racing gearbox. These parts increase mass and create losses. The disadvantage of lack of comfort is not an issue on race cars because it is not a competitive factor. Much more important is the ease of adapting gear ratios to engine and track conditions. In production gearboxes, oil is generally not even changed in the course of a vehicle's life. Gear geometry in racing transmissions is designed with shaft deflections in mind for the highest transmission efficiencies. Series gearboxes must be able to be manufactured and tested in large numbers using standard tools, and gear meshing must be quiet. This is one of the reasons why hypoid gearing is often used for axle drives, even though it has disadvantages in terms of transmission efficiency due to a high degree of longitudinal sliding in the rolling motion.

8.7.4 Suspension

The biggest differences between production and racing vehicles are probably in the area of suspension. This is not surprising, since the suspension has the greatest influence on handling. Cars are designed with comfort and safety in mind. Great compromises have to be made in the process. The vehicles are operated without modification – apart from summer and winter tyres – all year round in all weather conditions with different load conditions, with and without roof superstructures, with and without trailers on different road surfaces. Furthermore, very few drivers check the inflation pressure of the tyres or even their settings such as toe-in or camber before setting off. In the case of racing vehicles, these are routine activities that can also make all the difference on the results list. Racing suspensions must therefore be easy to adjust and within the required range. Production tyres should deliver long mileage with low rolling resistance and excellent wet grip (a contradiction that constantly challenges tyre developers). Noise emissions and rolling comfort are also considered and evaluated. Racing tires seem to have an easier time in this regard. They are supposed to have consistently high grip in circumferential and lateral directions and not change their behaviour. The short service life is accepted and sometimes even deliberately used by the organisers of racing series to increase the excitement for the spectators. Compared to their production counterparts, racing tyres are only partially vulcanised and only cure during operation. The rim width of production wheels is made as narrow as possible. Racing tyres, on the other hand, are mounted on wheels whose rim width is 1–2 inches wider than the tyre. This noticeably increases tire volume and lateral stiffness. The offsets of the standard wheels primarily take into account the balanced load on the wheel bearing pairs when driving straight ahead. In the case of racing vehicles, camber stiffness is decisive in this context. Figure 8.3 illustrates how corresponding compromises are made in the suspension designs. If – as in many passenger cars – the focus is on high tyre mileage, the static camber of 0° is selected and the tyres are fully utilised when driving straight ahead (top row). When cornering, the low camber is disadvantageous; the tyre cannot transmit the maximum possible lateral force. For sportier cars and sports cars, the compromise can be shifted in favour of cornering (middle row). For racing cars, the compromise is to the detriment of straight-line driving (bottom row). Tyre mileage is not an issue and for most tracks the importance of lateral dynamics far outweighs longitudinal dynamics (ratio up to 4:1). Tire utilization at maximum lateral acceleration is paramount.

Even though driving safety is the primary concern, there is still a classic conflict of objectives when it comes to spring/damper tuning in production vehicles. On the one hand, ride comfort should be high (soft spring, low damper effect) and on the other hand, wheel load fluctuation should be as small as possible (high damper force), Fig. 8.4. A solution to this conflict of objectives is provided by (technically complex) variable dampers.

To increase driving safety, the suspension joints on passenger cars are deliberately designed to be flexible. When lateral or circumferential forces are applied, the corresponding wheel "steers" in such a way that understeering behaviour results. Not

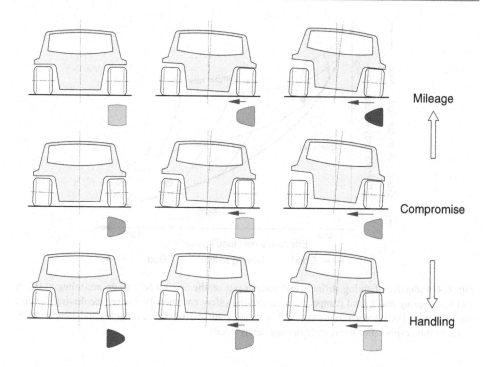

Fig. 8.3 Types of chassis designs, according to [7]. Three designs each are shown as a view of a wagon from behind when driving straight ahead (left), cornering (centre) and extreme cornering (right). In addition, the tyre contact patch shape is shown under the right-hand wheel or the wheel on the outside of the corner. The ideal contact patch utilization is shown in green. In this case the maximum tyre surface is evenly loaded on the road. In combination with lateral slip and missing or unfavourable camber angle, the worst utilisation of the tyre results, the tyre contact patch is strongly trapezoidal (red). Between these extremes you find the middle contact patch utilization (yellow). Depending on the primary goal (mileage or handling), different camber angles are accepted or aimed for, and thus different utilization or wear of the tire

only can racing drivers not use this elastokinematic wheel suspension, it also no longer functions at all in the saturation range[7] of the tires. Racing suspensions are therefore designed with almost play-free, low-friction joints. The racing driver assumes that the vehicle does what he expects through his steering inputs. The essential difference between production and racing suspensions becomes apparent when a sports car is converted to GT3 use. In [9] it is reported that the suspension of the sports car has been simplified for racing use. Not least so that problems on the track could be dealt with more quickly. For similar reasons, the adaptive dampers of the production car were replaced, but the control arms

---

[7] If the lateral force no longer changes via the slip angle, a steering movement of this tyre also no longer brings about a change in the lateral force.

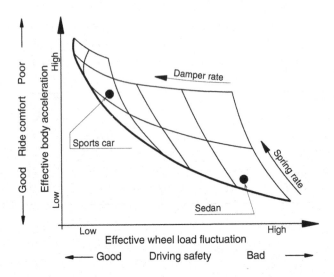

Fig. 8.4 Influence of spring stiffness and damper rate on the driving behaviour, according to [8]. A stiff body spring and a hard damper increase driving safety enormously, but reduce driving comfort due to the high body accelerations caused. For racing vehicles, the choice of spring and damper rate is easier in this respect. For them, only driving safety counts

remained the same. On the GT3 car, the front tires are wider and the weight distribution is more balanced than on the base car. The race car is about 100 kg lighter.

In general, racing vehicles are tuned more neutrally than production vehicles, for which (dynamically stable) understeering behaviour is recommended for safety reasons.

The disturbance force lever arm – the normal distance of the wheel center or the contact point from the steering axis – together with circumferential forces on the front tire (changes in rolling resistance, fluctuations in braking force, . . .) cause disturbance torques around the steering axis that the driver perceives at the steering wheel. In comfort-oriented passenger cars, this disturbance information is kept as small as possible, even if it does contain a certain amount of useful information. In sporty vehicles and racing cars, this information should be more pronounced.

References

1. Jenkins, M., et al.: Performance at the Limit. Business Lessons from Formula 1 Motor Racing. Cambridge University Press, Cambridge (2005)
2. Steiner, M.: Serienmodelle profitieren vom Rennsport-know-how. ATZ Jubiläumsausgabe 120 Jahre ATZ. **03**, 138–142 (2018)
3. Reuter, B. (ed.): Motorsport-Management. Grundlagen – Prozesse – Visionen. Springer Gabler, Berlin (2018)

4. Kunzmann, S.: Elektronische Reifendrucküberwachung mittels batterieloser Transpondertechnologie. In: Krappel, A. (ed.) Rennsport und Serie – Gemeinsamkeiten und gegenseitige Beeinflussung, pp. 183–197. Expert, Renningen (2003)

5. Ziegahn, K.-F.: Umweltschutz und Nachhaltigkeit im Motorsport. In: Reuter, B. (ed.) Motorsport-Management. Grundlagen – Prozesse – Visionen, pp. 311–333. Springer Gabler, Berlin (2018)

6. Indra, F., Tholl, M.: Der 3,0-l-Opel-Rennmotor für die Internationale Deutsche Tourenwagenmeisterschaft. MTZ. **52**(9), 454 ff (1991)

7. Serra, L., Andre, F.: Suspension systems: Optimising the Tyre contact patch. AutoTechnol. **4**, 66–68 (2001)

8. Krimmel, H., et al.: Elektronische Vernetzung von Antriebsstrang und Fahrwerk. ATZ. **5**, 368–375 (2006)

9. Scoltock, J.: McLaren MP4-12C GT3. Autom. Eng. **Jul.–Aug**, 8 f. Caspian Media, London (2011)

References

Appendix A

Glossary

1D simulation:	One-dimensional charge exchange calculation for pre-optimization of pipe lengths, tank volumes and valve timing of an internal combustion engine. Based on one-dimensional unsteady, compressible filament flow (acoustic theory), the engine is simulated as a system of pipes and tanks and wave travel times are determined. The torque curve over the speed can be determined by means of the cylinder filling that occurs. With this method, cam profiles, valve timing, intake manifold lengths, distributor volumes, duct geometries, exhaust pipe lengths and muffler designs can be pre-optimized without an existing test bench engine. Well-known software tools are available from AVL, Gamma Technologies, LMS, Lotus or Ricardo, among others.
ABS:	*Antilock braking system.* A control system in the hydraulic circuit of brake systems reduces the pressure in the brake line applied by the driver via the brake pedal as soon as a wheel threatens to lock. This requires, among other things, sensors that record the wheel speeds and compare them with a setpoint calculated from the deceleration. The main function of an ABS is to maintain the steerability of a vehicle. Locking wheels cannot build up usable lateral forces, which can lead to a loss of stability, especially on the rear axle.
	If different friction values occur on the left and right (μ-split), the driver must countersteer when braking. ABS can support the driver in this by building up braking force more slowly on the front wheel with more grip (yaw moment control). The yaw moment acting on the vehicle thus also builds up more slowly and there is

M. Trzesniowski, *Complete vehicle*,
https://doi.org/10.1007/978-3-658-39667-1

more time to countersteer. This inevitably increases the braking distance. In addition, the rear axle is controlled according to the wheel on the low friction value side (select low) [1].

An extension of the ABS control system is ABSplus or CBC (Cornering Brake Control). Here, the system detects the driving situation – in particular cornering – by means of the wheel speeds and regulates the braking forces on the individual wheels accordingly in order to keep the vehicle on track.

Acceleration:	Is the rate of change of velocity over time. In purely physical terms, it can be positive or negative, i.e. the speed increases or decreases. In the case of vehicles, however, one generally speaks of acceleration and deceleration.
ACO (Automobile Club de L'Ouest):	Automobile club that has been organizing the 24 Hours of Le Mans since 1923 and issues the regulations for the vehicles eligible to start. It also sets the regulations for the former European Le Mans Series (ELMS) and the American Le Mans Series (ALMS).
Actuated transmission (*shift by wire*):	Manually operated manual transmissions have a mechanical connection (linkage, cables) between the shift lever and the actual actuating device on the transmission housing. If the actual shifting process is carried out via actuators (pneumatic or hydraulic cylinders, electric motors, …), shifting can be initiated by the driver at the push of a button or by the on-board computer (automated transmission).
ALMS:	Abbreviation for American Le Mans Series. In this American racing series, the same regulations apply as in the famous 24-hour Le Mans race. However, the races are shorter and last between 2:45 and 12 hours.
Anisotropy:	Directional dependence of certain material properties, e.g. modulus of elasticity, strength. The opposite behaviour is called isotropic.
Boundary *layer*:	If air flows around a stationary body with a favourable flow, the air follows the contour of this body the closer to the surface the air layer under consideration is located. Due to friction effects, an air flow slows down the closer it gets to the surface of a stationary body. Thus a static to slow flow is formed at the surface of the body, the thickness of which increases towards the end of the body, the so-called boundary layer. Depending on the shape of the body and pressure conditions, this B. detaches from the surface of the body with increasing thickness and turbulence after a certain distance of flow along it. Outside this B. the friction can be

	neglected and the velocity of the particles increases with the wall distance.

Buckling: Type of failure of slender, bar-shaped components that transmit compressive forces. Compared to an ideal load that only presses the bar, imperfections occur in reality that lead to additional bending of the component. If the compressive force becomes too great, the bar deflects laterally in the middle and fails due to the excessive bending stress.

CAD: Abbreviation for Computer Aided Design. Components and their assemblies are designed three-dimensionally with the aid of suitable software. Clearances and movement spaces can be controlled more easily than on the drawing board, and numerical simulations (strength, flow investigations, . . .) can also be carried out. Some of the data can be used directly for the production of real components. *See also*: Rapid Prototyping.

CAN: Abbreviation for Controller Area Network. A two-core cable harness used instead of many lines to transmit signals in vehicles. It is a serial bus system in which messages from all participants (ABS control unit, engine control unit, sensors, actuators, . . .) can be sent or received one after the other. The CAN controller controls this sequence and sets the priorities if several signals are to be sent simultaneously. The wiring harness in a vehicle with CAN is much shorter than in a conventional system and the number of plug connections is halved.

CART: Abbreviation for Championship Auto Racing Teams. American formula series that was contested in oval stadiums and on road courses. The 2.6-liter V8 engines were powered by methanol and accelerated the single-seaters to 400 km/h. 2003 bankruptcy. Afterwards new start as ChampCar. Champ Car has since (early 2008) initially merged with IRL to form a formula series for financial reasons, and shortly afterwards became officially insolvent.

CFD (*computational fluid dynamics*): Similar to the finite element method (FEM), the geometry to be investigated is broken down into small areas ("grids)" for which the equations describing the flow are solved numerically. Depending on the equation used (potential equation, Euler equation or Navier-Stokes equation) and computer performance, even hydrodynamic boundary layers, turbulence and flow separation can be determined.

CFRP: *Carbon-fibre-reinforced* plastic *carbon-fibre-reinforced-plastic.* A composite material in which carbon fibre fabrics are impregnated

	with reactive resins and processed in several superimposed layers to form moulded parts or with internal honeycomb cores to form sandwich constructions. The targeted arrangement of the aligned fibres makes it possible to influence the mechanical component behaviour in the desired way.
Chassis:	This term is widely used from chassis to body. Probably not least because a clear separation into different assemblies cannot be made in every case. In this book, it is intended to mean the actual, load-bearing structure of a vehicle, to which wheel suspension, powertrain and bodywork components are attached. Accordingly, another term for C. is frame. In most passenger cars, the body is of self-supporting design and thus the bodywork, frame and floor panel are combined to form a structural unit. A clear assignment of the terms to one component each is therefore not possible in this case.
Coefficient of friction μ:	Value determined by tests to calculate the frictional force between two bodies. The C. depends, among other things, on the material pairing.
Compression ratio ε:	The C. of an engine is the ratio of maximum and minimum cylinder volume. The largest volume results when the piston is at bottom dead center. This volume is therefore the displacement of a cylinder plus the so-called compression volume. The smallest volume is enclosed by the piston at top dead centre. This volume represents the compression volume. The compression volume is made up of the combustion chamber volume and other components that result from the piston crown shape.
Concept:	First phase in a design process. In this phase, possible solutions for sub-functions of the overall system are sought and assembled into an effective structure. This phase is followed by the design phase.
Coordinate system:	Of the common, vehicle-fixed coordinate systems, the following is used in this book in accordance with DIN ISO 8855 (was DIN 70 000) and ISO 4130: The coordinate origin is the intersection of the vehicle longitudinal center plane with the front axle. The trihedron is aligned to this as follows. The positive x-axis points in the direction of travel, the y-axis points to the left and the z-axis points upwards.

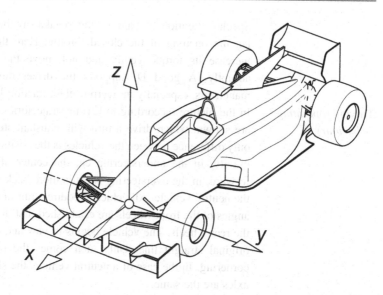

Degree of freedom (DOF):	A DOF. is a defined change of position of a rigid body according to a unique and reproducible function. A body has six DOF in space (three translations and three rotations). The machine elements that enable such DOF are called joints. A spherical joint offers three (rotational) DOF. All translations (the possible three translations) are locked. The piston rod of a damper leg is a rotational joint. It has two DOF: A translation (compression/deflection) and a rotation (rotation around the piston rod axis).
Design position (*reference standing height*):	Certain position of a vehicle in relation to the roadway, which is used as a basis for the design of running gears. Usually, the car is ready to drive with a half-full tank and the driver on board. Based on this position, the car can bounce in and out or pitch and roll. All nominal dimensions, e.g. for ground clearance, ground distance, king-pin inclination, caster, etc., are thus obtained in D.
Differential construction:	Design principle in which a functional carrier (component) is broken down into several parts. Each part can then be optimised for its partial function, e.g. multi-part wheels. The opposite is represented by the integral design.
Draft (*embodiment design*):	Phase of design activity in which the proposed solutions literally take shape. The search for solutions before the layout is the concept (conceiving) phase.
Drag:	Force acting on moving bodies due to the fact that they displace air and that the air rubs against the surface of the body.
Drivability:	For the human driver, a linear, predictable response of a system to his input is best. This is also the case with the accelerator pedal: A good D. means that the engine delivers as much torque as the driver expects based on his foot movement when he accelerates.

Special attention is paid to the breakaway behaviour (*tip-in*), i.e. the opening of the closed throttle. Here the engine should increase its torque gently and not move the vehicle forward abruptly. A good D. supports the driver during acceleration maneuvers, especially in overpowered, traction-limited vehicles.

Driving (*operating behaviour*):

In the picture (according to [2]) the trajectories of three vehicles are shown, which drive a turn with constant steering angle. The only difference between the vehicles is the position of the centre of gravity. In the understeering car the center of gravity is more forward, in the oversteering car it is more backward compared to the neutral vehicle. All vehicles require a run-in period where slip angles of the front wheels are established first, followed by slip of the rear wheels. The vehicle begins to yaw and deviates from the original straight line. Only then comes the phase of constant cornering. In the case of a neutral vehicle, the slip angles of both axles are the same.

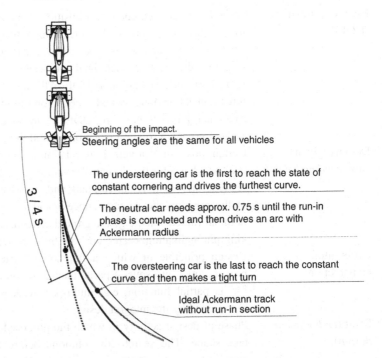

Beginning of the impact.
Steering angles are the same for all vehicles

The understeering car is the first to reach the state of constant cornering and drives the furthest curve.

The neutral car needs approx. 0.75 s until the run-in phase is completed and then drives an arc with Ackermann radius

The oversteering car is the last to reach the constant curve and then makes a tight turn

Ideal Ackermann track without run-in section

DTM:

Abbreviation for Deutsche Tourenwagen Masters (*German Touringcar Championship*). Touring car series whose vehicles must be based on production cars with at least four seats. The engines must be four-stroke gasoline engines with eight cylinders

	in a V arrangement with 90°. The engine capacity is limited to 4 litres.
Dynamic rolling radius:	The distance between the centre of the wheel and the contact patch surface is smaller for a stationary wheel than for a rolling wheel (static tyre radius). Depending on the tyre design and the wheel speed, the distance increases with increasing speed. The dyn. R. as measured value is calculated from the measured rolling circumference of a tire at 60 km/h.
Elongation at rupture	Relative elongation of a test bar at which fracture occurs. The E. is thus a measure of the toughness of a material. The higher the E., the more favourable is the fracture behaviour of a material, because failure is announced slowly.
ESP:	*Electronic stability programme.* Control system that influences driving stability. Sensors record the driving condition of the vehicle, in particular the yaw movement and the steering reaction of the driver. If the state of the car deviates from the calculated target state, the system intervenes by applying brakes to individual wheels or influencing the engine management system. ESP has a stabilising effect, for example, in the event of panic evasive manoeuvres, corners taken too fast or tyre blow-outs.
FIA:	Federation Internationale de l'Automobile. World automobile federation with headquarters in Paris. Issues the international sporting code and is thus also the supreme motorsport authority.
Filling pressure (*inflation pressure*):	The pressure difference between a tyre and the ambient pressure. The F. is usually measured on a cold tyre. If, for example, the air pressure is 1 bar[1] and the absolute pressure in the tyre is 2.5 bar, the inflation pressure is 1.5 bar. This is also referred to as overpressure.
Finite element method (FEM):	Stress calculation of components with numerical methods by a computer. In this process, the component is broken down into (thousands!) of finite elements and each element is calculated according to the laws of mechanics. These approximation methods also allow the stress calculation of parts of complex geometry and load, which cannot be calculated with formulas.
GFRP (*glass-fibre-reinforced plastics*):	Glass fibre reinforced plastic. Plastics reinforced with glass fibres in the form of mats, fabrics and strands of parallel threads to increase strength. GFRP parts are used as bodywork parts, wings, moulded parts.

---

[1] 1 bar = 100 kPa. Although the valid SI unit for pressure is Pascal (Pa), the book uses the unit bar, which is more "handy" in practice.

Glass transition temperature:	In plastics, a characteristic change in behaviour occurs when a certain temperature is reached. Below this so-called G., the oscillatory movements of the macromolecules come to a standstill and the materials become brittle. On further cooling, they reach a glassy-hard state. In the case of tires, the greater the difference between the G. of the rubber compound and the operating temperature, the softer the rubber becomes and the more friction it builds up.
Ground clearance	Distance between the vehicle underbody and the road surface. A distinction must be made between this and the ride height.
Haptics:	H. is the study of haptic perception. Haptic perception is the active sensing of size, contours, surface texture, weight, etc. of an object through the sense of touch.
Heat exchanger:	Structure in which heat is transferred from one liquid or solid substance at a higher temperature to another at a lower inlet temperature without the two substances being mixed with each other. Depending on the media involved, a distinction is made between, for example, water/air or air/air heat exchangers for the charge air cooling of a turbocharged engine.
High voltage (HV):	Electrical voltages greater than 60 V DC or 25 V AC are referred to as high voltage in the vehicle sector. This term is used to distinguish this range from "high voltage" in industrial standardization, which also has completely different numerical values. HV cables and connectors are identified by orange insulation.
IMSA:	International Motor Sports Association. International motorsport authority that runs the American Le Mans races, for example.
IndyCar Series:	Formerly IRL (Indy Racing League): Organizer of the Indianapolis 500 miles (Indy 500 on Memorial Day, May 30) and other races under the same rules on oval tracks. The cars are single-seaters and were powered by methanol-fuelled V8 engines with a displacement of 3.5 litres. Ethanol is now used as fuel and the engines are 2.2-liter V6 biturbo engines. The cost of the vehicles is limited by the regulations.
Integral construction:	Design principle in which an attempt is made to accommodate all the functions that a component must fulfil in one component. This eliminates the need for weight-increasing and strength-reducing joints. One example of this is sideshafts made from one piece with integrated tripod journals. The opposite is the differential design.
Isotropic:	The material properties are the same in all directions. The opposite behaviour is called anisotropic.
Knocking:	In a gasoline engine, a limit to the increase in compression results from (partially audible) knock at full load. Knocking is an uncontrolled sequence of a combustion initiated by the spark plug.

Especially towards the end of a knocking combustion, high pressure peaks occur which propagate at the speed of sound in the combustion chamber and damage the piston crown, seal surrounds and cylinder head. Therefore, continuous knocking must be avoided at all costs. This is achieved, among other things, by fuel additives, setting a rich fuel-air mixture, reducing the ignition angle, reducing the boost pressure, cooling the intake air, shaping the combustion chamber and targeted cooling of problematic combustion chamber areas (spark plug seat, exhaust valve seat rings).

Labisator, balance spring, Z-bar: Z-shaped connecting rod of the wheels of an axle. In contrast to a stabilizer bar, this arrangement reduces the wheel load differences and thus increases the grip on this axle. The self-steering behavior is thus influenced in exactly the opposite direction compared to the stabilizer.

Laminar flow: The flow runs in superimposed layers that do not mix. This means that there are no cross flows (turbulence).

LMS Le Mans Series: Is a racing series held according to the rules of the famous 24-hour endurance race at Le Mans. The races are usually held over 1000 km. Several drivers are entered per vehicle due to the duration of the race.

Load collective: In general, the load on a component is not constant over time, but changes irregularly. A drive shaft, for example, is subjected to extremely high loads during start-up and after a gear change, but almost no loads at all when braking and driving through a turn. However, simplified representations of loads (forces, moments) are required for the design of components. In test series (e.g. driving through a certain course), loads are therefore recorded and evaluated over time. In such evaluations, among other things, the load heights that occurred and their frequency (temporal proportion, number of load changes) are determined. The figure shows how a load collective is created from a load course.

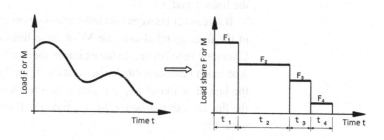

Mass inertia (first Newton's axiom): For a body to change its direction of motion or velocity, a force must act on it. This force is proportional to acceleration and mass (Newton's second axiom), $F = m \cdot a$.

Mean effective pressure $p_{m,e}$:

During an operating cycle of an internal combustion engine, the pressure in the combustion chamber changes. The mean pressure is a calculated comparative variable. It is an imaginary constant pressure that would perform the same work on the crankshaft as the actual periodically changing pressure in the course of an operating cycle.

Mixture formation:

The task of mixture formation in an engine is to produce an ignitable and combustible air-fuel mixture under all operating conditions. These mixtures only burn satisfactorily in a narrow mixture range. If the air content increases (lean mixture), fuel consumption decreases until combustion misfires increase and the running limit is reached. If the fuel content increases (rich mixture), the engine power increases until the fuel can no longer be completely burned due to a lack of oxygen.

Moment of inertia J_{polar}:

The M. in a rotation is a measure of the resistance to changes in angular velocity and is thus comparable to the mass in a translation. The M. depends on the distribution of the mass in relation to the axis of rotation. The further away mass components are from the axis of rotation, the greater the M.

Momentary pole (*instanteneous centre*):

Every movement between two rigid bodies can be described by a rotation around an instantaneous (=momentary) axis of rotation (= instantaneous pole). The location of the M. is therefore also the location at which no velocity exists between the bodies under consideration. The specification of the M. in coupling gears is done by the combination of the related links. In the picture a four-link gear is shown. If links are mounted in the (fixed) frame 1, the bearing point is taken as the M., i.e. in the example joints 12 and 14 for links 2 and 4. If links under consideration are not directly coupled, the M. can be determined by knowing two velocity vectors belonging to the rigid body. So here the pole for the links 1 and 3.

If forces act between two links, the position of the line of action of the force in relation to the M. of these links determines which kinematic state occurs. In the example, the force F_{31} (force on link 3 of link 1) causes a clockwise rotation. If the M. 13 would lie on the line of action of F_{31}, the gear would remain in equilibrium. If the line of action is below M. 13, link 3 will rotate counterclockwise. [3]

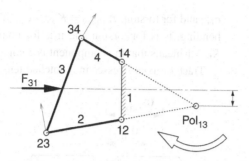

Monoposto (*single-seater*):	Single-seater racing car in which the driver's seat is located on the longitudinal centre plane of the vehicle.
NACA air inlet (*NACA air duct*):	(National Advisory Council for Aeronautics). Design of an air shaft according to the recommendations of the NACA.
NASCAR:	Abbreviation of National Association for Stock Car Auto Racing Inc. This is the rules authority for the NASCAR Sprint Cup Series (was 2004–2007 Nextel Cup Series, before that Winston Cup), a popular racing series in North America, the majority of which is run on oval tracks in stadiums. It represents the highest level racing series in the United States. The cars look like production cars on the outside, but consist of a tubular space frame and until 2011 were powered by carburetor engines driving a rigid axle on trailing arms via a cardan shaft. Since 2012, gasoline injection (Multipoint Fuel Injection MPFI) has been used.
Natural frequency:	An oscillating structure performs an oscillation (= a periodic movement around the rest position) after a single impulse left to itself. The frequency occurring in this process is the natural frequency. If such a structure is excited with a frequency equal or nearly equal to the natural frequency, the oscillation amplitudes become maximum (resonance).
Notch factor:	The stress on a component at a point is determined by calculating the mechanical stresses (bending stresses σ, torsion stresses τ etc.). In conventional calculation methods, the so-called nominal stresses are first determined, which result from the cross-section at the notch bottom of the unnotched component and the load. (In contrast, numerical methods exist that allow the approximate calculation of the stress curve, see Finite Element Method). The component is subjected to higher stresses at notch locations. The local stresses at the notch base are considerably greater than the nominal stresses. The notch factor K_f indicates by how much the maximum stresses become greater than the nominal stresses under dynamic, i.e. time-varying, loading. For bending, $\sigma_{b,\max} = K_{fb} \cdot$

$\sigma_{b,n}$ and for torsion, $\tau_{,ts\ max} = K_{,f\ ts} \cdot \tau_{ts,n}$. Where subscript b is for bending, ts is for torsion and n is for nominal. Thus, a value of $K_f = 1$ means that the component is completely notch insensitive. Trace of axial stresses in a notched tension bar.

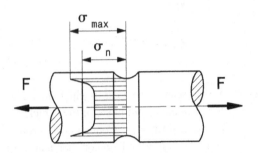

O-arrangement (*back to back a.*):	Two angular contact ball bearings or tapered roller bearings can be arranged in mirror image. If two bearings are fitted in such a way that the pressure lines point outwards (i.e. as in "O"), this is referred to as an O arrangement of bearings. If the pressure lines point towards each other, this is known as an X arrangement.
Octane number:	A parameter for the anti-knock properties of a fuel. The higher the octane number, the more resistant the fuel is to knocking. Two different methods are used to determine the octane number: The engine method (engine octane number MON) and the research method (research octane number RON).
Oversteer, *AE loose*:	See driving behaviour
Percentile:	Division of a population (normal distribution) into 100 sections. Here statistical division of the dimensions of the human body. This is used to design cockpits and passenger compartments that are suitable for a large proportion of the population. Thus, in car design, the 95% man and the 5% woman cover 90% of the total population. That is, only 5% of men are taller and only 5% of women are shorter than the percentiles used.
Pitching:	See vehicle movements
Planar moment of inertia:	Mathematical quantity that follows from the geometry of a cross-section. The MoI. is required in the strength calculation for bending stress on components.
Pressure angle:	At this angle, the force is transmitted from the outer ring and inner ring in a rolling bearing. The greatest load carrying capacity for a bearing is obtained when the contact angle coincides with the angle of the external bearing force.

Prototype:	Racing car of a certain category and group, which is only produced in small numbers or as a one-off.
Quality control:	The control of the load (and thus, in the case of constant load, of the speed) takes place in diesel engines by controlling the fuel supply to the combustion chamber. The engine draws in the combustion air unthrottled. This results in the desired air-fuel mass ratio in the combustion chamber solely by changing the fuel quantity.
Quantity control:	In gasoline engines with conventional mixture formation (carburetor, intake manifold injection), the load (and thus the speed at constant load) is controlled via throttle elements (throttle valve, slide valve). In the partial load range, the air or mixture quantity supply to the engine is changed by throttling the intake cross-section. At full load, the entire cross-section is released.
Raid, Rally Raid:	This generic term covers endurance races that are held cross-country in open terrain, primarily in desert regions. The basic course of the competition is similar to that of a rally, i.e. the vehicles drive from a starting point to a specific destination.
Rally:	These are competitions that are held on sections of road that are closed off for the duration of the competition. The road surface can be asphalt or similar, gravel, but also snow and ice. Each vehicle usually drives the route alone. A characteristic feature of R. is that a co-driver announces the course to the driver.
Rapid Prototyping (3D Printing, Additive Manufacturing):	This includes all processes with the help of which real models can be created directly from 3D CAD information. Some of these processes work like a printer that prints out three-dimensional plastic bodies. Depending on the process and the purpose, these models can be demonstration objects, test parts, casting models or molds. The aim is to quickly arrive at a functional (prototype) part (name!) based on CAD data.
Reynolds number Re:	Is a dimensionless similarity ratio in fluid mechanics. It compares the inertia forces with the friction forces in a fluid. In a wind tunnel test with a scaled-down vehicle model, the R. values of the model and the original must be the same in order to obtain comparable flow fields and thus useful measurement results.

Rockwell hardness:	Indication of the hardness of a material. Is determined by the permanent penetration depth of an indenter (cone, ball) into the workpiece.
Roll:	See vehicle movements
Ride height:	Is the distance of any point fixed to the vehicle from the road. During set-up, a certain ground clearance is assumed as a reference value and the car is set higher or lower. The ground clearance is therefore only a metrological simplification for determining the ground clearance.
Rubber:	Collective term for rubber-based elastomers (a subgroup of plastics). The actual rubber is obtained from the thickened sap (latex) of the rubber tree by sulphur treatment (so-called vulcanisation, leading to wide-meshed cross-linking of the molecules). In addition to this natural rubber, there is also synthetically produced rubber. The best known representative is Buna, which is produced by polymerizing butadiene.
	For elastomers, the service temperature is higher than the glass transition temperature. For the other plastics (thermoplastics and thermosets) it is exactly the opposite.
Self-steering properties:	(See also Driving Behaviour.) In the limit range of the drivable lateral acceleration, the vehicle rotates about its vertical axis in a different way than corresponds to the steering angle during pure rolling of the tyre. The lateral forces increase differently at the front and rear axle (more precisely at each individual wheel). However, lateral forces are only transmitted to the rubber-tyred wheel if it rolls at an angle to its plane (slip). If the slip angle on the front axle of a vehicle increases faster than on the rear axle, the car "pushes" out of the curve via the front wheels. The driver has to turn in more than he would have to if the car was just rolling (understeering S.). The opposite behaviour is called oversteer. The behaviour of a vehicle with (approximately) equally increasing slip angles at all wheels is called neutral. However, a given vehicle need not exhibit the same self-steering behavior over the entire drivable limit range. In addition to vehicles that exhibit constant behavior, there are also those that understeer at low lateral accelerations, but switch to oversteering behavior at higher lateral accelerations, and vice versa. In addition, there is the not inconsiderable influence of circumferential forces on the drive wheels,

	especially at high engine outputs. For example, a rear-wheel drive vehicle that is neutral when rolling will oversteer when accelerating strongly because the drive forces cause the tires to become laterally softer.
Sequential shifting:	A type of gear change in a manual transmission in which the individual gears are only engaged one after the other (sequentially). The driver merely has to make a simple movement. Motorcycle transmissions are an example of this. In contrast, common passenger car manual transmissions have an H-shift, where any gear can be engaged with a compound movement.
Shear modulus (slip modulus) G_{shear}:	Material constant determined by shear tests on test specimens. For many materials, the ratio between shear stress and angular distortion remains the same under shear loading. This ratio is the S.
Simulation:	Simulations are used to calculate the effects of complex physical relationships, usually over time. For this purpose, the system to be investigated is first represented in simplified form by a model. This model is then described mathematically by a system of equations. With the help of a computer, this system of equations is solved (usually by numerical approximation methods). The results are then displayed (visualized) as graphics or animations. Simulations allow many changes to be made to the system under investigation in a short time, which would either not be possible or too expensive on the real object in isolation. Among other things, the driving behaviour of a car with different tyres, axle loads, centre of gravity heights, downforce, etc. on different routes (which must of course have been recorded three-dimensionally for this purpose) is simulated. Because of the simplifications made, a simulation does not represent reality exactly, but it does provide qualitative information about the variables influencing the system under investigation. By comparison with measured test results, models are tested for their usefulness and subsequently improved.
Spring rate:	Specification of the spring stiffness. If the behaviour of a spring is plotted in a force/displacement diagram, the spring characteristic curve is obtained. The slope of the characteristic curve is the spring rate c_{Sp}. The spring rate does not have to be constant, but can change during compression. If the spring becomes stiffer during compression (the line becomes steeper), this is called progressive behaviour. The opposite behaviour is called degressive. The characteristic curve flattens out and the spring becomes increasingly softer when loaded.

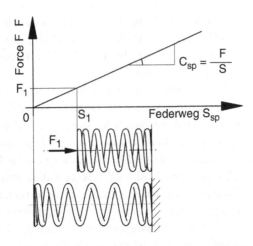

Stress:	An external load (force, moment, torque) causes a stress state in the material structure of a component. This stress state is the stress. It is recorded by (technical) stresses (tensile stress, compressive stress, shear stress, ...).
Stroke/bore ratio:	The ratio of the piston stroke s to the cylinder bore B in a reciprocating engine. Based on the appearance of a cylinder from the side, a distinction is made between square (stroke = bore), undersquare or long-stroke (stroke > bore) and oversquare or short-stroke (stroke < bore) engine designs. The figure shows schematically a short-stroke (a) and a long-stroke (b) design of a crank mechanism.

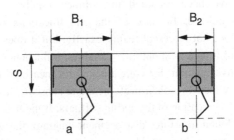

Tensile strength R_m:	Material characteristic value determined in a tensile test. It results from the quotient of the maximum force during the test and the cross-section of the test bar before the test. The T. is included in many material abbreviations.
Tension stress:	If a component is loaded by external forces and/or moments or if it is hindered in its thermal expansion, a stress occurs in the interior.

This stress is recorded mathematically by mechanical stresses, e.g. in N/mm². If the stress at a point in the component exceeds a material-dependent characteristic value, failure (cracking, flow, ...) occurs at this point.

Tribology: The study of the interaction of friction, lubrication and wear. If relative motion occurs between bodies, this leads to loss of energy (friction) and material removal (wear).

Turbulent flow: Is a flow form in which cross flows and turbulence occur in different sizes and directions.

Tyre contact patch: The contact area of a tyre. All forces between the tyre, and therefore the vehicle and the road, are transmitted via this surface.

Understeer, AE push: See driving behaviour

Vehicle coordinate system (*axis system*): See coordinate system

Vehicle level: See ground clearance

Vehicle motion: A vehicle – like any rigid body – has six degrees of freedom in space. The possible individual movements (displacements and rotations) about the three main axes are designated as follows:

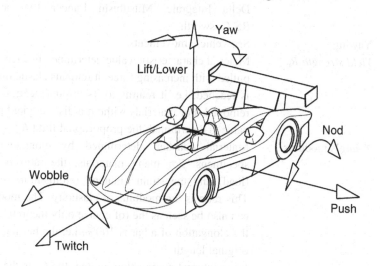

Along the transverse axis: Push (*drift*).

Along the vertical axis: Raise or lower (*heave*).

Around the transverse axis: *Pitch.*

Around the vertical axis:	Yaw.
Displacements (Translations):	Along the longitudinal axis: twitch (*jerk*).
Rotations (Rotations):	Around the longitudinal axis: roll (tilt). If a vehicle drives on a roadway, the movements are a combination of the possible individual movements and result from the given movements of the roadway and the driver's influence by steering.
Wheel frequency:	Natural frequency of an oscillating wheel connected to the car body by spring and movable links.
WRC – *World Rally Car*:	Rally car based on a generous set of regulations that do not stipulate a minimum number of cars built. The minimum weight is 1230 kg. The number of cylinders in the engines is limited to eight. The displacement depends on the number of valves and the supercharging method. Other rally cars belong to group A and N. For these cars it is required that 2500 basic models are built within one year. To Group A we owe such road cars as the Lancia Delta Integrale, Mitsubishi Lancer Evo and Ford Escort RS-Cosworth.
Yawing:	See vehicle movements
Yield strength R_e:	Material characteristic value determined in a tensile test. If a bar is pulled with increasing force, it remains elastic until the yield point is reached, i.e. it returns to its original length when the load is removed. For materials without a distinct yield point, a substitute value is determined, the proportional limit $R_{p0,2}$.
Young's modulus:	Material constant determined by elongation tests on test specimens. For many materials, the ratio between the stress (load) and the strain obtained (elongation) remains the same. This ratio is the modulus of elasticity. The modulus of elasticity can also be seen as the (of course only theoretical) stress at which the elongation of a bar is 100%, i.e. the bar has reached twice its original length.
Charging λ_a efficiency:	In an internal combustion engine, the C. is the ratio of the fresh charge supplied (this is everything that flows through the air filter) to the charge mass theoretically possible in the cylinder. Thus, the C. is not equal to the degree of delivery. Due to scavenging losses in the charge exchange top dead center, for example, fresh charge can be lost via the exhaust tract. This loss is taken into account in the C., but not in the degree of delivery. In this example, the C. would be greater than the degree of delivery if the mass

Volumetric efficiency: λ_1

supplied is greater than the theoretically possible mass. The C. is easier to measure than the degree of delivery.

In an internal combustion engine, the V. is the ratio of the charge mass actually in the cylinder after completion of the charge exchange compared to the charge mass theoretically possible in the cylinder ($=$ swept volume times air density). The V. is less than 1 in naturally aspirated engines. As the flow velocity (speed) increases, the losses increase due to throttling in the lines and valves. This is partly compensated or even overcompensated by gas dynamic effects at certain speeds.

Air-fuel ratio λ:

The air-fuel mixture in the engine ignites and burns satisfactorily only within a certain mixture range. For gasoline, this ratio is about 14.7:1, i.e. 14.7 kg of air are required for complete combustion of 1 kg of fuel (stoichiometric mixture).

The air number λ compares this theoretical demand with the actual mixture present.

$\lambda = \frac{\text{existing mixture}}{\text{stoichiometric mixture}}$. $\lambda = 1$ means that there is a stoichiometric mixture in the combustion chamber. $\lambda < 1$ means there is a lack of air (rich mixture). $\lambda > 1$ means there is excess air (lean mixture).

Listed below are the differences between corresponding American (AE) and British terms (BE) for some common parts:

Component	American	British
Side shaft	Axle shaft	Half shaft
Drive shaft	Driveshaft	Prop shaft
Wheelhouse	Fender	Wheel arch
(Engine) hood	Hood	Bonnet
Oversteer	Loose	Oversteer
Bevel gearbox	Ring & pinion	Crown wheel & pinion
Understeer	Tight (push)	Understeer
Trunk	Trunk	Boot
Shock absorber	Shock absorber	Damper
Torsion stabilizer	Sway bar	Anti roll bar
Gurney bar	Wicker	Gurney
Windscreen	Windshield	Windscreen

Different racing classes also use different names for what is essentially the same component:

- Triangular wishbone: A-arm/wishbone, control arm
- Wheel carrier: spindle, knuckle (touring car)/upright (monoposto)
- tie rod: tie rod/toe link

References

1. Breuer, B., Bill, K.-H. (Hrsg.): Bremsenhandbuch, 1. Aufl. GWV Fachverlage/Vieweg Verlag, Wiesbaden (2003)
2. Milliken, W.F.: Chassis Design: Principles and Analysis. Society of Automotive Engineers, Warrendale (2002)
3. Neumann, R., Hanke, U.: Eliminierung unerwünschter Bewegungen mittels geeigneter Momentanpolkonfiguration. Konstruktion, Heft 4, S. 75–77. Springer, Berlin (2005)

Printed in the United States
by Baker & Taylor Publisher Services

Printed in the United States
by Baker & Taylor Publisher Services